MITOCHONDRIA

ACADEMIC PRESS RAPID MANUSCRIPT REPRODUCTION

MITOCHONDRIA
Bioenergetics, Biogenesis and Membrane Structure

Edited by

Lester Packer

Membrane Bioenergetics Group
Lawrence Berkeley Laboratory
University of California and the
Department of Physiology-Anatomy
Berkeley, California

Armando Gómez-Puyou

Departamento de Biología
Universidad Nacional Autónoma de México
México

ACADEMIC PRESS, INC. New York San Francisco London 1976
A Subsidiary of Harcourt Brace Jovanovich, Publishers

ACADEMIC PRESS, INC.
111 Fifth Avenue, New York, New York 10003

United Kingdom Edition published by
ACADEMIC PRESS, INC. (LONDON) LTD.
24/28 Oval Road, London NW1

Library of Congress Cataloging in Publication Data

Main entry under title:

Mitochondria: Bioenergetics, Biogenesis and Membrane Structure

    Papers from a symposium held at the Instituto de biolo-
gía, Universidad Nacional Autonoma de Mexico, in June 1975
       Bibliography: p.
       Includes index.
       1.   Mitochondria-Congresses.   I.   Packer, Lester.
II.   Gómez-Puyou, Armando.   III.   Mexico(City).
Universidad Nacional.   Instituto de biología.
[DNLM: 1.   Mitochondria—Congresses.   QH581.2
M684 1975]
QH603.M5M55   574.8'734   75-13077
ISBN 0–12–543460–X

CONTENTS

LIST OF CONTRIBUTORS   ix
PREFACE   xiii

PART I. BIOENERGETICS

Monovalent Cation Transport by Mitochondria   3
  Gerald P. Brierley

The Proton Pump in the Yeast Cell Membrane   21
  Antonio Peña

Kinetic and Thermodynamic Aspects of Mitochondrial   31
Calcium Transport
  A. Scarpa

Mitochondrial Calcium Transport and Calcium Binding Proteins   47
  Ernesto Carafoli

Comparison of Electroneutral and Electrogenic Anion Transport   61
in Mitochondria
  Kathryn F. LaNoue and Marc E. Tischler

Mitochondrial Metabolism and Cell Regulation   79
  J.R. Williamson

Kinetic and Binding Properties of the ADP/ATP Carrier as a   109
Function of the Carrier Environment
  P.V. Vignais, G.J.M. Lauquin and P.M. Vignais

The Adenine Nucleotide Transport of Mitochondria   127
  M. Klingenberg

Correlation of Mitochondrial Swelling and Localization of   151
Malate Dehydrogenase Activity
  A. Rendón and L. Packer

On the Problem of Site Specific Agents in Oxidative   155
Phosphorylation. The Action of Octylguanidine and K$^+$
  M. Tuena de Gómez-Puyou, M. Beigel, and A. Gómez-Puyou

Electron Transfer and Energy Coupling at the NADH-
Ubiquinone Segment of Mitochondria   167
  Sergio Estrada-O and Carlos Gómez-Lojero

Energy Coupling in Reconstituted Complexes from   183
Mitochondria
  Peter C. Hinkle

PART II. BIOGENESIS

Application of Continuous Culture in the Study of   195
Mitochondrial Membrane Biogenesis in Yeast
  A.W. Linnane, S. Marzuki, G.S. Cobon and P.D. Crowfoot

Mitochondrial Assembly: Attempts at Resolution of   213
Complex Functions
  Henry R. Mahler

Cytoplasmic and Nuclear Mutations Affecting   241
Mitochondrial Functions
  Alexander Tzagoloff

The Control of Mitochondrial Membrane Formation   251
  Diana S. Beattie

Biochemical Genetic Studies of Oxidative Phosphorylation   265
  David E. Griffiths

Binding and Uptake of Cationic Dyes to a Nuclear   275
Ethidium Bromide Resistant Mutant of the Petite-
Negative Yeast *Kluyveromyces lactis*
  Aurora Brunner

PART III. MEMBRANE STRUCTURE

Spin Labeling the Hydrocarbon Phase of Biological   291
Membranes
  Alec D. Keith

NMR Studies on the Precise Localization of Drugs   303
in the Membrane. Characterization of the Reaction Site
  Jorge Cerbón

The Use of Apolar Azides to Measure Penetration of   315
Proteins into the Membrane Lipid Bilayer
  Amira Klip and Carlos Gitler

CONTENTS

Nanosecond Fluorescence Spectroscopy of 327
Biological Membranes
  P.A. George Fortes

The Structure of the Mitochondrial Inner 349
Membrane: Chemical Modification Studies
  Harold M. Tinberg and Lester Packer

Fluidity of Mitochondrial Lipids 367
  P.M. Vignais

The Interaction of Ionic Detergents with 381
Submitochondrial Membranes
  R.J. Mehlhorn

Light Increases the Ion Permeability of 389
Rhodopsin-Phospholipid Bilayer Vesicles
  A. Darszon and M. Montal

SUBJECT INDEX 403

LIST OF CONTRIBUTORS

Diana S. Beattie
    Department of Biochemistry, Mt. Sinai School of Medicine, The City
    University of New York, New York 10029

M. Beigel
    Departamento de Biologia Experimental, Instituto de Biologia,
    Universidad Nacional Autonoma de Mexico, Mexico 20, D.F., Mexico

Gerald P. Brierley
    Department of Physiological Chemistry, College of Medicine, Ohio State
    University, Columbus, Ohio 43210

Aurora Brunner
    Departamento de Biologia Experimental, Instituto de Biologia,
    Universidad Nacional Autonoma de Mexico, Mexico 20, D.F., Mexico

Ernesto Carafoli
    Laboratorium fur Biochemie, Eidgenossische Technische Hochschule,
    Zurich, Switzerland

Jorge Cerbón
    Departamento de Bioquimica, Centro de Investigacion y de Estudios
    Avanzados, Instituto Politecnico Nacional, Mexico 14, D.F., Mexico

G.S. Cobon
    Department of Biochemistry, Monash University, Clayton, Victoria 3168,
    Australia

P.D. Crowfoot
    Department of Biochemistry, Monash University, Clayton, Victoria 3168,
    Australia

A. Darszon
    Departamento de Bioquimica, Centro de Investigacion y de Estudios
    Avanzados, Instituto Politecnico Nacional, Mexico 14, D.F., Mexico

Sergio Estrada-O
    Departamento de Bioquimica, Centro de Investigacion y de Estudios
    Avanzados, Instituto Politecnico Nacional, Mexico 14, D.F., Mexico

P.A. George Fortes
Department of Biology, University of California, San Diego, La Jolla, California 92037

Carlos Gitler
Departamento de Bioquimica, Centro de Investigacion y de Estudios Avanzados, Instituto Politecnico Nacional, Mexico 14, D.F., Mexico

Carlos Gómez-Lojero
Departamento de Bioquimica, Centro de Investigacion y de Estudios Avanzados, Instituto Politecnico Nacional, Mexico 14, D.F., Mexico

A. Gómez-Puyou
Departamento de Biologia Experimental, Instituto de Biologia, Universidad Nacional Autonoma de Mexico, Mexico 20, D.F., Mexico

M. Tuena de Gómez-Puyou
Departamento de Biologia Experimental, Instituto de Biologia, Universidad Nacional Autonoma de Mexico, Mexico 20, D.F., Mexico

David E. Griffiths
Department of Molecular Sciences, University of Warwick, Coventry CV4 7AL, England

Peter C. Hinkle
Department of Biochemistry and Molecular and Cell Biology, Cornell University, Ithaca, New York 14853

Alec D. Keith
Department of Biophysics, Pennsylvania State University, University Park, Pennsylvania 16802

M. Klingenberg
Institut fur Physiologische Chemie und Physikalische Biochemie, Universitat Munchen, 8000 Munchen 15, Germany

Amira Klip
Departamento de Bioquimica, Centro de Investigacion y de Estudios Avanzados, Instituto Politecnico Nacional, Mexico 14, D.F., Mexico

Kathryn F. LaNoue
Department of Physiology, Milton S. Hershey Medical Center, Pennsylvania State University, Hershey, Pennsylvania 17033

A.W. Linnane
Department of Biochemistry, Monash University, Clayton, Victoria 3168, Australia

G.J.M. Lauquin
Laboratoire de Biochimie, Dept. de Recherche Fondamentale, Centre d'Etudes Nucleaires, 38041 Grenoble CEDEX, France

Henry R. Mahler
Department of Chemistry, Indiana University, Bloomington, Indiana 47401

S. Marzuki
Department of Biochemistry, Monash University, Clayton, Victoria 3168, Australia

R.J. Mehlhorn
Membrane Bioenergetics Group, Energy and Environment Division, Lawrence Berkeley Laboratory, University of California, Berkeley, Berkeley, California 94720

M. Montal
Departamento de Bioquimica, Centro de Investigacion y de Estudios Avanzados, Instituto Politecnico Nacional, Mexico 14, D.F., Mexico

Lester Packer
Membrane Bioenergetics Group, Energy and Environment Division, Lawrence Berkeley Laboratory, University of California, Berkeley, Berkeley, California 94720

Antonio Peña
Departamento de Biologia Experimental, Instituto de Biologia, Universidad Nacional Autonoma de Mexico, Mexico 20, D.F., Mexico

A. Rendón
Department of Physiology-Anatomy, University of California, Berkeley, Berkeley, California 94720
On Leave From: Centre de Neurochimie, C.N.R.S., Strasbourg 6700, France

A. Scarpa
Johnson Research Foundation, University of Pennsylvania, Philadelphia, Pennsylvania 19174

Harold M. Tinberg
Physiology Research Laboratory, Veterans Administration Hospital, Martinez, California 94553

Marc E. Tischler
Department of Physiology, Milton S. Hershey Medical Center, Pennsylvania State University, Hershey, Pennsylvania 17033

Alexander Tzagoloff
Public Health Research, Institute of the City of New York, New York, New York 10016

P.M. Vignais
Laboratoire de Biochimie, Dept. de Recherche Fondamentale, Centre d'Etudes Nucleaires, 38041 Grenoble CEDEX, France

P.V. Vignais

Laboratoire de Biochimie, Dept. de Recherche Fondamentale, Centre d'Etudes Nucleaires, 38041 Grenoble CEDEX, France

J.R. Williamson

Johnson Research Foundation, University of Pennsylvania, Philadelphia, Pennsylvania, 19174

PREFACE

The chapters in this book comprise the main lectures of an advanced course on the biogenesis, structure, and function of mitochondria that was sponsored by the Instituto de Biologia of the Universidad Nacional Autonoma de México in June, 1975. The lectures were imparted at the Instituto de Biología and at the Instituto de Investigaciones Biomédicas at the UNAM and at the Centro de Investigaciones y Estudios Avanzados, IPN in Mexico City.

It is now clear that only the joint efforts of researchers of a wide variety of disciplines will lead to a full understanding of mitochondria. The book is divided into three sections. Parts I and II cover many of the important approaches to the study of mitochondrial bioenergetics and biogenesis. These chapters are written to provide students of bioenergetics and other fields with an adequate background of the subject as well as with some of the more recent findings of the authors' laboratories. In Part III, membrane structure is emphasized.

During the two weeks of this course, it was recognized that a satisfactory comprehension of the molecular events that occur in mitochondria appears to be in sight. The basic questions are now well defined, and diffent and powerful methodologies are being applied to explore the various facets of the same problem. Therefore, it would seem that the answers to these questions will be the consequence of further experimentation, and in some cases of an improvement of present day methodology. In other words, a proper understanding of the molecular events that occur in mitochondria is now on the horizon.

The Editors are very much indebted to Dr. José Laguna for his encouragement and support in the developing of this course. We are also grateful to Dr. Antonio Peña for his collaboration in the organization of the meeting. We thank Ms. Yolanda Díaz D'Castro for her assistance in solving many of the administrative problems. We are also indebted to Ms. Sharon Anderson for faithfully and carefully assisting with the preparation of the chapters for publication, for editorial assistance, and for interfacing the contents with the publishers. We thank Dr. Alvaro Rendon for assisting in communications between the Editors and for his counsel with the assistance of Rosa Laura Oropeza-Rendón on the indexing of the book. The Editors also acknowledge the financial support of the Organization of American States, the Mexican Society of Biochemistry and the U.S. Bioenergetics Group.

Lester Packer
Armando Gómez-Puyou

PART I
BIOENERGETICS

A substantial amount of effort has been expended in the past in attempts to ascertain the molecular mechanism of oxidative phosphorylation. While a completely satisfactory explanation for this process has not yet emerged, it has become apparent that phosphorylation is intimately related to numerous transport processes which take place across the inner membrane of the mitochondrion. Therefore, a complete picture of mitochondrial energy coupling must include not only a molecular mechanism for ATP formation, but also a mechanistic description of the transport reactions. In addition, the role of the transport reactions in metabolic regulation and communication between the mitochondrial and extra-mitochondrial compartments of the cell must be specified if we are to understand how mitochondrial oxidative phosphorylation is integrated into cellular metabolism

These objectives are currently being pursued by a number of laboratories and the variety of experimental approaches reflects the appeal of these questions to investigators in widely different disciplines. The chapters in this section illustrate the progress which has been made in establishing the characteristics of the transport reactions, in isolation and characterization of both transport and phosphorylation components, and in reconstitution of reactions from purified components.

MONOVALENT CATION TRANSPORT BY MITOCHONDRIA

Gerald P. Brierley

Although there is still no general consensus as to the mechanism of oxidative phosphorylation, it has become increasingly clear that the movement of ions across the mitochondrial membrane is intimately associated with the process of energy coupling. Mitochondria exchange substrate anions, inorganic phosphate (Pi), and adenine nucleotides across the inner membrane (Fig. 1) on a series of exchange-diffusion carriers [1] and the anion gradients thus produced, in conjunction with proton and cation movements, combine to provide a capacity for production of both pH gradients and potential differences between the mitochondrial matrix and extramitochondrial compartments [2,3]. If these gradients are not indeed the primary mode of energy coupling as suggested by Mitchell [2], they are at least in close communication with the primary coupling process. Early studies recognized that mitochondria could retain certain ions against concentration gradients and expend energy to accumulate others (see Ref. [4] for a review of the earlier literature), but recent intense interest in mitochondrial ion transport has stemmed largely from the predictions of the Mitchell hypothesis [2,3] and from a series of observations on mitochondrial divalent cation accumulation [4]. These studies revealed that impressive amounts of Ca^{+2}, Mn^{+2}, and Mg^{+2} could be accumulated at rapid rates and that the reaction was supported by an intermediate state or condition prior to the production of ATP in the phosphorylation process. A capacity for extensive uptake of monovalent cations upon the addition of inducing agents (ionophores), such as valinomycin, was demonstrated shortly thereafter (see [5] for a review), and much recent interest has centered on the high rates of monovalent cation uptake which result in the presence of these reagents. Rapid uptake of monovalent cations in the absence of ionophores can also be obtained when mitochondria are suspended in isotonic acetate or phosphate salts [6]. In addition to these ion uptake reactions, which are characterized by deposition of phosphate salts of divalent cations and by osmotic swelling when soluble ion pairs are accumulated, swollen mitochondria have been shown to contract and extrude ions in an energy-dependent reaction under a number of conditions [6-8].

The present state of monovalent cation transport in isolated mitochondria can be resolved into the following three questions: (1) What is the mechanism by which monovalent cations are accumulated? (2) By what mechanism are accumulated ions extruded, and what controls the switch from accumulation to extrusion which is apparent under a variety of conditions? (3) What is the physiological role of mitochondrial cation movements? In the present chapter evidence will be developed which suggests that cations move into the mitochondrion in response to a primary pH gradient and a secondary membrane potential (negative interior). With the exception of Ca^{+2}, which appears to enter electrophoretically by means of a charged carrier (Fig. 1), there seems to be no firm evidence that carriers are involved in the uptake of Mg^{+2} or monovalent cations, and, indeed, it appears that a low permeability to these components is essential for normal mitochondrial function. Evidence will also be presented which is consistent with the participation of an electroneutral cation/H+ exchanger in the process of energy-dependent ion extrusion and contraction. One could speculate that an interplay between electrophoretic cation influx and electroneutral extrusion could provide the mitochondrion with a controllable mechanism for adjusting the size of its internal anion pools and for osmotic volume control [8].

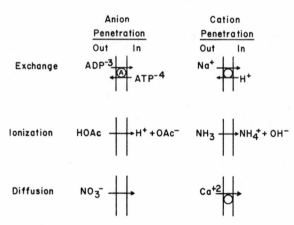

Fig. 1. *Possible modes of entry of anions and cations into the mitochondrion.*

METHODS

Ion uptake and extrusion can be followed by direct analysis of centrifuged or filtered mitochondria, but the completeness of the separation from the suspending medium and the metabolic condition of the organelles is often open to question. If the ion to be measured is rather permeable, obviously washing procedures to remove suspending medium will result in loses of accumulated ion. For this reason centrifugation incubation methods [1], usually involving sedimentation of the particles through an insulating layer of silicone oil, have proven useful for direct analysis and estimation of cation flux by isotope distribution [9]. If such a procedure also employs an acid layer to denature the mitochondria, questions as to the state of aeration of the pellet, etc. can also be minimized [1]. A correction for the amount of suspending medium which is carried down with centrifuged mitochondria can be made using tritiated water to get the total pellet water volume and ^{14}C-sucrose which will equilibrate with extramitochondrial water and the solvent in the intermembrane space, but which is excluded by the inner membrane. Ion-sensitive electrodes [5] have been very helpful in cases in which a vigorous uptake of $K+$ from a solution containing less than 5mM $K+$ can be induced by an ionophore, such as valinomycin. One of the most convenient and generally useful methods for ion uptake and extrusion is the use of light scattering or absorbance records to follow osmotic swelling and contraction as will be discussed in the next section.

EXPERIMENTAL

Mitochondria respond as osmometers to changes in osmotic pressure exerted by non-penetrating solutes in the suspending medium. It is now generally accepted that the outer membrane is permeable to all solutes of low molecular weight and that osmotic response is at the level of the inner membrane. Non-electrolytes penetrate the inner membrane as a function of their oil/water partitition coefficients and produce osmotic swelling. As first pointed out by Chappell and Crofts [10], osmotic swelling in the presence of electrolytes will occur only when both an anion and a cation can enter the matrix compartment so that the internal osmolarity can be increased without inducing a large diffusion potential. Osmotic swelling in electrolytes can be either energy-independent (passive) or energy-dependent, and rates of ion movement can be calculated from rates of swelling and contraction [7]. In the absence of energy, osmotic swelling studies indicate that the entrance of anions into the matrix can occur by one of three different

5

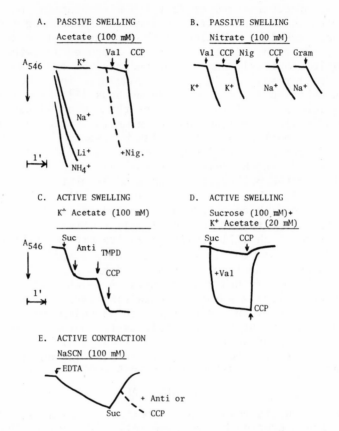

Fig. 2. *Swelling and contraction of isolated mitochondria.*
Isolated beef heart mitochondria were treated with rotenone
and suspended in the following media at 0.5 mg/ml, 25°C,
buffered at pH 7.1 with 2 mM Tris: A. Indicated acetate salts
(100 mM) B. Indicated nitrate salts (100 mM) C. K+ acetate
(100 mM) D. Sucrose (100 mM) plus K+ acetate (20 mM) E. NaSCN
(100 mM). A circular cuvette equipped with a water jacket and
magnetic stirrer was mounted on an Eppendorf photometer and
absorbance at 546 nm was recorded. This arrangement permits
convenient simultaneous recording of O_2 uptake and pH with an
oxygen and a combination glass electrode respectively. Where
indicated, the following additions were made: succinate, 2 mM;
CCP (m-chlorocarbonylcyanide phenylhydrazone), 5 x 10^{-7}M;
valinomycin, 1 x 10^{-7}M; nigericin, 1 x 10^{-7}M; gramicidin,
2 x 10^{-7}M; antimycin 0.5 nmoles/mg; TMPD (tetramethylphenylene-
diamine plus ascorbate), 0.05 mM + 2 mM; EDTA, 2 mM.

mechanisms (Fig. 1). These are (a) electrically neutral (malate^{-2}/succinate^{-2}) or electrogenic (ATP^{-4}/ADP^{-3}) exchange of anions on one of the exchange-diffusion carriers which have been identified in the membrane [1,11], (b) simple diffusion of certain anions, such as SCN$^-$ and NO$_3$-, through the membrane, and (c) ionization of a permeant free acid, such as acetic or lactic acid, in the matrix compartment. A corresponding analysis of cation entrance pathways in the absence of energy shows evidence for an electroneutral exchange of Na+/H+ and for the generation of NH$_4$+ and other similar cations in the matrix as a result of ionization of a permeant free base (NH$_3$). Whereas there are indications that Ca++ can penetrate the membrane in the absence of energy by a carrier-mediated, electrophoretic pathway [12], passive swelling studies have established that K+ and similar monovalent cations do not penetrate the non-energized membrane rapidly enough to support swelling by either exchange or electrophoretic diffusion [6,8].

Typical experiments showing passive swelling of heart mitochondria in acetate and nitrate salts are reproduced in Fig. 2A and 2B respectively. Swelling in acetate salts appears to depend on penetration of free acetic acid, ionization to produce H+ and acetate$^-$ (as shown in Fig. 1), and then either exchange of the proton for a cation on an endogenous exchanger as in the case of Na+ and Li+, or by providing a pathway for cation penetration (valinomycin in the case of K+) and H+ exit (uncoupler, CCP in Fig. 2A). In ammonium acetate, swelling is rapid since NH$_3$ and HOAc provide a permeant, self-neutralizing solute pair which rapidly builds up NH$_4$+ and acetate$^-$ in the interior. Swelling in nitrate or thiocyanate salts depends on diffusion of NO$_3^-$ into the matrix and the electrophoretic permeability of the membrane to cations. Thus the charged valinomycin-K+ complex produces a maximum rate of swelling in K+ nitrate (Fig. 2B) whereas the mobile cation for H+ exchanger nigericin promotes swelling in K+ nitrate only when a source of protons in the matrix is supplied by addition of a proton-carrying uncoupler (CCP).

Energy-Dependent Swelling - In contrast to the results for passive swelling, mitochondria suspended in 0.1M acetate or phosphate salts of K+ and other cations take up ions and swell extensively when the mitochondria are energized by addition of substrate or ATP (Fig. 2C). The reaction occurs in the absence of reagents which modify permeability, is sensitive to uncouplers, inhibitors of respiration, and to oligomycin when exogenous ATP is used as the energy source. The rate of swelling and ion uptake is proportional to respiration rate and to the number of phosphorylation sites

7

traversed by electron transport [13]. In addition to K+,
all cations tested including Mg++, choline+, Tris+, tetra-
methylammonium+ (TMA+), and other large cations, all of
which appear to be impermeable in the absence of energy,
are capable of supporting swelling and are accumulated in
the presence of respiration. Accumulation of K+ is more
efficient than a cation such as TMA+, however, from the
standpoint of O_2 or ATP utilization. Ion accumulation requires the presence of either a weak acid, such as acetate,
or phosphate as the anion. Phosphate accumulation appears
to involve the phosphate transporter.

Sucrose and other non-penetrating solutes inhibit the
swelling reaction, but apparently only by providing an
osmotic pressure in opposition to the energy dependent salt
accumulation. In media which contain sucrose, therefore,
significant ion uptake and swelling are obtained only when
the reaction is magnified by addition of an ionophore, such
as valinomycin, or some other modification of the membrane
which serves to increase the energy-dependent ion accumulation (Fig. 2D). In the presence of sucrose, passive contraction and extrusion of accumulated ions due to the osmotic
pressure of the non-permeant disaccharide occurs when the
energy supply is interupted (Fig. 2D), and the rate of
contraction under these conditions can be taken as an
indication of the passive permeability of the membrane to
the accumulated ions [6]. In addition to this passive,
osmotic contraction, numerous cases of energy-dependent
contraction and extrusion of ions have been reported, and
an example of this phenomenon is reproduced in Fig. 2E.
In this case a continued supply of energy is necessary as
accumulated ions are extruded against an apparent concentration gradient. Na+ is more effective than K+ in supporting
energy-dependent contraction and ions such as TMA+ and cho-
line+ are completely unreactive. Permeant anions are also
required for this reaction and follow the Hofmeister
series in their effectiveness [7].

DISCUSSION

Possible Mechanisms for Monovalent Cation Uptake - The
availability of the antibiotics valinomycin, gramicidin,
and nigericin provides us with natural models for the
behavior of an electrophoretic cation carrier, a cation-
selective pore, and a mobile cation exchanger, respectively.
The properties of each of these compounds have been established in a number of natural and artificial membranes
(see [5,14,15] for reviews, for example) and it would seem
that a comparison of these properties with those of mitochondrial monovalent cation movement may permit a choice as

to the mechanism of the cation transport in unmodified mitochondrial membranes.

The passive swelling studies (Fig. 2) show no indication
of an overt K+ carrier of the valinomycin type present in
the non-energized membrane (since no spontaneous swelling
occurs in K+ salts of permeant anions). These data also
show that, if K+/H occurs in analogy to the observed Na+/H+
exchange, this exchange does not occur in non-energized mitochondria to an extent sufficient to support osmotic swelling
in K+ acetate.

The studies summarized here have established that K+
gains access to the matrix with relative ease in energized
mitochondria, however. In cases in which a strong driving
force is provided (i.e. a high concentration of acetate or
phosphate) the energized entrance of K+ is rapid and does
not require the addition of an exogenous ionophore. The
question arises as to exactly how this rapid entrance of K+
under these conditions is accomplished. Is the entrance of
K+ mediated by an endogenous carrier, in analogy to the
uptake of Ca^{+2} (Fig. 1)? If there is a K+ carrier, does it
provide an electrophoretic pathway, an exchange pathway, or
does it promote electrogenic movement of K+ by a K+ pump?
If such a carrier is present, why is it cryptic under
passive conditions? If K+ entrance is not carrier-mediated,
how does K+ traverse the membrane and why is the uptake
favored to such a large extent in the energized mitochondrion? These and other questions cannot be resolved with
certainty at present, but some of the arguments can be
summarized using the diagrams shown in Fig. 3.

Model (A)- Passive cation distribution in response to pH-
dependent anion gradients [2,3,6,16]. In the model shown
in Fig. 3A the primary event is the formation of a respiration-dependent pH gradient (either directly by the electron
transport system or by a proton pump). This is followed by
the exchange of anions for internal OH⁻ on a series of inter-
locked anion carriers which establishes a negative charge
in the matrix compartment, or alteratively by penetration of
a free acid (such as HOAc), which is formally equivalent to
A^-/OH^- exchange and which produces the same elevated concentration of anions and negative charge in the interior. The
final event in this model is an electrophoretic distribution
of cations past the permeability barrier to compensate this
negative charge. It is implicit in this model that the
process is limited by the low permeability of the membrane
to cation under normal conditions. When the interior potential becomes large enough, (as in 0.1 M acetate or phosphate media) the permeability barrier to cation penetration

9

will be overcome and these reactions will produce sufficient osmotic pressure to cause swelling of the matrix.

Model (B)- Electrogenic cation accumulation driven by a cation pump which utilizes an energy-conserving intermediate (or condition). Passive distribution of anions occurs in this model in response to the positive interior potential [17].

Model (C)- The third general class of reactions suggests that neither an electogenic H+ nor an electrogenic cation pump is involved, but rather an electroneutral cation for H+ exchanger which ejects H+ and takes up cations. This model suggests that metabolism alters the affinity of the carrier for H+ and thereby drives the reaction [7].

The energy-dependent swelling data are readily explained by the anion gradient model (A in Fig. 3). Since this model requires no specific transporter or exchanger for cations, it predicts that cations other than K+ (such as TMA+) would support swelling, provided that a sufficient membrane potential were available to overcome the limited electrophoretic permeability which is normally encountered. In Mechanisms B and C of Fig. 3, and any other carrier-mediated mechanism, the cation exchanger or pump would have to have an exceptionally broad specificity in order to accomodate unusual cations, such as choline+ and TMA+, and to account for the results. Cation pump models (Fig. 3B) predict that swelling should be favored by permeant anions rather than by permeant acids, and this is not the case [6,8]. In line with the data, however, mechanism (A) and (C) both predict that a permeant acid (HOAc) would favor the swelling reaction. In addition to this swelling data, which is compatible with the Model in Fig. 3A, studies of ion uptake and uncoupling in submitochondrial particles (SMP) are best explained by an electrogenic extrusion of H+ into the interior of the particle [14,18], which has polarity opposite to that of the intact mitochondrion. Thus, in a K+ medium, SMP are uncoupled by nigericin + valinomycin, but not by nigericin alone, and in an ammonium medium, SMP are uncoupled by the combination of NH_3 penetration and valino-mycin-induced loss of interior NH_4^+. These observations favor electrophoretic cation movement (mechanism A) and argue against electroneutral exchange of cations for H+ (mechanism C of Fig. 3).

One of the most fundamental points in favor of Mechanism (A) would appear to be the generalization that reagents which increase the passive permeability of the membrane to cations also markedly activate the energy-dependent ion accumulation reaction [6,8]. It is clear from the model in

A. Electrogenic H⁺ Pump

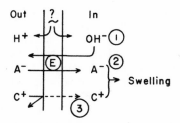

B. Electrogenic Cation Pump

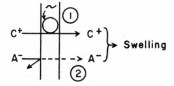

C. Electroneutral Cation H⁺/Exchange

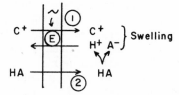

Fig. 3. Three general models for the mechanism of mitochondrial ion accumulation. In (A) an electrogenic H⁺ pump (or electron transport directly as suggested by the Chemiosmotic Model) provides the driving force. Anion exchangers (designated E) convert the pH gradient into a membrane potential (negative interior) and electrophoretic cation flow follows when sufficient driving potential builds up. Penetration of HA is formally equivalent to the A⁻/OH⁻ exchange shown. In (B) an electrogenic cation pump utilizes the high-energy intermediate (or condition) to establish a membrane potential (interior positive) and any net anion uptake is an electrophoretic response to this potential. In (C) energy is used to activate an electroneutral cation/H⁺ exchanger which creates a trans-membrane pH gradient. Penetration of HA or OH⁻ for A⁻ exchange will then produce net accumulation of anions as shown.

11

Fig. 3A that an increased permeability to cations would permit more rapid ion accumulation, so long as impermeability to H+, OH⁻, and the accumulated anion were maintained. Mechanisms (B) and (C) predict that increased passive permeability to K+ would favor leaking of K+ out of the matrix in opposition to a pump, be it electrogenic or electrically neutral. Reagents which promote the passive, transmembrane distribution of cations and which also activate the energy-dependent uptake of cations by mitochondria include: (a) cation-specific, charged, mobile ionophores, such as valinomycin, (b) pore-forming antibiotics, such as gramicidin, (c) certain heavy metals which appear to alter the protein portion of the membrane, such as Zn^{+2}, Pb^{+2}, Cu^{+2}, and Hg^{+2}, and (d) organic mercurial reagents which appear to modify specific thiol groups in the membrane, such as p-chloromercuriphenyl sulfonate. Removal of membrane Mg^{+2} with a chelator, such as EDTA, also promotes both passive ion flow and increased energy-dependent ion uptake.

The electrogenic H+ or anion-distribution model (Fig. 3A) can also be modified to take into account possible respiration-dependent conformational changes or a respiration-dependent change in membrane polarity which might contribute to the observed increase in permeability to monovalent cations seen in respiring mitochondria [6,8]. Such alterations would also help explain the increased effectiveness of certain heavy metals and mercurials in the induction of passive permeability when reacted with respiring as opposed to non-energized mitochondria [6,8], and the behavior of certain fluorescent probes and other ligands (as reviewed by Azzone and Massari [7]). It should be mentioned that none of the evidence which favors the electrophoretic equilibration of K+ into energized mitochondria has eliminated the possibility that the penetration of the cation is mediated by an endogenous ionophore. If such a component were present in the membrane it would have to be inactivated or otherwise masked under passive swelling conditions, accomodate a large variety of cations in the energized condition, and be highly specific for K+ in mercurial treated membranes to account for the experimental observations. It should be noted that "voltage gated" changes in permeability such as those seen with alamethicin [15] may provide a model for such a latent endogenous ionophore.

Evidence for Mitochondrial Cation for H+ Exchange - Whereas the arguments just presented are in line with the suggestion that monovalent cations distribute across the mitochondrial membrane in response to a primary H+ gradient and secondary anion gradients, much evidence suggests that a Na+/H+ and a much less effective K+/H+ exchange system may also be present. It is now well established that the introduction of a pulse of oxygen into a suspension of anaerobic mitochondria results in a cycle of proton ejection into the medium followed by re-equilibration of the pH. Mitchell [2] has interpreted this phenomenon as showing the production of H+ in the outer phase by the respiratory chain and the subsequent decay of the resulting pH gradient. However, a number of authors have noted that large H+ displacements are seen only under conditions in which the movement of Ca^{+2} or K+ can occur as well, and the observed H+ ejection could be visualized as an H+/cation+ antiport reaction.

An exchange of Na+ for H+ across the mitochondrial membrane was first postulated by Mitchell and Moyle [19] who observed that when K+ was replaced by Na+ there was an increased decay of the ΔpH produced by an O_2 pulse in anaerobic mitochondria. A lower activity of K+/H+ exchange was also suggested by the conductance data. Additional support for the existence of a Na+/H+ exchange activity came from studies of swelling in isotonic acetate salts in which it was observed that Na^+OAc supported rapid passive osmotic swelling and K+ did not (Fig. 2A). As just discussed, this phenomenon can be explained by penetration of HOAc into the matrix and subsequent exchange of Na+ in for H+ out, with the net accumulation of Na^+OAc^- in the matrix.

Cockrell [20] found indications that a K+/H+ exchange, which was much like nigericin, could support respiration-dependent cation uptake by SMP (A-particles). Douglas and Cockrell [21] compared swelling of rat liver mitochondria with ion uptake by SMP to establish the cation descrimination ratio (Na+:K+) of the endogenous cation/H+ exchanger. It was found that the reaction favored Na+ by as much as 50:1, was pH dependent, and both the flux and the cation descrimination were highly sensitive to Mg++.

Estrada-O [22] has also invoked a Na+>K+/H+ exchange to explain (a) the different ionic selectivity patterns obtained with the antibiotic beuvericin when mitochondrial membranes are compared to SMP and to phospholipid bilayers and (b) the substrate dependency of the ion selectivity in mitochondria.

Papa et al. [18] studying H+ uptake and extrusion following aeration of anaerobic submitochondrial particles found that the efflux of H+ could be resolved kinetically into two parallel first order processes. Of these, the fast reaction appeared to involve Na+/H+ exchange, showed

marked preference for Na+ over K+, and was inhibited by
Li+ and choline+ and activated by dibucaine. The slow
appearance of H+ was ascribed to H+ diffusion across the
membrane, showed no cation dependence, and was stimulated
by uncouplers. It was also concluded that the respiration-
dependent H+ movement (inward in SMP) is electrogenic and
does not require cation counter flow, and that the resulting
membrane potential is coupled to flow of permeant anions
and cations. These and subsequent observations [18] seem
to indicate that the primary event in energy-coupling is
associated with electrogenic H+ ejection, but that a
cation for H+ exchange system is also present and functions
under some experimental conditions.

At present it seems reasonably certain that some component in the mitochondrial membrane promotes the exchange
of cations for H+ across the membrane and that, as usually
measured, Na+ is greatly favored over K+ in this reaction.
The physiological significance, if any, of this exchange
system is completely unknown, however. The arguments just
presented above make it unlikely that such an exchanger
would participate in energy-linked cation uptake by mitochondria, but the possibility that it is involved in energy-
dependent contraction and ion extrusion seems attractive
[8]. A comparison of an osmotic mechanism for contraction
which utilizes the cation/H+ exchanger with a contractile
model is shown in Fig. 4. We feel that the available evidence favors such an osmotic mechanism for contraction
(see [23] also). It seems possible, therefore that both
electrophoretic cation permeability and cation+/H+
exchange may be involved in mitochondrial ion movements, but
that the cation uptake is electrophoretic (K+>Na+) whereas
cation extrusion occurs by the exchange mechanism (Na+>K+).
Alteration of the membrane can result in retention or amplification of a K+ discrimination over Na+, as in the case of
mercurial reaction [24] and certain heavy metal interactions
[6,8] or can result in a generalized increase in permeability to monovalent cations. It appears that many K+-spe-
cific responses, like the mercurial-dependent ATPase which
we have reported [24] have as their basis the induction of
a greatly preferential permeability to K+ over Na+. This
suggests that some endogenous component may provide a pathway for electrophoretic penetration of K+ and Na+ which
has the opposite cation discrimination to that shown by the
exchanger. Such a discrimination would be distinct from
that introduced by an exogenous ion carrier, such as valinomycin, and might correspond to the endogenous ionophores proposed by Southard et al. [25]. However, as will be dis-

A. Osmotic Contraction

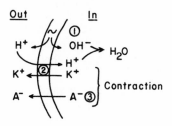

B. Contractile Mechanism

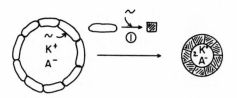

*Fig. 4. Osmotic vs. contractile mechanism for energy-dependent con-
traction of swollen mitochondria. The osmotic mechanism (A)
suggests that respiration produces a pH gradient (H⁺ outside),
that cation (out) for H⁺ (in) exchange occurs on the mitochondrial cation/H⁺ exchanger to produce water in the interior with
a decrease in net positive charge, and finally that electrophoretic anion egress results in osmotic contraction. This
mechanism assumes that the rate of cation extrusion on the exchanger is greater than the rate of cation leak into the matrix,
that electrophoretic H⁺ permeability is low, and that there is
sufficient permeability to internal anions to permit their
electrophoretic exit. The contractile mechanism (B) suggests
that a conformational change in membrane components occurs as a
result of energization and that cations, anions and water are
extruded as the membrane contracts. This mechanism assumes
that the membrane does not provide a permeability barrier to
the exit of either internal cations or anions.*

cussed below, cation discrimination has been observed in model systems which contain no specific transporting components and the issue of whether or not an endogneous ionophore for electrophoretic uptake of monovalent cations is present in the mitochondrial membrane is still open.

Endogenous ionophores for electrophoretic cation movement - The uptake of Ca^{+2} by mitochondria seems to be mediated by an endogenous electrophoretic carrier which has been characterized with respect to its kinetic properties, its inhibitor profile, and its distribution in comparative studies [26,27]. Considerable progress has been made in isolation of Ca^{+2}-binding components which exhibit properties consistent with these requirements of the endogenous component [28].

The situation with regard to the uptake of Mg^{+2} and monovalent cations is much less clear, however. As mentioned above the evidence that electrophoretic carriers for these cations are present is inconclusive.

Blondin and coworkers [25,29], however, have isolated an incompletely characterized peptide from heart mitochondria which induces swelling in NaCl and KCl and which they feel is a mitochondrial ionophore which promotes electrophoretic permeability of Na+ and K+ rather than Na+ or K+/H+ exchange. The material is obtained in reasonable amounts only after treatment with a mercurial (fluorescein mercuric acetate), is extracted from freeze dried mitochondria into ethanol, and purified by elution from silicic acid and alumina. An evaluation of the significance of this ionophore must await its further characterization, since the assays involved are rather non-specific and the identical responses can be obtained by mercurials alone (as we have documented in some detail, Ref. 24 and citations therein). It would seem that the appropriate inhibitor and kinetic criteria are not yet available to permit the distinction between the activity of a specific monovalent cation ionophore and a number of non-specific promotors of permeability (such as fatty acids or surfactants, for example).

In addition, studies with model systems indicate that considerable cation discrimination can be demonstrated in systems which contain no peptide ionophores. Cation discrimination and the self-diffusion rates for monovalent cations in phospholipid vesicles depend on the type of phospholipid involved, the degree of unsaturation of the side chains, and whether or not components such as cholesterol or proteins such as cytochrome c have been incorporated into the liposomes. The relationship of membrane proteins to the various possible lipid domains of the membrane also seems to affect permeability. For example, the Na+ permeability of liposomes seems to reach a maximum at the midpoint of a solid to liquid crystaline phase transition in liposomes [30]. It was suggested that boundary regions between the lipid domains were responsible for the permeability maximum at the transition temperature. It would appear that similar perturbations of membrane barrier function might result from chemical alterations of either the lipid or protein portions of the membrane, as well as temperature changes, and that the mercurial, heavy metal, and pH effects previously discussed could all have a common mechanism for increasing the permeability of the mitochondrial membrane [6,8]. In this regard, Trauble and Eibl [31] have concluded that small changes in the ionic environment (pH, monovalent, and divalent cations) can induce major alterations in the structure of lipid bilayers with monovalent cations stabilizing more fluid membranes.

These and other studies suggest that, if one views the membrane as a fluid mosiac [32], the options available for ion pathways across the membrane may be considerably more varied than those which are consistent with more rigid membrane models. The suggestion of Trauble [33] that mobile structural defects or "kinks" may permit rapid diffusion of small molecules across a membrane would be one such possibility. The simple diffusion of ions through the membrane in response to increased electrical potential or as a result of chemical modification of lipids or proteins would seem to be quite reasonable under these circumstances and would make invoking the presence of endogenous ionophores superfulous. It would seem that additional information on the temperature sensitivity, cation selectivity, and kinetic properties of electrophoretic cation permeability as opposed to cation/H+ exchange reactions would be of great value in the resolution of these questions. Such a program is currently under way in our laboratory.

Future Prospectives - Resolution of the mechanism of cation uptake will require additional information on the permeability of cations as a function of metabolic state. In order to resolve the question of whether or not a cryptic carrier participates in K+ or Na+ uptake (or Mg^{+2} for that matter) it appears that a precise evaluation of passive permeability coefficients will be necessary, as well as a complete kinetic characterization of ion uptake processes, both in the presence and the absence of energy. If the process is carrier-mediated, specific inhibitors should eventually show up. A similar characterization of the cation for H+ exchange reaction should be useful in

establishing its chemical nature, distribution with respect
to respiratory components, and possible inhibitor characteristics. Once the characterization of monovalent cation
movements in these terms has been completed, one could
logically proceed toward isolation of possible carriers.

At present it seems likely that monovalent cations
have both an electrophoretic (K+>Na+) and an exchange pathway (Na+>K+) available in the mitochondrial membrane. It
also seems likely that the electrophoretic pathway is the
more important in ion uptake and swelling reactions and that
the cation movement is in response to a primary electrogenic
H+ extrusion and secondary anion gradients. There is no
clear cut evidence that electrophoretic movement of K+ or
Na+ is carrier-mediated and experiments with model systems
suggest that considerable cation discrimination would be
possible in the absence of specific transporters. The
exchange pathway seems more likely to be involved in energy-
dependent contraction of swollen mitochondria and this
reaction does seem to require that an endogenous ionophore
of the nigericin type be postulated. One might speculate
that the two opposing processes of electrophoretic cation
uptake with osmotic swelling and exchange-dependent cation
extrusion and contraction might represent a volume control
mechanism for the mitochondrion within the cell which would
permit the organelle to maintain its functional integrity in
the presence of changes in ionic environment. To test this
postulate, further information as to cation permeability of
mitochondria in situ must be obtained. Possible control and
regulation of mitochondrial reactions by divalent and monovalent cations can be implicated by in vitro experiments
with purified enzyme systems, and would then require verifi-
cation at a higher level of organization.

REFERENCES

1. Klingenberg, M. (1970) Essays in Biochem. 6, 119-159.
2. Mitchell, P. (1968) Chemiosmotic Coupling and Energy
   Transduction, Glynn Res. Ltd. Bodmin, Cornwall.
3. Greville, G.D. (1969) Cur. Topics Bioenergetics 3, 1-78.
4. Lehninger, A.L., Carafoli, E., and Rossi, C.S. (1967)
   Adv. Enzymol. 29, 259-320.
5. Pressman, B.C. (1970) in Racker, E. (Editor) Membranes
   of Mitochondria and Chloroplasts, Van Nostrand, N.Y.,
   p. 213-250.
6. Brierley, G.P. (1974) Ann. N.Y. Acad. Sci. 227, 398-411.
7. Azzone, G.F. and Massari, S. (1973) Biochim. Biophys.
   Acta 301, 195-226.

8. Brierley, G.P. (1975) Molecular Cellular Biochem., in press.
9. Johnson, J. and Pressman, B.C. (1969) Arch. Biochem. Biophys. 132, 139-145.
10. Chappell, J.B. and Crofts, A.R. (1966) in Tager, J.M., Papa, S., Quagliariello, E., and Slater, E.C. (Editors) Regulation of Metabolic Processes in Mitochondria, Elsevier, Amsterdam, p. 293-316.
11. Meijer, A.J. and Van Dam, K. (1974) Biochim. Biophys. Acta 346, 213-244.
12. Lehninger, A.L. (1972) in Schultz, J. (Editor) Molecular Basis of Electron Transport, Academic Press, N.Y., p. 133-146.
13. Brierley, G.P., Jurkowitz, M., Merola, A.J., and Scott, K.M., (1972) Arch. Biochem. Biophys. 152, 744-754.
14. Chance, B. and Montal, M., (1971) Cur. Topics Membranes and Transport 2, 99-156.
15. Mueller, P. and Rudin, D.O., (1969) Cur. Topics Bioenergetics 3, 157-249.
16. Rottenberg, H., (1972) J. Membrane Biol. 11, 117-137.
17. Harris, E.J. and Pressman, B.C., (1969) Biochim. Biophys. Acta 172, 66-70.
18. Papa, S., Guerrieri, F., and Lorusso, M., (1974) in Ernster, L., Estabrook, R.W., and Slater, E.C. (Editors) Dynamics of Energy-Transducing Membranes, Elsevier, Amsterdam, p. 417-432.
19. Mitchell, P. and Moyle, J. (1967) Biochem. J. 105, 1147-1162.
20. Cockrell, R.S. (1973) J. Biol. Chem. 248, 6828-6833.
21. Douglass, M.G. and Cockrell, R.S. (1974) J. Biol. Chem. 249, 5464-5471.
22. Estrada-O, S. (1974) in Estrada-O, S. and Gitler, C. (Editors) Perspectives in Membrane Biology, Academic Press, N.Y., p. 281-302.
23. Nicholls, D.G. (1974) Eur. J. Biochem. 240, 2729-2748.
24. Brierley, G.P., Jurkowitz, M, and Scott, K.M. (1973) Arch. Biochem. Biophys. 159, 742-756.
25. Southard, J.H., Blondin, G.A., and Green, D.E. (1974) J. Biol. Chem. 249, 678-681.
26. Lehninger, A.L. (1974) Circulation Res. Supl. III, 34, 83-88.
27. Carafoli, E. (1973) Biochimie 55, 755-762.
28. Carafoli, E. and Sottocasa, G.L. (1974) in Ernster, L., Estabrook, R.W. and Slater, E.C. (Editors) Dynamics of Energy-Transducing Membranes, Elsevier, Amsterdam, p. 455-469.

29. Blondin, G.A. (1974) Ann. N.Y. Acad. Sci. 227, 392-397.
30. Papahadjopoulos, D., Jacobson, K., Nir, S., and Isoe, T. (1973) Biochim. Biophys. Acta 311, 330-348.
31. Trauble, H. and Eible, H. (1974) Proc. Natl. Acad. Sci. U.S. 71, 214-218.
32. Singer, S.J. (1971) in Rothfield, L.I. Structure and Function of Biological Membranes, Academic Press, N.Y., p. 145-222.
33. Trauble, H. (1971) J. Membrane Biol. 4, 193-208.

THE PROTON PUMP IN THE YEAST CELL MEMBRANE

Antonio Peña

INTRODUCTION

The studies on K^+ transport in yeast were initiated very long ago (1,2) and it was by the experiments of Conway and Duggan (3) that the existence of a carrier-mediated transport was first postulated. Studies of the phenomenon by the same group of Conway and that of Rothstein (cf. 1,2) described a system in which K^+ was taken up by yeast in an exchange for H^+ which was produced during the metabolism of the cell. It was evident from the beginning that K^+ ions had to enter the cell against a concentration gradient, and even more, if they had to be exchanged by H^+, it was most times necessary to expend energy to expulse H^+ from the cell also. In summary, the process of K^+ uptake was one which required a substantial amount of energy to be driven. Conway and Kernan postulated the existence of a transport system in which protons were generated by a carrier connected to the metabolism of the cell by means of the redox carriers (4). This system was supposedly capable of expulsing protons, and at the same time could carry out the exchange for K^+. However, most of the evidence for this scheme was indirect, and based mainly on the effects of several redox dyes on both K^+ transport and the redox potential of yeast cells measured directly with a platinum electrode. This scheme represented the first attempt to give a logical explanation to the transport of K^+ into the cell, and also to the connection required with the metabolism of the cell to provide for the energy required in this endergonic process. This model in fact was partly responsible for the elaboration of the elements of Mitchell's chemiosmotic hypothesis (5).

The transport of cations in many other cells, as observed at the level of the plasma membrane, seems to be based on the existence of a Na^+ pump, which by using the energy of ATP can expulse Na^+ ions and take up K^+. So far no example has been described of a eukariotic cell in which a H^+ pump has been described in detail in the plasma membrane. Proton pumps have been described in mitochondria (6-9), bacteria (10) and chloroplasts (11). Yeast cells seem an interesting subject, due to the fact that they seem to possess a proton pump in the plasma membrane, in contrast to the Na^+ pump commonly found in other eukariotic cells. The following is an attempt to give a summary of the data that support the existence of such a translocator in yeast.

The Effects of K^+ on Yeast Metabolism: Our studies were
started by the analysis of a phenomenon previously reported:
the acceleration of glycolysis by the presence of K^+ in the
incubation medium of yeast described by Lasnitski and Szorenyi (12) and further studied by Rothstein and Demis (13).
Our first studies (14,15) were directed to the analysis of
the relationship between metabolic rate and the presence of
K^+ in the external medium of the cell. The presence of K^+
in the medium could modify metabolism essentially by one of
three reasons: first, the cation could be taken up and
modify the activity of enzymes or enzyme systems sensitive
to the cation; second, it could modify the metabolic rate
because of the energy required for its uptake, and third, it
could alter metabolism by virtue of secondary effects resultant from the entrance of K^+; the changes of pH resultant
from the expulsion of protons, for instance.

The studies carried out showed that the instantaneous
addition of K^+ to the incubation medium at low pH (4.0)
with glucose as substrate, produced a series of changes in
the levels of the intermediates of glycolysis (15) that at
the end could be explained by a decrease of the levels of
ATP, which was accompanied by the increase of the levels of
ADP and Pi. The studies carried out in aerobiosis, with
ethanol as substrate showed the same type of changes in the
levels of ADP, ATP and Pi, which could mean that the addition
of K^+ produces the immediate stimulation of an ATPase activity (the shortest period for the measurement of the levels
of adenine nucleotides after the addition of K^+ was 30 sec).
This increase could explain the increased rate of both
respiration and fermentation that the addition of K^+ produces in yeast (14). The facts besides agreed with the
hypothesis that K^+ uptake is an endergonic process, especially at low pH values in which protons also have to be
expulsed against a concentration gradient. The link between
transport and metabolism however, seemed to be ATP. Experiments were carried out in which the effect of K^+ was measured also on the levels of NADH. If the K^+ transport
system was driven by redox power, the levels of reduced
carriers should decrease by the addition of K^+ to the external medium. The experiments with glucose as substrate (14)
showed that the addition of K^+ not only did not produce a
decreased level of NADH, but an immediate increase was observed. With ethanol as substrate (15), the addition of K^+
to the medium did not produce any immediate change in the
levels of NADH. The experimental data then are more in
agreement with the existence of an ATPase as the link between metabolism and transport than with a redox system.

Many studies have been carried out in which it has been shown that the uptake of K^+ in yeast is accompanied by the expulsion of the same amount of H^+ (cf. 1,2). Thus it seems that the uptake of K^+ occurs as a H^+/K^+ exchange. The mechanism of this exchange as proposed by Conway's group (4,16) is more or less explained in the scheme of Fig. 1. Two essential points emerge from this diagram: first, the system is driven by redox power, and second, it is one molecule; the one which carries out the exchange, i.e., the translocation of both H^+ and K^+. Since the studies by Rothstein and Demis on the effects of K^+ on yeast fermentation, it became clear that the stimulation of fermentation that occurred at the low pH values did not take place at high pH values, and that high pH values per se produced an important stimulation of fermentation. Our studies (15) confirmed these data and showed that the same is true for respiratory rate with ethanol as substrate. Since the effect on metabolic rate produced by K^+ could not be produced at high pH, the possibility existed that both K^+ and the high pH had a common site of action, or that a high pH could block the effect of K^+. These facts prompted us to the realization of the following experiments.

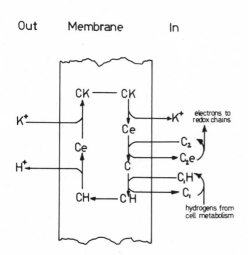

Out   Membrane   In

K^+

H^+

CK —— CK
Ce
K^+ electrons to redox chains
Ce
C
C_2
C_2e
C_1H
C_1
CH ←— CH
hydrogens from cell metabolism

Fig. 1. Schematic representation of the conception of Conway et al. (17) of the K^+/H^+ exchange mechanism in yeast.

Effects of pH on Fermentation and Respiration in Yeast: The studies mentioned before showed that the increase of the pH of the incubation medium produced an increase in the rate of fermentation as reported by Rothstein and Demis (cf.13) and a significantly increased rate of respiration as well. The analysis of the changes of the glycolytic intermediaries (17) produced by the sudden increase of the pH of the incubation medium showed approximately the same type of changes observed upon the addition of K^+. In these experiments it was found that the changes observed in the levels of the glycolytic intermediates could be explained by the stimulation of an ATPase activity which produced an increased level of both

23

ADP and Pi and a decrease of the ATP concentration.

On the other hand, yeast incubated at low pH values (4.0) does not produce any change in the pH of the medium unless K^+ is added. When incubated at high pH values yeast cells produce an important amount of acid even in the absence of K^+. This could be explained if the cells had a proton pump that could work at high pH values independently of the presence of cations which exchange the H^+ ions. This proton pump could be inhibited by high H^+ concentrations of the medium and under these conditions it could act only if there was a cation to exchange the protons. At the high pH values in the absence of K^+ the pump might expulse protons accompanied by anions. In fact Conway and Brady (18) reported that the expulsion of some anions, succinate for instance, is decreased when there is K^+ in the medium of incubation. Other experiments showed that when yeast cells are incubated at high pH values there is no change in the levels of ATP, ADP and Pi upon the addition of K^+ (cf. 15) and vice versa when the pH of the medium is increased suddenly from 4.0 to 7.5 and K^+ has been added previously to the medium, no change of the levels of the same intermediates takes place (cf. 17). Even if there is an indirect mechanism involved, both the presence of K^+ and the high pH value produce the same effects on metabolism on one hand and both conditions give rise to the expulsion of protons on the other. These data are also in agreement with the idea of the existence of a proton pump in the yeast cell and seem to indicate that it is a system that can function independently of K^+ transport.

Measurements were made of the changes of the internal pH of the cells under different conditions to find out if the protons expulsed actually came from the inside of the cell. The results showed (cf. 17) that the increased pH of the medium, in a process which required a substrate, actually produced an increased pH of the interior of the cell whether in the presence or in the absence of K^+ in the medium. As previously reported by Conway and Brady (cf. 18), the sole addition of K^+ to the incubation medium at low pH also produced an increase of internal pH. Up to here the results point to two main characteristics of the K^+ transport system; first, it seems that in the case of the plasma membrane of the yeast cell there exists an ATPase system which functions as a proton pump and this system can be inhibited by large concentrations of H^+ in the medium but it can be stimulated by the presence of K^+. At low pH it seems to be coupled to, but independent of, the uptake of K^+ since at high pH values the efflux of H^+ can work in the absence of K^+. These conclusions make our model different from the one proposed by Conway, both in the direct energy source from the cell and the fact that we postulate the independence of the proton pump and the K^+ transport system (Fig. 2).

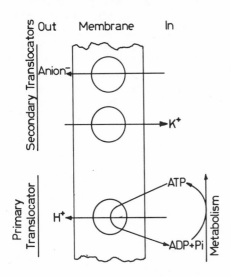

Fig. 2. Our actual conception of the K^+/H^+ exchange systems in yeast.

Studies with Inhibitors of K^+ Transport: Pressman in 1963 (19) found that alkylguanidines inhibit respiration and phosphorylation linked to site I in mitochondria without finding an explanation to the data. Hille in 1971 (20) pointed out certain similarities of guanidine to Na^+ in relation with the Na^+ channel in the nerve. Since then several reports have appeared that seem to explain the effects of guanidine derivatives on the basis of certain similarities of the guanido group with different cations (21,22,23). In yeast we thought that guanidine could be used to study the relationship between K^+ and H^+ movements; it was to be expected that guanidine and alkylguanidines would inhibit K^+ transport and it was then left to see if H^+ efflux was also affected. The experiments showed that guanidine in fact inhibits K^+ transport but at rather high concentrations (40 mM); octylguanidine on the other hand produces an inhibition at much lower concentrations. In a kinetic analysis of the inhibition it was found that alkylguanidines inhibit K^+ transport and the inhibition is of the competitive type (24). When the effect of the alkylguanidines was measured on the ability of yeast cells to expulse protons, no change in this activity was observed (24). These results seem to indicate that an inhibitor exists that can block completely the K^+ transport activity at high concentrations without affecting the expulsion of protons. With ethidium bromide, which in some aspects behaves as a hydrophobic cation (25) and inhibits the uptake of K^+ by yeast, the effect on the expulsion of protons was not one of inhibition but stimulated it to some extent. Ethidium produces an inhibition that perhaps is of a more pure competitive type on on the uptake of K^+ than that of alkylguanidines and has no

inhibitory effect on the expulsion of protons. Ethidium
seems to be taken up by the cells instead of K^+ (cf. 25).

The inhibition of K^+ uptake by cationic molecules was
further explored by the use of a series of cationic dyes.
As shown in Fig. 3, this inhibition can be produced with all
the molecules employed and also with the substances employed,
practically no change in the rate of H^+ expulsion was observed. This significant degree of inhibition of K^+ uptake without change of H^+ expulsion rates would be explained with
difficulty if it was a single molecule carrier, the one involved in the movements of both ions across the yeast cell
membrane.

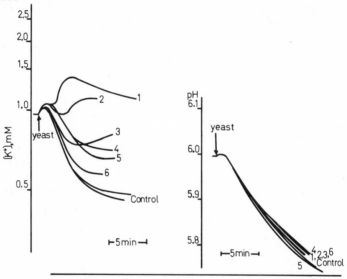

*Fig. 3. Effect of different cationic dyes on K^+ uptake and
H^+ expulsion in yeast. The concentrations of K^+ and H^+ were
measured with electrodes (27). The incubation mixture was
the following: 10 mM maleate-triethanolamine buffer, pH 6.0;
50 mM glucose; 1 mM KCL; yeast 250 mg net weight; final
volume 10.0 ml; temperature 30°. To this mixture it was
added: curve 1, 100 µM methylene blue; 2, 50 µM acriflavine;
3, 50 µM neutral red; 4, 50 µM methyl violet; 5, 100 µM
brilliant cresyl blue; 6, 100 µM ethidium bromide.*

Another approach to study the relative inhibitions of both
K^+ and H^+ movements consisted of employing alkylguanidines in
which the carbon chain was varied in length. It was found
(Fig. 4) that the longer the alkyl chain the greater the
effect on K^+ transport. The relative concentrations required to obtain similar inhibitions on K^+ uptake show

that the potency of the alkylguanidines increases approximately four fold for each two carbon atoms added. The data seem to indicate that the cationic head occupies the site of the cation and the alkyl chain acts as an anchor in the hydrophobic portion of the membrane. In spite of the practically total inhibition of K^+ uptake by different concentrations of alkylguanidines, no effect could be shown on H^+ expulsion which agrees with the independence of the system for H^+ expulsion and K^+ uptake.

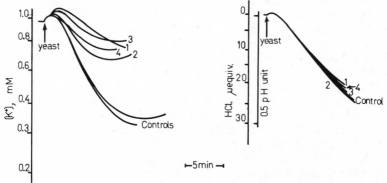

Fig. 4. *Effect of the length of the side chain of alkylguanidines on the inhibition of K^+ uptake and H^+ expulsion in yeast. Experimental conditions were the same as for Fig. 3, but tracings 1 to 4 were obtained in the presence of 20 mM ethylguanidine, 4 mM butylguanidine, 1 mM hexylguanidine and 0.25 mM octylguanidine, respectively.*

Effects of Uncouplers: Uncouplers have long been known as inhibitors of K^+ uptake in yeast (26). In principle, according to Mitchell's hypothesis, this would agree perfectly with the conception of transport systems driven by proton gradients. However, it was necessary to show in yeast that uncouplers do not act directly on the H^+/K^+ exchange system but that they can act as proposed (as proton conductors) and by that mechanism alter the coupling between the proton gradients and the transport of K^+. In a series of experiments carried out it was shown that uncouplers actually can increase the influx of protons previously expulsed by the cells (27). Furthermore, it was shown that in the absence of energy under adequate conditions of pH and K^+ concentration gradients, uncouplers can stimulate instead of inhibiting the efflux of protons and the uptake of $^{86}Rb^+$ as long as both gradients and not only one are favorable to the movements. This was taken to mean that there is a coupling between both H^+ and K^+ movements but passive movements of

27

H^+, i.e., apparently independent from the proton pump, can produce the uptake of K^+ ($^{86}Rb^+$), because this is a system coupled to but independent of the proton pump.

DISCUSSION

The experiments here revised relative to the movements of H^+ and K^+ in yeast are relevant because they show the existence of a proton pump in the plasma membrane which seems to be the basis for the transport of at least a few other ions. The data seem to be clear to indicate that there exists a coupling between H^+ and K^+ fluxes and both systems probably are independent from one another. The studies by other workers have shown that phosphate, for instance, is transported faster if K^+ with a substrate has been included previously in the incubation medium (28). In mitochondria a system has been postulated for the uptake of Pi according to which this anion is accumulated by virtue of an exchange for OH^- ions (29). It would not be surprising that the uptake of Pi in yeast occurred according to the same or a similar mechanism. One important support of this view is the fact that monovalent organic acids (acetic, propionic, butyric) inhibit the uptake of Pi (30,31) and it is known that these acids penetrate in the undissociated form into the yeast cell (32) and can thus acidify the interior of the cell causing the decrease of OH^- available for the exchange of Pi. The uptake of K^+ in an exchange for H^+ actually gives rise to an increase of the internal pH (cf. 18) which would favor the uptake of Pi according to the same mechanism as in mitochondria. Monocarboxylic acids, like acetic, distribute across the yeast membrane according to the pH differences between the outside and the inside of the cell; the uptake of K^+ which results in the increase of internal pH and decrease of external pH also favors the accumulation of monovalent acids (31,32). As a further consequence of this, the presence of organic acids, by reversing the increase of the internal pH produced during the uptake of K^+, can help to allow the uptake of larger amounts of the monovalent cation (27,34). In the same sense point the results of Eddy and Nowacki (35) who have demonstrated some relationship between the uptake of K^+ and that of several amino acids.

Although the results just mentioned do not permit establishment of direct links between the uptake of K^+ and that of many other ions nor would allow conclusions as to their mechanisms, they could be good indications to support the dependence of these systems on the uptake of K^+. Since K^+ requires the operation of a H^+ pump to be taken up by the yeast cell, this means that this might be the primary driving force to the uptake of a much larger number of ions besides K^+.

The proton pump in the yeast plasma membrane, as it is in mitochondria, could be conceived as the main primary translocator capable of creating the electrochemical potential necessary for the uptake of numerous other ions by the further action of the basic type of symport and antiport systems described by Mitchell (cf. 15).

REFERENCES

1. Conway, E.J. and O'Malley, E. (1946) Biochem. J. 40, 59.
2. Rothstein, A. and Enns, L.H. (1946) J. Cell. Comp. Physiol. 28, 231.
3. Conway, E.J. and Duggan, F. (1958) Biochem. J. 69, 265.
4. Conway, E.J. and Kernan, R.P. (1955) Biochem. J. 61, 32.
5. Mitchell, P. (1966) Chemiosmotic Coupling and Energy Transduction, Glynn Research, Bodmin, Cornwall, England.
6. Mitchell, P. and Moyle, J. (1967) in Biochemistry of Mitochondria (Slater, E.C., Kaniuga, Z. and Wojtczac, L., eds.) pp. 53-91, Academic Press, New York.
7. Mitchell, P. and Moyle, J. (1965) Nature 208, 1205.
8. Greville, G.D. (1969) in Current Topics in Bioenergetics (Sanadi, R., ed.) pp. 1-77, Academic Press, New York.
9. Papa, S., Guerrieri, F. and Lorusso, M. (1974) in Dynamics of Energy Transducing Membranes (Estabrook, R.W. and Slater, E.C., eds.) pp. 417-432, Elsevier, Amsterdam.
10. Harold, F. (1972) Bacteriol. Revs. 36, 172.
11. Jagendorf, A.T. (1967) Fed. Proc. 20, 1361.
12. Lasnitski, A. and Szorenyi, E. (1935) Biochem. J. 29, 580.
13. Rothstein, A. and Demis, C. (1953) Arch. Biochem. Biophys. 44, 18.
14. Peña, A., Cinco, G., García, A., Gómez-Puyou, A. and Tuena Gómez-Puyou, M. (1967) Biochim. Biophys. Acta. 158, 673.
15. Peña, A., Cinco, G., Gómez-Puyou, A. and Tuena Gómez-Puyou, M. (1969) Biochim. Biophys. Acta. 180, 1.
16. Conway, E.J., Brady, T.G. and Carton, E. (1950) Biochem. J. 47, 369.
17. Peña, A., Cinco, G., Gómez-Puyou, A. and Tuena Gómez-Puyou, M. (1972) Arch. Biochem. Biophys. 153, 413.
18. Conway, E.J. and Brady, T.G. (1950) Biochem. J. 47, 360.

19. Pressman, B.C. (1963) J. Biol. Chem. 238, 401.
20. Hille, B. (1971) J. Gen. Physiol. 58, 599.
21. Papa, S., Tuena Gómez-Puyou, M. and Gómez-Puyou, A. (1975) Eur. J. Biochem., in press.
22. Estrada-O.S. (1974) in Perspectives in Membrane Biology (Estrada-O.S. and Gitler, C., eds.) pp. 281-302, Academic Press, New York.
23. Davidoff, F. (1973) New Eng. J. Med. 289, 141.
24. Peña, A. (1973) FEBS Letters 34, 117.
25. Peña, A. and Ramírez, G. (1975) J. Membrane Biol., in press.
26. Conway, E.J. (1955) Intern. Rev. Cytol. 4, 377.
27. Peña, A. (1975) Arch. Biochem. Biophys. 167, 397.
28. Goodman, J. and Rothstein, A. (1975) J. Gen. Physiol. 40, 915.
29. Chappell, J.B. and Crofts, A.R. (1966) in Regulation of Metabolic Processes in Mitochondria, BBA Library, Vol. 7, pp. 293-316 (Tager, J.M., Papa, S., Quagliariello, E. and Slater, E.C., eds.) Elsevier, Amsterdam.
30. Samson, F.E., Katz, A.M. and Harris, D.L. (1955) Arch. Biochem. Biophys. 54, 406.
31. Borst-Pauwels, G.W.F.H. and Dobbelmann, J. (1972) Biochim. Biophys. Acta. 290, 348.
32. Conway, E.J. and Downey, M. (1950) Biochem. J. 47, 347.
33. Conway, E.J. and Downey, M. (1950) Biochem. J. 47, 355.
34. Ryan, H., Ryan, J.P. and O'Connor, W.H. (1971) Biochem. J. 125, 1081.
35. Eddy, A.A. and Nowacky, J.A. (1971) Biochem. J. 122, 701.

KINETIC AND THERMODYNAMIC ASPECTS OF MITOCHONDRIAL CALCIUM TRANSPORT

A. Scarpa

INTRODUCTION

Ca^{++} transport by isolated mitochondria has been studied in great detail during the last three decades. It is well established that Ca^{++} accumulation by mitochondria is a energy dependent process in which the energy can be derived either from electron transport or ATP hydrolysis (1-2). The existence of a specific Ca^{++} carrier in mitochondria has also been suggested by binding studies (3) and by kinetic studies in the presence of inhibitors (4), but the validity of these data has recently been questioned (5-6). Since a comprehensive review of Ca^{++} transport in mitochondria appears in this book (7), the contents of this chapter will be mostly confined to two aspects of Ca^{++} uptake by mitochondria: a) the velocity and the affinity of mitochondrial Ca^{++} transport and b) the coupling between metabolic reaction of mito-chondria and the energy-consuming reaction of Ca^{++} transport. The clarification of these aspects is not only required for the elucidation of the mitochondrial Ca^{++} transport mechanism, but is at the core of understanding the physio-logical significance of mitochondrial Ca^{++} uptake in vivo and the overall regulation of intracellular Ca^{++}.

Kinetic Measurements of Ca^{++} Transport

Most of the contradiction and confusion existing in the literature on the velocity and affinity of mitochondrial Ca^{++} transport are probably attributable to the different techniques used to measure Ca^{++} transport and to the inter-pretation of the results obtained. The rate and the apparent K_m of the energy-dependent Ca^{++} uptake by isolated mito-chondria have been measured in several ways; by measuring $^{45}Ca^{++}$ distribution in mitochondria and supernatants (2,8) or by measuring events occurring during Ca^{++} accumulation such as increased rate of O_2 uptake or H^+ release (1), or the shift in the redox state of respiratory pigments (1,9). The most common technique, which under controlled conditions possess both sensitivity and reproducibility, is the measure-ment of $^{45}Ca^{++}$ distribution. However, this method does not permit to discriminate between Ca^{++} transport and Ca^{++} exchange and, in addition, is seldom applicable to kinetic measurement since it requires the separation of mitochondria

31

from the surrounding media. Measurements of shift in redox states of cytochromes or H^+ movements, which accompanies Ca^{++} uptake, are indirect events not always stoichiometrically coupled to Ca^{++} transport (10).

An alternative approach is the kinetic measurements of Ca^{++} disappearance from a reaction mixture containing mitochondria. Ca^{++} sensitive electrodes have intrinsic limitations in response times and in sensitivity, but sensitive measurement of ionized Ca^{++} concentration can be followed kinetically by recording the spectral changes of Ca^{++} indicators.

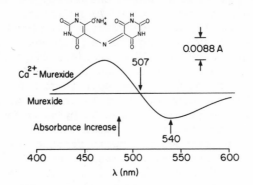

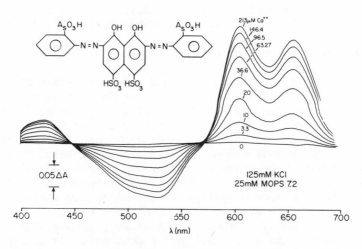

Fig. 1A and B. Differential spectra of murexide versus murexide and calcium (A) and of arsenazo III versus arsenazo III and calcium (B). The cuvettes contained 100 mM KCl, 10 mM Tris HCl (pH 7.2) and 50 μM indicator.

32

The differential absorbance spectra of two metallochromic
indicators that can be used to measure Ca^{++} transport by
mitochondria are shown in Figure 1. Figure 1A shows that
the addition of Ca^{++} to murexide (ammonium purpurate) produces a decrease in Δ_A at 540 nm and an increase in Δ_A at
470 nm with an isosbestic point at 507 nm. Murexide, which
has been extensively used as an indicator of Ca^{++} transients
across a variety of cells and cell fractions, has a fast
complex formation rate with Ca^{++} (τ = < 3 μsec), little
affinity for Ca^{++} (K_D = 4 mM), no binding to nor side effects
on functions of mitochondrial membranes [11].

Arsenazo III (2,2'-[1,8-dihydroxy-3,6-bisulfo-2,7-
naphtalene-bis (azo)] dibenzene arsonic acid) is a new
indicator of ionized Ca^{++} which is several times more
sensitive than murexide. The differential spectra of
arsenazo III and various amounts of Ca^{++} versus arsenazo
alone are shown in Figure 1B. Although various cations
produces spectral changes of arsenazo, at 675-685 nm changes
in $[Ca^{++}]$ can be measured without the interference of Mg^{++}
or other cations. The arsenazo-calcium complex has a Δ_ε of
21 (cm $\cdot$ moles)$^{-1}$, a K_D of 20 μM and a halftime for complex
formation of 2.8 msec [12]. With respect to murexide,
arsenazo has the advantage of a greater sensitivity and a
spectral change with Ca^{++} at around 700 nm, where the interference by cytochromes is minimal.

The absorbance changes of these indicators, which are
within certain ranges linear function of changes in ionized
Ca^{++}, are measured at two close λ pairs by differential
wavelength spectroscopy, so that aspecific absorbance changes
due to mitochondrial swelling or shrinking or change in
redox state of cytochromes can be minimized or abolished.
This can be achieved by a dual wavelength spectrophotometer
[13] or by a multifilter time sharing rotating wheel [14],
which both electronically subtract the absorbance of the
"measure" wavelength (absorption maximum of the dye-Ca^{++}
complex) from the absorbance of the "reference" wavelength
(isosbestic point or absorption minimum of dye-Ca^{++} complex).

Owing to their fast complex formation with Ca^{++}, these
indicators make possible the recording of the initial
velocities of Ca^{++} transport when rapid mixing devices [15]
are used to start the reaction. Figure 2 illustrates the
stopped flow apparatus connected with the double beam spectrophotometer used for the measurements of initial velocity
of Ca^{++} uptake by mitochondria. The large syringe of the
flow apparatus contained the reaction mixture together with
murexide and the mitochondria. The smaller syringe contained
$CaCl_2$. The plungers of both syringes were pneumatically

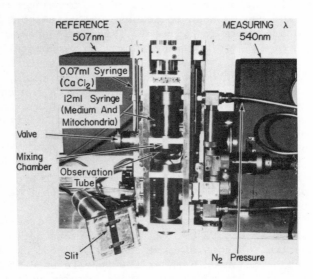

Fig. 2. Dual wavelength spectrophotometer and stopped flow apparatus used for the measurements of initial velocities of Ca^{++} uptake.

driven by high pressure N_2 and $CaCl_2$ could be mixed with the reaction mixture in less than 2 msec. The photomultiplier tube, located after the mixing chamber, observed the changes in absorbance of murexide at 540-507 nm from a few msecs to several minutes after mixing. One of the most important features of this stopped flow apparatus is the presence of two syringes of unequal volume. This proves to be of particular importance in optical measurements in the presence of turbid solutions and makes possible addition of Ca^{++} to reaction mixtures with negligible dilution of the system as a whole.

The Velocity and the Affinity of Mitochondrial Ca^{++} Transport

Figure 3 shows recording of an experiment measuring the rate of energy-dependent Ca^{++} uptake by rat liver mitochondria as obtained with a stopped flow apparatus and dual wavelength technique. Addition of 7.5 µM $CaCl_2$ to the reaction mixture containing murexide and mitochondria produced an abrupt decrease in absorbance at 540 nm, due to the formation of calcium-murexide complex with absorbs less than murexide alone. The formation of this complex was completed within the mixing time and was followed by a

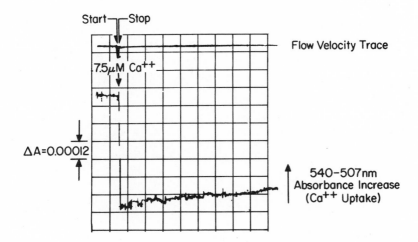

Fig. 3. The kinetic of the energy-dependent Ca^{++} uptake by rat liver mitochondria. The large syringe of the stop-ped flow apparatus contained 0.15 M sucrose, 0.075 M KCl, 2 mM K_2HPO_4, 3 mM MOPS (morpholinopropane sulfonate) (pH 7.2), 2 mM $MgCl_2$, 4 mM Na succinate, 3 μM rotenone, 40 μM murexide and 3.6 mg mitochondrial protein/ml. The reaction was started by the discharge of 7.5 μM $CaCl_2$ through the small syringe. Temperature 25°.

slower increase in absorbance related to the energy-dependent Ca^{++} uptake by mitochondria. Measurements at longer times (not shown) indicate that all the Ca^{++} added is taken up by mitochondria. In this way, the initial velocity of the energy-dependent Ca^{++} uptake by rat liver mitochondria was measured after addition of various Ca^{++} concentrations.

Figure 4 shows that the initial velocity of Ca^{++} uptake was low at low concentrations of Ca^{++} and increased in a sigmoidal fashion to 8 nmoles Ca^{++}/sec/mg protein at 200 μM Ca^{++}. The Ca^{++} concentration at which half-maximal activation of Ca^{++} uptake occurred was about 50 μM. Qualitatively similar results were obtained with other mammalian mitochondria and by varying the composition of the medium (protein concentration, presence of various permeable anions, etc.) (17).

Reliable values of K_m and rates of Ca^{++} transport by mitochondria are important for understanding the physiological role of mitochondrial Ca^{++} transport in vivo during contraction and secretion. In particular, these values are necessary to assess the proposed role of mitochondria as

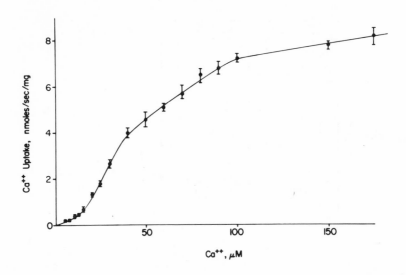

*Fig. 4. Initial velocities of Ca^{++} uptake by rat liver
mitochondria at different Ca^{++} concentrations. The
experimental conditions were similar to that of
Figure 3 except for the protein concentration which was
1.9 mg/ml. The reaction was started by discharging in
the reaction mixture the amounts of Ca^{++} indicated in
the abscissa. Initial velocity refers to the amount of
Ca^{++} taken up by mitochondria during the first 200 msec.
Every point represents the average of three different
experiments obtained within a few hours with the same
preparation of mitochondria. From Vinogradov and
Scarpa (16).*

controllers of the beat-to-beat Ca^{++} cycle of cardiac muscle,
by taking up and releasing during the cycle the amount of
Ca^{++} necessary for physiological needs (18,19).

Figure 5 compares the specific activity of Ca^{++} transport by sarcoplasmic reticulum and mitochondria isolated from
guinea pig hearts. Similar reaction mixtures were used for
both experiments and arsenazo III, which is a more sensitive
indicator of ionized Ca^{++}, was used for measuring Ca^{++}
uptake. Maximal rate of Ca^{++} transport by cardiac sarcoplasmic reticulum was already attained at 2.5 μM Ca^{++},
whereas only half-maximal rate of Ca^{++} transport by cardiac
mitochondria was observed at 50 μM Ca^{++}. Due to these
differences in apparent K_m values, the rate of Ca^{++} uptake

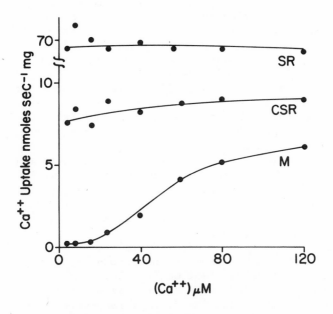

Fig. 5. Initial velocity of Ca^{++} uptake by mitochondria (M), by sarcoplasmic reticulum of guinea pig heart (CRS) and by sarcoplasmic reticulum from rabbit hind white skeletal muscle. Isolation procedures and experimental conditions are reported in reference 20 except that 30 μM arsenazo was used as a $[Ca^{++}]$ indicator. Protein concentrations were: M = 2.8 mg/ml; CRS = 3.1 mg/ml and SR = 1.1 mg/ml.

by cardiac mitochondria is significantly lower than that of cardiac sarcoplasmic reticulum, especially at Ca^{++} concentrations below 20 μM. Furthermore, the specific activity of Ca^{++} uptake by cardiac sarcoplasmic reticulum is probably an underestimated value, since sarcoplasmic reticulum isolated from cardiac muscle is notoriously impure, labile and less active (21). A better preparation, if available, should probably yield values close to that of sarcoplasmic reticulum from white skeletal muscle (top trace).

In our experimental conditions, 1 to 5 millimolar $MgCl_2$ was present in order to minimize Ca^{++} complex formation with components of the reaction mixture and Ca^{++} binding to aspecific binding sites in mitochondria. It has recently been reported that in the absence of Mg^{++} the apparent K_m for Ca^{++} uptake by cardiac mitochondria was shifted from 40 μM to around 5 μM. This low value for apparent K_m of Ca^{++} transport is still far too high with respect to the concentration of Ca^{++} free in cardiac cells to render the mitochondrial Ca^{++} transport effective in regulating the beat-to-beat Ca^{++} cycle of the heart. Furthermore, this low value may be of little physiological significance, due to the large cytosolic content of Mg^{++} in cardiac cells.

The high K_m values for Ca^{++} uptake and the effect due to cooperativity, make Ca^{++} uptake by mitochondria very inefficient at the cytosolic concentration of Ca^{++} (18,19). Based on the reported values of Ca^{++} uptake by mitochondria at 10 μM Ca^{++}, the amount of mitochondria present in mammalian cardiac tissue (80 mg/g wet tissue) and the average time of relaxation for mammalian heart (200 msec), it can be calculated that the mitochondria contained in 1 g of cardiac muscle can take up only 1 nmole Ca^{++} during the relaxation time of mammalian myocardium (17). Since this value is 1-2 orders of magnitude lower than the amount of Ca^{++} that should be sequestered in vivo to obtain relaxation (18,19), Ca^{++} uptake by mitochondria seems to be completely inadequate in controlling the beat-to-beat Ca^{++} cycle.

On the other hand, mitochondria seem to play a significant role in controlling the overall long-term intracellular Ca^{++} homeostasis and seem particularly effective in sequestering large amounts of Ca^{++} when the cytosolic Ca^{++} concentration rises in response to physiological or pathological events. This regulation of cytosolic Ca^{++} by mitochondria may be significantly different in other kinds of muscle such as smooth muscle and   frog heart, where the mitochondria have a higher affinity and capacity for Ca^{++} (17), and where the overall control of Ca^{++} homeostasis has different features.

Ca^{++} Transport and Energy-Coupling

Several models have been proposed in the literature to describe the coupling between energy-transducing metabolic reactions of mitochondria and the energy-utilizing reactions leading to the accumulation of Ca^{++} against concentration gradients. Two of the models which have generated much interest and controversy are the chemical intermediate hypothesis (22,23) and the chemiosmotic hypothesis (24), which can be simplified as follows:

(1)
$$\text{Respiratory chain} \rightleftarrows X \backsim I \rightleftarrows \text{ATP synthesis}$$
$$\text{Cation translocation}$$
$$\longleftrightarrow$$
$$H^+ \text{ translocation}$$

Respiratory chain $\rightleftharpoons$ H^+ translocation $\rightleftharpoons$ ATP synthesis

(2)

Cation translocation

An excellent review on these and other hypotheses has been written by Greville (25), and the relationship with ion transport has been discussed in detail by Chance and Montal (26) and by Lehninger (27).

According to the chemical intermediate hypothesis (scheme 1), the primary event is the formation of a high-energy intermediate, $X \smallsmile I$, which drives ATP synthesis and the Ca^{++} pump. Ca^{++} transport should be coupled with electron flow and the direction of the ion movement depends on the direction of the electron flow through the respiratory chain. In this model, H^+ movement is secondary to Ca^{++} translocation. Therefore, the mitochondrial membrane should be permeable to protons and impermeable to Ca^{++}, which will be transported by the operation of a specific carrier coupled to $X \smallsmile I$ and the respiratory chain.

In the chemiosmotic hypothesis (scheme 2) and its various modifications, the primary event is the translocation of H^+ across the mitochondrial membrane with consequent development of an electrochemical H^+ gradient (or proton motive force) consisting of two components: a concentration gradient (ΔpH) and a membrane potential attributable to charge transfer across the membrane ($\Delta \Psi$). It has been proposed that this gradient is responsible for an electro-phoretic cation movement across the mitochondrial membrane and that the intra-extramitochondrial cation distribution is an equilibrium distribution in which the cation concentration gradient is balanced by a membrane potential according to the Nerst equation. In the case of monovalent cation distribution, these predictions have recently been confirmed experimentally. The membrane of isolated mitochondria is impermeable to K^+ and Rb^+ but become permeable to these ions in the presence of valinomycin, an uncharged ionophore which produces electrophoretic transport of Rb^+ or K^+ across artificial or natural membranes (28). Under these conditions it has been shown that the distribution of intra-extra-mitochondrial (Rb^+) varies in the presence of various energy-coupling agents, and it has been proposed that high intra-mitochondrial Rb^+ gradients are held by the existence of a membrane potential, negative on the inside (29). Therefore, a comparison of Rb^+ distribution in the presence of valino-mycin, with Ca^{++} distribution under various metabolic conditions can provide information concerning the relation-ship between membrane potential and Ca^{++} transport.

Intra-extramitochondrial Ca^{++} and Rb^+ concentration ratios under steady-state conditions were obtained by measurements of Ca^{++} and Rb^+ in the mitochondria and in the surrounding medium by spectrophotometric and radiochemical techniques, and the matrix space of mitochondria was calculated as the space of ^3H_2O corrected for the non-matrix space as measured by $[^{14}C]$ sucrose (30). External Ca^{++} binding to aspecific binding sites in mitochondria was minimized by the presence of Mg^{++} and intra-mitochondrial binding and external precipitation was prevented in the presence of acetate. Ca^{++} distribution ratios were calculated from the experiments of Figure 6 where Ca^{++} uptake by mito-

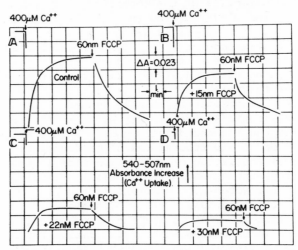

Fig. 6. *Steady-state levels of Ca^{++} uptake at various concentrations of uncoupler. The reaction mixture was composed of 200 mM sucrose, 5 mM sodium succinate, 1 μM rotenone, 3 mM $MgCl_2$; 1 mM sodium acetate, 10 mM MOPS (pH 7.2), 5 mM KCl, 0.5 μg/ml of valinomycin and 2.5 mg mitochondrial protein/ml. In B, C and D, the indicated concentrations of FCCP were added prior to the addition of calcium. From Rottenberg and Scarpa (30).*

chondria was measured by the changes in absorbance of murexide. The energy-dependent Ca^{++} release induced by the uncoupler FCCP are shown in Figure 6A. In Figure 6B, C and D, increasing concentrations of FCCP were added before the addition of Ca^{++}, resulting in decreased steady-state levels of Ca^{++} accumulation. The results of this experiment, and parallel radiochemical analysis of matrix water and ^{86}Rb distribution in mitochondrial pellets are summarized in

TABLE I

Comparison of the Intra-Extramitochondrial Distribution of Calcium and Rubidium

	$\dfrac{[^{86}Rb]_{in}}{[^{86}Rb]_{out}}$	Log Ratio	$\Delta\Psi^a$	$\dfrac{[Ca]_{in}}{[Ca]_{out}}$	Log Ratio	$\Delta\Psi^b$
Control	13.3	1.24	67	230	2.36	71
+15 nM FCCP	5.87	0.77	46	95.2	1.98	59
+22 nM FCCP	5.20	0.72	43	41.8	1.62	49
+30 nM FCCP	4.01	0.60	36	12.6	1.10	33

The reaction mixture was similar to that of Figure 6, except that ^{86}Rb and $^{3}H_2O$ and 400 μM calcium were added to the mitochondrial suspension before precipitation. Mitochondrial preparation and concentration were identical to that of the experiment of Figure 5. Determination of matrix water volume was carried out in a parallel experiment under the same experimental conditions in the presence of $[^{14}C]$ sucrose and $^{3}H_2O$. $\Delta\Psi^a = (RT/F)\ln([Rb]_{in}/[Rb]_{out})$. $\Delta\Psi^b = (RT/2F)\ln[Ca]_{in}/[Ca]_{out})$.

Table I. By increasing FCCP concentration, the intra-extramitochondrial Ca^{++} ratio falls from 230 to 12.6 whereas the rubidium ratio, which was always smaller, decreased from 13.3 to 4. The two ratios are approximately related by the equation: log $([Ca^{++}]_{in}/[Ca^{++}]_{out})$ = 2 log $([Rb^+]_{in}/[Rb^+]_{out})$. Under the assumption that, in the presence of valinomycin, the rubidium ratio is controlled by the membrane potential in accordance with the relationship $\Delta\Psi = (RT/ZF)\ln(C_{in}/C_{out})$, membrane potential can be calculated from Rb^+ distribution. The same relationship should hold true if calcium transport is a simple electrogenic process. Since in this case Z = 2, for the same membrane potential the log of the calcium ratio should be twice the log of the rubidium ratio. This prediction was verified experimentally. Table I shows the membrane potential as calculated from calcium distribution (in which Z = 2). This also indicates that the net charge transfer for Ca^{++} accumulated is 2.

The indication that Ca^{++} uptake by mitochondria is an electrophoretic process driven by membrane potential is also

consistent with the effect of various ionophores on mitochondrial Ca^{++} uptake and release. Figure 7 shows several

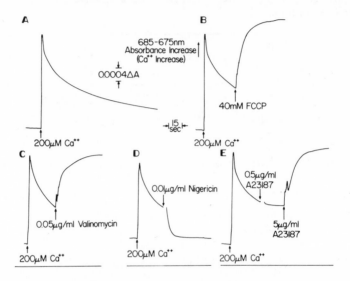

Fig. 7. *The effect of various ionophores on Ca^{++} transport.*
The reaction mixture contained 50 mM KCl, 100 mM
sucrose, 10 mM MOPS (pH 7.2), 5 mM Na succinate, 1 μM
rotenone, 3 mM $MgCl_2$, 20 μM arsenazo and 1.1 mg/ml
mitochondrial protein.

traces of energy-dependent Ca^{++} uptake by rat liver mitochondria obtained by recording the changes in absorbance of
the dye arsenazo III. Ca^{++} was added to a reaction mixture
containing small amounts of mitochondrial protein and no
permeant anions, so that steady-state levels of Ca^{++} accumulation were obtained when about half of the Ca^{++} added was
taken up by mitochondria (Figure 7A). The addition of the
uncoupling agent FCCP completely releases the accumulated
Ca^{++} (Figure 7B). Upon addition of smaller concentrations
of the uncoupler FCCP (not shown) there was a well-defined
steady-state level of Ca^{++} accumulation. Valinomycin (in
the presence of KCl) also releases the Ca^{++} accumulation and
this effect was saturated at very low valinomycin concentrations (Figure 7C). Figure 7D shows that low concentrations
of nigericin produce additional accumulation of Ca^{++}, so
that higher steady-state levels of Ca^{++} accumulation are
obtained. Figure 7E shows that the divalent ionophore
A23187 has no effect on steady-state Ca^{++} levels at con-

centrations which were effective in translocating Ca^{++} across the lipid phase of mitochondria (31). Ca^{++} release was observed only at higher concentrations of A23187.

The presence of these ionophores induces a stimulation of either Ca^{++} uptake or Ca^{++} release in a way consistent with the known effect of these compounds on mitochondrial ΔpH or transmembrane potential. Release of the accumulated Ca^{++} is expected in the presence of FCCP or valinomycin + K^+, since either compound is able to collapse a membrane potential; FCCP, by acting as a proton conductor abolishes the transmembrane H^+ gradient; valinomycin promotes massive electrophoretic transport of K^+ inside the mitochondrial membrane, thereby collapsing the existing membrane potential negative inside. Nigericin, a monocarboxylic acid which can transport K^+ in an unprotonated form and H^+ in a protonated form, mediates the electroneutral H^+-K^+ exchange across the mitochondrial membrane. The effect of nigericin in stimulating Ca^{++} uptake by mitochondria can be due to a decrease of ΔpH, an increase of $\Delta\Psi$, or partial conversion of ΔpH to $\Delta\Psi$. Finally, the addition of a small amount of A23187 to energized mitochondria containing Ca^{++} produces an exchange of $^{45}Ca^{++}$ and an increase in respiration, but not a release of the accumulated Ca^{++}. Since A23187 catalyzes a neutral Ca^{++}-$2H^+$ or Ca^{++}-Mg^{++} exchange (31), the efflux of Ca^{++}-ionophores across the hydrophobic phase of mitochondria will be neutral and will not affect membrane potential. As long as the efflux of Ca^{++} mediated by the ionophore is limited and neutral, mitochondrial uptake of Ca^{++} increases to maintain the concentration gradients of Ca^{++} that can be held by membrane potential.

Although not yet conclusive, these results are interpreted as evidence that Ca^{++} uptake in mitochondria is an electrophoretic process driven by a transmembrane potential with a net charge transfer of 2. Ca^{++} transport can be accounted for either by a mitochondrial membrane highly permeable to Ca^{++} (32) or by the operation of an uncharged Ca^{++} carrier not coupled with electron flow and with no compulsory exchange with other ions. Such a mechanism of Ca^{++} transport accomodates and offers a unified explanation for events occurring simultaneously with Ca^{++} transport such as increase of oxygen consumption and H^+ release, as well as for Ca^{++}/H^+ stoichiometries and ADP/Ca^{++} competition.

Control of mitochondrial Ca^{++} transport in vivo through membrane potential suggests that energized mitochondria within a cell may contain large amounts of Ca^{++} in the matrix space. Moreover, since little change in membrane potential occurs among the various physiological states of mitochondria (i.e., state 4 - state 3 transition) (29), Ca^{++} transport by mitochondria in vivo seems to be more adequate for slow and large adjustments rather than for rapid and repetitive adjustments of cytosolic Ca^{++}. These points are susceptible to experimental tests, and experiments are in progress where Ca^{++} uptake by mitochondria can be measured in situ within the cytosol of large cells.

Acknowledgements

Some of these experiments were carried out in collaboration with Drs. A. Vinogradov and H. Rottenberg. Many thanks are due to Nancy Jones and to Ken Ray for the preparation of the manuscript and figures. This study was supported by grants 73-743 and HL-15835 from NIH. A. Scarpa is an investigator of the American Heart Association

REFERENCES

1. Chance, B. (1965) J. Biol. Chem. 243, 2729.

2. Lehninger, A.L., Carafoli, E. and Rossi, C.S. (1967) Adv. Enzymol. 29, 259.

3. Reynafarje, B. and Lehninger, A.L. (1969) J. Biol. Chem. 244, 584.

4. Mela, L. (1968) Arch. Biochem. Biophys. 123, 286.

5. Akerman, K.E., Saris, N.E.L. and Järvisalo, J.O. (1974) Biochem. Biophys. Res. Commun. 58, 501.

6. Reed, K.G. and Bygrave, F.L. (1974) Biochem. J. 138, 239.

7. Carafoli, E., this volume.

8. Bygrave, F.L., Reed, K.G. and Spencer, T. (1971) Nature New Biol. 230, 89.

9. Carafoli, E. and Azzi, A. (1972) Experientia 28, 906.

10. Scarpa, A. (1974) in Calcium Transport in Contraction and Secretion (Carafoli, E., Clementi, F. and Margreth, A., Eds.) North-Holland Publishing Co., Amsterdam, in press.

11. Scarpa, A. (1972) Methods Enzymol. 24, 343.

12. Scarpa, A., Abstr. 10th Intern. Congr. Biochem., Hamburg, 1976, in press.

13. Chance, B. (1972) Methods Enzymol. 24, 322.

14. Chance, B., Legallais, V., Sorge, J. and Graham, N. (1975) Anal. Biochem. 66, 498.

15. Chance, B. (1973) *in* Techniques of Chemistry, Vol. 6, part 2 (Hammes, G. and Weissberger, A., Eds.) pp. 5-62, Wiley and Sons, London.

16. Vinogradov, A. and Scarpa, A. (1973) 248, 5527.

17. Scarpa, A. and Graziotti, P. (1973) J. Gen. Physiol. 62, 756.

18. Ebashi, S., Endo, M. and Ohtsuki, I. (1969) Quart. Rev. Biophys. 2, 351.

19. Katz, A.M. (1970) Physiol. Rev. 50, 63.

20. Scarpa, A. and Williamson, J.R. (1973) *in* Calcium Binding Proteins (Drabikowski, W., Strzeleka, H. and Carafoli, E., Eds.) pp. 547-585, Elsevier, Amsterdam.

21. Inesi, G., Ebashi, S. and Watanabe, S. (1964) Amer. J. Physiol. 207, 1339.

22. Chance, B., Lee, C.P. and Mela, L. (1967) Fed. Proc. Fed. Amer. Soc. Exp. Biol. 26, 1341.

23. Slater, E.C. (1967) Eur. J. Biochem. 1, 317.

24. Mitchell, P. (1966) *in* Chemiosmotic Coupling in Oxidative and Photosynthetic Phosphorylation, Glynn Research, Bodmin, Cornwall, England.

25. Greville, G.D. (1969) Curr. Top. Bioenerg. 3, 1.

26. Chance, B. and Montal, M. (1971) Curr. Top. Membr. Transp. 2, 99.

27. Lehninger, A.L. (1973) *in* The Molecular Basis of Electron Transport (Shultz, J. and Cameron, B.F., Eds.) pp. 259-273, Academic Press, New York.

28. Pressman, B.C. (1968) Fed. Proc. Fed. Amer. Soc. Exp. Biol. 27, 1283.

29. Rottenberg, H. (1973) J. Mem. Biol. 11, 117.

30. Rottenberg, H. and Scarpa, A. (1974) Biochem. 13, 4811.

31. Reed, P.W. and Lardy, H.A. (1972) J. Biol. Chem. 247, 6970.

32. Selwyn, M.S., Dawson, A.P. and Dunnett, S.J. (1970) FEBS Lett. 10, 1.

MITOCHONDRIAL CALCIUM TRANSPORT AND CALCIUM BINDING PROTEINS*

Ernesto Carafoli

It is now well known (1-3) that mitochondria from most eukaryotic cells have the ability to accumulate Ca^{2+} against concentration gradients. The transport process can be driven by respiratory or ATP energy, and is inhibited by agents which abolish the energy supply (respiratory inhibitors when the process is energized by respiration, oligomycin when the energy source is exogenous ATP). It is also inhibited specifically by two inhibitors which act at very low concentrations, La^{3+} (and other lanthanides) and ruthenium red (4-7). The commonly accepted interpretation of these observations is shown in Fig. 1. The nature of the [∿] which drives the uptake is unspecified in the figure: experiments which are relevant to this point have recently been carried out by Rottenberg and Scarpa (8), who have compared the distribution of Ca^{2+} between mitochondria and medium with that of Rb^+ in valinomycin-treated mitochondria. Since the latter is known to be determined by the membrane potential maintained by respiration, the experiments of Rottenberg and Scarpa (8) have provided convincing evidence that the membrane potential is the driving force also for the uptake of Ca^{2+}.

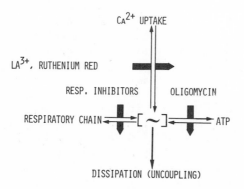

Fig. 1. Ca^{2+} transport and energy transformations in mitochondria.

*Parts of the original work described in this article have been supported by the financial contribution of the Swiss National Funds (grant # 3-1720-73).

Other interesting characteristics of the transport process are listed in Table I. The maximal amount of Ca^{2+} which can be taken up by mitochondria varies between 100 and 150 nmoles per mg of protein. This amount can be increased several fold if a permeant anion also penetrates into mitochondria to precipitate Ca^{2+} inside as an insoluble salt (phosphate), or to dilute it in the matrix by favoring the osmotic penetration of water (acetate). One question which has not been solved yet, is whether the Ca^{2+} which is taken up in the absence of permeant anions remains free in the intramitochondrial water, or whether it is bound to fixed anionic sites on the inner side of the inner membrane. The aforementioned results by Rottenberg and Scarpa would favor the first possibility, whereas older experiments on submitochondrial vesicles by Gear et al. (9) would favor the second. Experiments of Chappell et al. (10) and Gunter and Puskin (11) in which Mn^{2+} has been used as a paramagnetic analogue for Ca^{2+}, also indicate that the cation is not free but bound. Table I also shows that the initial rate of Ca^{2+} uptake (measured at a concentration of Ca^{2+} near the K_m) corresponds to about 2.0 nmoles of Ca^{2+} per mg of protein per sec. This figure has been recently measured directly in our laboratory using heart mitochondria (Sigel et al., in preparation), and is lower than previous figures (estimated indirectly, or extrapolated from other measurements) which are found in the literature (12,13). Values of initial velocity, not very different from those directly measured in our laboratory, have recently been measured by Vinogradov and Scarpa (14) using liver mitochondria and the spectrophotometric murexide method. The affinity of the energy-linked Ca^{2+} transport system for Ca^{2+} is also mentioned in Table I. This is obviously a very important parameter, since many Ca^{2+}-dependent cell reactions, which could conceivably be influenced by mitochondria, are regulated by Ca^{2+} in the µM concentration range. Most authors, using different experimental approaches, have measured K_m's of the order of 1-10 µM (13,15,16, Sigel et al., in preparation). A recent study of Scarpa and Graziotti (17) on heart mitochondria, however, indicates a K_m in excess of 100 µM, a figure which would clearly rule out mitochondria as significant participants in the metabolic regulation of cell Ca^{2+}. The reason for the abnormally high K_m measured by Scarpa and Graziotti became clear after it was discovered (18-20) that Mg^{2+} is a powerful inhibitor of the energy-linked uptake of Ca^{2+}, specifically in heart mitochondria. The reaction media used by Scarpa and Graziotti contained Mg^{2+}, which evidently depressed very significantly the ability of mitochondria to take up Ca^{2+}. The observation that Mg^{2+} inhibits the uptake of Ca^{2+} by heart mitochondria may on the other hand be very significant from a physiological standpoint, since it has been estimated that the cytosol of heart cells contains about 0.9mM free Mg^{2+} (21). Thus, although it is now clear that heart mitochondria under optimal conditions may take up Ca^{2+} with an affinity corresponding to a K_m between 1 and 10 μM, they may not be able to accumulate Ca^{2+} with this efficiency under the conditions prevailing in the cell. Mg^{2+} may thus be an efficient means to modulate the ability of mitochondria to remove Ca^{2+} from the cytosol.

TABLE I

Characteristics of Ca^{2+} transport in mitochondria

Total Capacity	150 nmoles/mg protein (no Pi) > 2 μmoles/mg protein (+ Pi)
Affinity	K_m < 10 μM
Rate of uptake	about 2.0 nmoles/mg protein/sec
Inhibition	Competitive inhibition by Sr^{2+} Inhibition by La^{3+} and ruthenium red

The other parameters of Ca^{2+} uptake listed in Table I (competitive inhibition by Sr^{2+}, inhibition by very low concentrations of La^{3+} and ruthenium red), indicate the presence of a specific Ca^{2+} carrier. In particular, the carbohydrate specificity of ruthenium red (22) suggests that the Ca^{2+} carrier may be a carbohydrate-containing compound, possibly a glycoprotein. In the last three years, attempts have been made in several laboratories to isolate the Ca^{2+} carrier (or at least to resolve the Ca^{2+} transport system into some of its components), and it is of interest that the isolated fractions when adequately characterized have been shown to contain carbohydrates. The first report of a mitochondrial fraction able to bind Ca^{2+} is that of Evtodienko et al. (23) who have isolated and partially purified from acetone powders of liver mitochondria, a protein which had a rather high affinity for Ca^{2+} (K_d, 1 μM) and was able to split ATP in a reaction which was stimulated by Ca^{2+}. The protein was not characterized in detail, and it is thus not known whether it contains carbohydrates. Carbohydrates are, on the other hand, present in a water-soluble protein which Lehninger (24) and later Gómez-Puyou et al. (25) liberated from liver mitochondria by osmotic shocks, and which has then been purified essentially by an ammonium sulphate fractionation procedure. The protein has a molecular weight of approximately 120,000, contains lipids and rather large amounts of phosphoprotein phosphorous, and binds Ca^{2+} at two classes of sites, one of which has a very high affinity for the cation. Very interestingly, the reaction of the protein with Ca^{2+} is strongly inhibited by La^{3+} at the same low concentrations which abolish the energy-linked inter-action of Ca^{2+} with mitochondria. Preliminary reports on Ca^{2+} binding, carbohydrate-containing fractions, have been published recently by three other laboratories. Kimura et al. (26) have isolated from adrenal cortex mitochondria a glycoprotein able to bind Ca^{2+}, but only with low affinity; Tashmukhamedov et al. (27) have purified from liver mitochondria a carbohydrate-containing lipid fraction which binds Ca^{2+} with high affinity, and which is inhibited by hexamine cobaltichloride, an inhibi-tor of Ca^{2+} transport in mitochondria. Utsumi and Oda (28) have extracted from liver mitochondria a water-soluble glyco-protein which contains lipids, has a molecular weight in excess of 70,000, and binds ruthenium red. This glycoprotein is able to remove the inhibition of Ca^{2+} transport when added to ruth-enium red-treated mitochondria. Among the Ca^{2+} binding carbo-hydrate-containing fractions which have been isolated from mitochondria, the best characterized is a glycoprotein purified by Carafoli, Sottocasa, and their coworkers (29-31). (See Table II for a summary of the different Ca^{2+} binding fractions which have been isolated from mitochondria). The glycoprotein is extracted from mitochondria by an osmotic shock procedure (or by exposing mitochondria to chaotropic agents) and is then purified by preparative polyacrylamide gel electrophoresis. It has a molecular weight between 30,000 and 33,000, contains about 10% carbohydrates and a variable amount of phospholipids. Among its interesting properties is the ability to bind Ca^{2+} at two classes of sites, in a reaction which is inhibited by La^{3+} and by ruthenium red. Also of interest is the absence of high-affinity Ca^{2+} binding sites in glycoproteins isolated from mitochondria which do not have the ability to actively transport Ca^{2+}. It has also been established that the glycoprotein is normally present in both mitochondrial membranes and/or the intermembrane space, but not in the matrix (32,33). Using this glycoprotein, Prestipino et al. (34) have attempted to recon-stitute the transport of Ca^{2+} in artificial lipid bilayer systems (black films). Upon addition of low amounts of glyco-protein to lecithin black films in the presence of Ca^{2+}, a marked increase in the electrical conductance could be observed (Table III), which could not be reproduced using a variety of commercially available glycoproteins, and which was inhibited by ruthenium red. In the same system, however, Prestipino et al. (34) could not measure Ca^{2+}-dependent Nernst potentials, an observation which is not compatible with a glycoprotein-mediated transport of Ca^{2+} across the lipid bilayer. This is supported by experiments carried out in our laboratory (P. Gazzotti, personal communication) on Ca^{2+}-loaded liposomes,

CALCIUM TRANSPORT AND BINDING PROTEINS

TABLE II

Mitochondrial Ca^{2+}-binding fractions

Author	Nature of fraction	Affinity for Ca^{2+}
Evtodienko et.al. (1971)	Protein	high
Lehninger (1971) Gómez-Puyou et al. (1972)	Glycoprotein	high
Kimura et al. (1972)	Glycoprotein	low
Sottocase et al. (1972)	Glycoprotein	high
Tashmukhamedov et al. (1972)	Glycolipid	high
Utsumi and Oda (1974)	Glycoprotein	-

TABLE III

Effect of a mitochondrial Ca^{2+} binding glycoprotein on the electrical conductance of lipid bilayers.

Addition	G_M $(\Omega^{-1} xcm^{-2} x10^{-8})$
Mitochondrial glycoprotein	10.1
Mitochondrial glycoprotein plus ruthenium red	0.31
Immunoglobulin G	0.09
Horseradish peroxidase	0.28
Ovalbumin	0.18

The preparation of the bilayer (black film) and the details of the measurement are described by Prestipino et al. (1974). The G_M is the difference between the electrical conductance measured 22 to 33 min (depending on the different glycoprotein) after the addition of the glycoproteins, and that measured immediately before their addition. The glycoproteins were added at equal concentrations to both sides of the bilayer (1.48 to 5 x 10^{-7} M). The bathing solutions contained 4 mM CaCl$_2$ and 1 mM Tris-Cl, pH 7.8. Ruthenium red was 1 μM.

where no increase in the efflux of Ca^{2+} could be observed upon exposure of the liposomes to the Ca^{2+}-binding glycoprotein. Taken at face value, these experiments indicate that the glycoprotein is not a transmembrane, mobile or immobile Ca^{2+} carrier, but rather may be a superficial (and specific) Ca^{2+} receptor, which may function in series with a transmembrane carrier, and which may become associated with the mitochondrial membrane under the influence of Ca^{2+}. The nature of the species which is responsible for the increase in the electricial conductance of the bilayer remains to be determined, but it is of interest that a Ca^{2+}-binding protein isolated from the nervous system of lower animals (35) has been found to promote the efflux of Rb^+ from Rb^+-loaded liposomes. If the association of the Ca^{2+}-binding glycoprotein with the mitochondrial membrane is controlled by Ca^{2+}, it could be expected that removal of Ca^{2+}, e.g., by EDTA, would result in the extraction of greater amounts of glycoprotein from the membrane. That this is, indeed, the case has been recently verified in the laboratory of Sottocasa (personal communication). It is also possible to suggest that other hydrophillic proteins from mitochondria and other organelles could become reversibly associated with, or dissociated from, the membrane environment under the influence of Ca^{2+}. Indications for this possibility can be found in the literature (36-43).

There is an additional aspect of the process of mitochondrial Ca^{2+} transport which may be considered, and this is its role in metabolic regulation. Some of the properties of the process are of particular interest from this standpoint, among them, the affinity for Ca^{2+}, the initial rate of its uptake, and the total capacity for Ca^{2+} accumulation. It is easy to calculate, for example, that the total capacity of mitochondria for Ca^{2+} storage in average cells correspond to the impressive figure of 8 to 15 µmoles of Ca^{2+} per g of tissue in the absence of permeant anions, and to more than 200 µmoles per g of tissue in the presence of phosphate. The latter condition is probably the closest to physiology, since in the cytosol of all cells phosphate is always available for mitochondrial uptake. Mitochondria thus represent a very important buffer system for cell Ca^{2+}. In epithelial cells, which do not possess other important Ca^{2+}-transporting organelles, mitochondria are by far the most important Ca^{2+} sinks.

Perhaps the most important point in a discussion on the possible role of mitochondria in the regulation of cell Ca^{2+} is, however, the reversibility of the process. Here, a distinction is clearly necessary between the case of mitochondria loaded with Ca^{2+} and phosphate, and that of mitochondria loaded with Ca^{2+} in the absence of complexing permeant anions. Clearly, the release of Ca^{2+} stored in mitochondria as an insoluble phosphate salt must be a slow process, probably without interest from the standpoint of metabolic regulation. On the other hand, it may be expected that Ca^{2+} stored in mitochondria in the absence of complexing anions is released rather rapidly.

As Fig. 2. shows, this is indeed the case: interruption of energy coupling with an uncoupler induces a slow release of the Ca^{2+} accumulated by mitochondria in the presence of phosphate, and a much more rapid release of the Ca^{2+} accumulated by mitochondria in the absence of complexing anions. It is interesting that not only the accumulated Ca^{2+}, but also the endogenous Ca^{2+} which is normally present in mitochondria is released rapidly by uncouplers, an observation that suggests that the endogenous Ca^{2+} pool is not maintained in mitochondria as a salt precipitate. Direct measurements of the initial rates of Ca^{2+} release would obviously be very important. Preliminary estimates in our laboratory (43) indicate that the rate of release is proportional to the amount of Ca^{2+} taken up by mitochondria. In addition to uncouplers, other agents which interrupt the energy flow, like respiratory inhibitors, can induce release of Ca^{2+} from mitochondria, and it has also been demonstrated that the efflux of Ca^{2+} down its concentration gradient

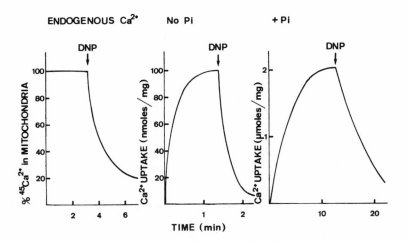

Fig. 2. *Reversibility of mitochondrial Ca^{2+} transport. Mitochondria were prepared from rat heart by the method of Pande and Blanchaer (44). For the experiments on endogenous Ca^{2+}, the rats were injected intraperitoneally with carrier-free $^{45}Ca^{2+}$, 5 min before sacrifice. The in vitro uptake of Ca^{2+} was measured by Millipore filtration in a medium containing 210 mM mannitol, 70 mM sucrose, 10 mM glutamate, 10 mM malate, 10 mM Tris-HCl, pH 7.4, 10 mg of mitochondrial protein, in a final volume of 4 ml at $25^{o}C$. The concentration of added $^{45}Ca^{2+}$ was 250 μM in the experiment without phosphate, and 3 mM in the experiment with phosphate (5 mM phosphate buffer, pH 7.4). In the experiment on endogenous Ca^{2+}, the medium contained the same components minus $^{45}Ca^{2+}$. Concentration of DNP (2,4-dinitrophenol) 0.1 mM.*

can be coupled to the synthesis of ATP in the absence of respiration (46). It has also been shown that two Ca^{2+} specific acid ionophores, X-537 A and A 23187, induce the rapid loss of the accumulated (Fig. 3), or endogenous (47) Ca^{2+} from mitochondria. It is evident, then, that the uptake of Ca^{2+} by mitochondria is a completely reversible process which can make large amounts of Ca^{2+} rapidly available to the cytosol. The problem at this point is that of identifying release-inducing agents which are likely to operate in vivo, i.e., "natural" release-inducing agents. Only recently emphasis has been put on this problem, but some encouraging progress is already evident. The

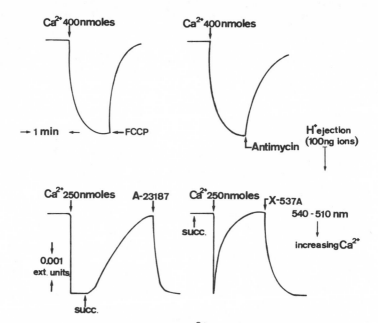

Fig. 3. Release of mitochondrial Ca^{2+} by interruption of the energy flow and by specific ionophores. Rat liver mitochondria were prepared by a conventional method. In experiment 1-2, movements of Ca^{2+} were followed by measuring the opposite movements of protons; in the last two experiments they were followed with the spectrophotometric murexide method. Reaction medium for experiments 1-2, 80 mM NaCl, 2 mM Tris-HCl, pH 7.4, 5 mM succinate and 5 mg mitochondrial protein (1.8 ml at 20°C). In experiments 3-4, the reaction medium contained (2.7 ml at 25°C) 210 mM mannitol, 10 mM sucrose, 10 mM Tris-HCl, pH 7.4, 14 μM murexide, 1 μM rotenone and 2.5 mg mitochondrial protein. Concentrations of inhibitors: FCCP (carbonyl cyanide p-trifluoromethoxyphenylhydrazone) 1 μM, antimycin A, 0.25 μg/mg of protein; a 23187, 2 μg/mg of protein; X 537 A, 2 μg/mg of protein.

recent report by Borle (48) that cyclic AMP could induce large losses of Ca^{2+} from mitochondria was greeted with considerable excitement. It has recently become clear, however, that the releasing effect of cyclic AMP is not easily reproducible and may depend considerably on the experimental conditions employed. In our laboratory, efforts to induce efflux of Ca^{2+} from mito-chondria by the cyclic nucleotide have been largely unsuccess-ful (Malmström and Chiesi, unpublished).

Two other "natural" agents have recently been found to re-lease Ca^{2+} from mitochondria: Na^+ and prostaglandins. Na^+ (47) has been found to induce release of Ca^{2+} specifically from heart mitochondria (the effect is negligible, or at least much less evident, in liver mitochondria). Among the inter-esting features of the Na^+ effect is the fact that it requires that mitochondria are loaded with minimal amounts of Ca^{2+}, or that they contain only the endogenous Ca^{2+} pool. In other words, the effect of Na^+ is seen optimally in mitochondria which are closest to the "physiological" situation. As Fig.4

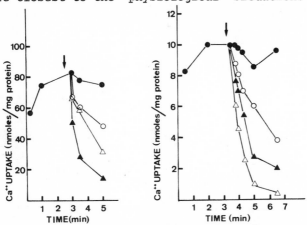

Fig. 4. Release of mitochondrial Ca^{2+} by prostaglandins (PG) and Na^+ in the presence of ruthenium red (RR). Rat liver mitochon-dria (prepared by a conventional method) were used for the PG experiment, rat heart mitochondria (see legend for Fig. 1) for the Na^+ experiment. PG experiment (left): mitochondria (12.5 mg protein) in 150 mM sucrose, 10 mM Na-succinate, 10 mM imidazole buffer, pH 6.4, and 500 μM $^{45}CaCl_2$ (final volume 5 ml, $25^{\circ}C$). At 2.5 min 10 μM RR was added followed by 10 μM PGA_1 (O—O), 20 μM PGA_1 (Δ—Δ), 50 μM PGA_1 (▲—▲) or EtOH (●—●). Na^+ experiment (right): heart mitochondria (5 mg protein) in 210 mM mannitol, 70 mM sucrose, 10 mM Tris-Cl, pH 7.4, 10 mM Tris-succinate and 50 nmoles of $^{45}CaCl_2$ (final volume 4 ml, $25^{\circ}C$). At 3 min, 2.5 μM RR followed by 5 mM (O—O), 10 mM (▲—▲), 50 mM (Δ—Δ) NaCl, or medium (●—●) were added. The movements of Ca^{2+} were followed by Millipore filtration.

TABLE IV

Ca^{2+} *Dependent Reactions in Cells*

Activation of Enzyme Systems

Glycogenolysis (phosphorylase b kinase)
Lipases and phospholipases
α-glycerophosphate dehydrogenase
Pyruvate dehydrogenase
Succinate oxidation
Synthesis of some phospholipids

Inhibition of Enzyme Systems

Pyruvate kinase
Synthesis of some phospholipids

Activation of Contractile and Motile Systems

Muscle myofibrils
Cilia and flagella
Microtubules and microfilaments
Cytoplasmic streaming
Pseudopod formation

Hormonal Regulation

Formation and/or function of cyclic AMP (GH, LH, TSH, MSH, PTH)
Release of insulin, steroids, vasopressin, catecholamines, thyroxine and progestrone

Membrane-linked Functions

Excitation-secretion coupling at nerve endings
Excitation-contraction coupling in muscles
Exocrine secretion (pancreas, salivary glands, and HCl in the stomach)
Aggregation of platelets
Action potential (nerve and muscle cells)
Na^+ and K^+-activated adenosine triphosphatases of several membranes
Tight junctions
Cell contact
Binding of prostaglandins to membranes

shows, the effect is seen already, if care is taken to prevent the re-accumulation of the lost Ca^{2+} at 5 mM Na^+. This is a rather low concentration, and it would thus be tempting to attribute physiological significance to the Na^+ effect. This would, however, be premature as several problems are still open, e.g., the rate of the Na^+-induced release, its tissue specificity, and the high concentrations of Na^+ necessary to induce release in the absence of agents that prevent the re-uptake of Ca^{2+}.

In our laboratory, we have recently found (49,50) that prostaglandins (PG) can induce loss of Ca^{2+} from mitochondria. The experiments, triggered by the suggestion of Kirtland and Baum (51) that one of the PG (PGE_1) could act as a natural mitochondrial Ca^{2+} ionophore, have led to the finding (50) that a variety of PG added to mitochondria after the accumulation of a pulse of Ca^{2+} induce a pronounced release. In Fig. 4, PGA_1 has been used and it can be seen that an extensive release is evident already after the addition of 10 μM PG. This is admittedly a very high concentration, and for this reason great caution is necessary in extrapolating these findings to the physiological situation. Indeed, we currently consider the effect of PG primarily as an interesting artificial model. We have also found that PG uncouples mitochondrial respiration, but only when added to mitochondria which have accumulated Ca^{2+} (50). In this respect, they differ from fatty acids, a category of uncouplers to which they could be superficially related.

If and when the "natural" agent that induces release of Ca^{2+} from mitochondria will be identified, the discussion of the role of mitochondria in the regulation of the intracellular Ca^{2+} will be based on much more solid ground. It will then be pertinent to consider mitochondria as candidates for the regulation of a great variety of cell reactions. Table IV lists the most important among them.

REFERENCES

1. Lehninger, A.L., Carafoli, E. and Rossi, C.S. (1967) Adv. Enzymol. 30, 259.
2. Carafoli, E. and Lehninger, A.L. (1971) Biochem. J. 122, 681.
3. Carafoli, E. and Rossi, C.S. (1971) Adv. Cytopharmacol. 1, 209.
4. Mela, L. (1968) Arch. Biochem. Biophys. 123, 286.
5. Lehninger, A.L. and Carafoli, E. (1971) Arch. Biochem. Biophys. 143, 506.
6. Moore, C. (1971) Biochem. Biophys. Res. Commun. 42, 298.

7. Vasington, D.F., Gazzotti, P., Tiozzo, R. and Carafoli, E. (1972) Biochim. Biophys. Acta 256, 43.
8. Rottenberg, H. and Scarpa, A. (1974) Biochem. 13, 4811.
9. Gear, A.R.L., Rossi, C.S., Reynafarje, B. and Lehninger, A.L. (1967) J. Biol. Chem. 242, 3403.
10. Chappell, J.B., Cohn, M. and Greville, G.D. (1963) in Energy-linked Functions of Mitochondria (Chance, B., ed.), p. 232, Academic Press, New York.
11. Gunter, T.D. and Puskin, J.S. (1972) Biophys. J. 12, 625.
12. Chance, B. (1965) J. Biol. Chem. 240, 2729.
13. Carafoli, E. and Azzi, A. (1972) Experientia 27, 906.
14. Vinogradov, A. and Scarpa, A. (1973) J. Biol. Chem. 248, 5527.
15. Spencer, T. and Bygrave, F.L. (1973) Bioenergetics 4, 347.
16. Reynafarje, B. (1974) in Myocardial Cell Damage (Fleckenstein, A., ed.), University Park Press, Baltimore, Md. (in press).
17. Scarpa, A. and Graziotti, P. (1974) J. Gen. Physiol. 62, 756.
18. Jacobus, W.E., Tiozzo, R., Lugli, G., Lehninger, A.L. and Carafoli, E. (1975) J. Biol. Chem., in press.
19. Sordahl, L.A. (1974) Arch. Biochem. Biophys. 167, 104.
20. Carafoli, E., Malmström, K. Capano, M., Sigel, E. and Crompton, M. (1975) in Calcium Transport in Contraction and Secretion (Carafoli, E., Clementi, F. Drabikowski, W. and Margreth, A., eds.), Elsevier-North Holland, (in press).
21. Endo, M. (1975) Proc. Jap. Acad., (in press).
22. Luft, J.H. (1972) Anat. Rec. 171, 347.
23. Evtodienko, Y.V., Peshkova, L.V. and Shchipakin, V.N. (1971) Ukrainian J. Biochem. 43, 98.
24. Lehninger, A.L. (1971) Biochem. Biophys. Res. Commun. 42, 312.
25. Gómez-Puyou, A., Tuena de Gómez-Puyou, M., Becker, G. and Lehninger, A.L. (1972) Biochem. Biophys. Res. Commun. 47, 814.
26. Kimura, T., Chu, J.W., Mukai, R., Ishizuka, I. and Yamakawa, T. (1972) Biochem. Biophsy. Res. Commun. 49, 1678.
27. Tashmukhamedov, B.A., Gagelgans, A.I., Mamatkulov, K.H. and Makhmudova, E.M. (1972) FEBS Letters 28, 239.
28. Utsumi, K. and Oda, T. (1974) in Organization of Energy-transducing Membranes (Nakao, K. and Packer, L., eds.), p. 265, University Park Press, Baltimore, Md.
29. Carafoli, E., Gazzotti, P., Vasington, F.D., Sottocasa, G.L., Sandri, G., Panfili, E. and de Bernard, B. (1972) in Biochemistry and Biophysics of Mitochondrial Membranes (Azzone, G.F., Carafoli, E., Lehninger, A.L., Quagliariello, E. and Siliprandi, N., eds.) p. 623, Academic

Press, New York.
30. Sottocasa, G.L., Sandri, G., Panfili, E., de Bernard, B., Gazzotti, P., Vasington, F.D. and Carafoli, E. (1972) Biochem. Biophys. Res. Commun. 47, 808.
31. Carafoli, E. and Sottocasa, G.L. (1974) in Dynamics of Energy-transducing Membranes (Ernester, L, Estabrook, R. and Slater, E.C., eds.) p. 455, Elsevier, Amsterdam.
32. Sottocasa, G.L., Sandri, G., Panfili, E., Gazzotti, P. and Carafoli, E. (1973) Ninth Intern. Congr. Biochem., Stockholm, Abstract 4f-12.
33. Melnick, R.L, Tinberg, H.M., Maguire, J. and Packer, L. (1973) Biochim. Biophys. Acta 311, 230.
34. Prestipino, G.F., Ceccarelli, D., Conti, F. and Carafoli, E. (1974) FEBS Letters 45, 99.
35. Calissano, P. and Bangham, A.D. (1971) Biochem. Biophys. Res. Commun. 43, 504.
36. Manery, J.F. (1966) Fed. Proc. 25, 1804.
37. Burger, S.P., Fujii, T. and Hanahan, D.J. (1968) Biochem. 7, 10.
38. Duggan, P.F. and Martonosi, A. (1970) J. Gen. Physiol. 56, 147.
39. Rorive, G., Nielson, R. and Kleinzeller, A. (1972) Biochim. Biophys. Acta 266, 376.
40. Reynolds, J.A. (1972) Ann. N.Y. Acad. Sci. 195, 75.
41. Gilbert, I.G.F. (1972) Eur. J. Cancer 8, 99.
42. Azzi, A. and Montecucco, C. (1974) in Membrane Proteins in Transport and Phosphorylation (Azzone, G.F., Klingenberg, M., Quagliariello, E. and Siliprandi, N., eds.), Academic Press, New York.
43. Carafoli, E. (1975) Mol. and Cell Biochem., (in press).
44. Pande, S.W. and Blanchaer, M.C. (1971) J. Biol. Chem. 246, 402.
45. Carafoli, E. (1975) in Basic Function of Cations in Myocardial Activity (Fleckenstein, A., ed.), University Park Press, Baltimore, Md., (in press).
46. Haaker, H., Berden, J.A., Kraayenhof, R., Katan, M., Van Dam, K. (1972) in Biochemistry and Biophysics of Mitochondrial Membranes (Azzone, G.F., Carafoli, E., Lehninger, A.L., Quagliariello, E. and Siliprandi, N., eds.) p. 239, Academic Press, New York.
47. Carafoli, E., Tiozzo, R., Lugli, G., Crovetti, F. and Kratzing, C. (1974) J. Mol. Cell. Cardio. 6, 631.
48. Borle, A.B. (1974) J. Membrane Biol. 16, 221.
49. Carafoli, E. Crovetti, F. and Ceccarelli, D. (1973) Arch. Biochem. Biophys. 154, 40.

50. Malmström, K. and Carafoli, E. (1975) Arch. Biochem. Biophys., in press.

51. Kirtland, S.J. and Baum, H. (1972) Nature (London) 236, 47.

COMPARISON OF ELECTRONEUTRAL AND ELECTROGENIC ANION TRANSPORT IN MITOCHONDRIA*

Kathryn F. LaNoue and Marc E. Tischler

The mitochondrial membrane surrounds an anatomical compartment in the cell whose major function is to transform
energy available from respiration into a form (ATP) usable
by the rest of the cell. In many mammalian organs mitochondrial enzymes catalyze certain biosynthetic reactions as
well. This matrix compartment contains the enzymes of the
citric acid cycle, and the initial rate controlling enzymes
for gluconeogenesis from pyruvate, lactate, and alanine. In
addition, many of the enzymes of the urea cycle are found
compartmented within the mitochondrial matrix, while the
enzymes of fatty acid β-oxidation, the enzymes of electron
transport, and the ATP synthesizing enzymes are part of the
membrane.

Since certain biosynthetic enzymes of the cell are found
only within the mitochondria, there is a cytosolic requirement for mitochondrial products other than ATP. Citrate
must be transported out of the matrix, where it is formed,
into the cytosol for fatty acid synthesis. Malate and
aspartate formed in the matrix are used in the cytosol as
the carbon source for gluconeogenesis, while aspartate and
citrulline are needed for urea synthesis.

Permeases for both substrates and products must exist
since most of the mitochondrial metabolites are hydrophylic
ions at neutral pH (1) and as such would not be expected to
penetrate a lipid bilayer in the absence of specific carrier
systems (2). In addition, mechanisms for concentrating the
substrates of mitochondrial enzymes within the mitochondrial
matrix would be of obvious value to the cell, as would facilitated transport of products out of the matrix. Since the
osmotically active membrane encloses a very small space, it
is necessary that the transport processes operate primarily
by exchange mechanisms to maintain a constant osmolarity
within the matrix.

Isolated mitochondria can maintain relatively large
concentration gradients of phosphate, acetate, pyruvate, and
the di- and tri-carboxylic acid intermediates of the citric
acid cycle. Originally, it was thought that these anions

*   Supported by grants: AHA-72-739 from the American Heart
Association, GM-22158-01 from the National Institutes of
Health, HE-14461, AM-15120 and AA-00292 from the U.S. Public
Health Service

were accumulated electrophoretically in response to an electrical potential gradient, positive inside and negative outside (3,4). The postulated potential gradient was estimated to be about 20 millivolts and was thought to be the result of an energy driven accumulation of cations, chiefly potassium.

However in recent years an impressive body of evidence has developed indicating that such is not the case. It was proposed by Mitchell (5,6) and later by Palmieri and Quagliariello (7) that anions accumulate in the mitochondrial matrix by electroneutral exchange with OH^- ions or by transport of the neutral acid form of the anion. The entry of anions in exchange for OH^- would be facilitated by H^+ on the outside of the membrane whereas the reverse exchange would be inhibited by OH^- inside. The gradient of any given anion at equilibrium would be expected to be a function of the Δ pH across the membrane according to the following equation (8):

(1) $$\log \frac{A_i}{A_e} = a + n \, \Delta \, pH$$

where A_i = internal anion concentration
A_e = external anion concentration
a = logarithm of the activity coefficient
n = number of charges on the anion

The original proposal of Mitchell (5) was a natural consequence of his scheme for oxidative phosphorylation in which flow of the electrons through the enzymes of the electron transport chain resulted in the development of an electrical potential gradient (negative inside) and a pH gradient (alkaline inside) across the membrane. These gradients are the result of an outward directed proton pump, and the proton motive force thus developed, defined as:

(2) $$\Delta \rho = \Delta \Psi - Z \, \Delta \, pH$$

where $\Delta \Psi$ = membrane potential
and Z = 2.3 RT/F

was proposed to be the driving force for ATP synthesis. Direct evidence has been obtained in other laboratories in recent years, supporting the Mitchell proposal that the Δ pH portion of the proton motive force provides the energy for anion accumulation in mitochondria. This evidence consists of the following observations:

1. In dilute solutions of permeant anions, mitochondria develop anion gradients according to equation 1 (8,9). The slope of the curve relating anion gradient to Δ pH has been shown to be equal to the charge (n) of the anion at neutral pH (8). Δ pH's have been varied by changing the pH of the external media (8-10) or by changing the pH of the

62

matrix with uncoupling agents or potassium ionophores (9,11).

2. Measurements of H^+ movements accompanying the accumulation of anions has shown the expected stoichiometry of H^+ entry to anion entry (12-14).

Prior to the proposal of Mitchell concerning the driving force for anion accumulation, a series of papers appeared from Chappell's laboratory (15,16) in which it was shown that many of the anion transport systems operated by very specific exchange processes. Thus, using techniques for mitochondrial swelling and for kinetic measurement of the state of reduction of the mitochondrial pyridine nucleotides, a number of separate carrier systems were identified. These included:

(1) A phosphate carrier specific for the exchange of phosphate for hydroxyl ions.

(2) A dicarboxylate carrier specific for the exchange of phosphate with malate, succinate, oxalacetate or malonate.

(3) An α-ketoglutarate carrier which catalyzes the specific exchange of α-ketoglutarate and malate.

(4) A tricarboxylate carrier which catalyzes the exchange of citrate or isocitrate with malate. It has been shown recently that phosphoenol pyruvate may also be transported on this carrier.

(5) A carrier which brings about the exchange of glutamate for aspartate.

(6) A glutamate carrier which catalyzes the exchange of glutamate for hydroxyl ions.

The original proposal, that all anions are accumulated by means of a direct proton symport or hydroxyl ion exchange, has been modified as a result of various studies of the different carrier systems. It now appears that the di and tri-carboxylates are accumulated in the matrix space in a manner proportional to the Δ pH due to indirect coupling with the transport of phosphate, which is coupled directly to proton movements.

This has been demonstrated most clearly by experiments (12, 17) in which mitochondria were loaded with citrate, malate, α-ketoglutarate, phosphate or pyruvate. Suspensions of loaded mitochondria were layered over silicone oil in a centrifuge tube, and the mitochondria were centrifuged at

0^0 through a second incubation media, then through a second
layer of heavier silicone oil and finally into perchloric
acid. The second incubation media, to which the mitochondria
were exposed for only a few seconds, contained no permeant
anions but was buffered at various pH's ranging from 6.5 to
8.5. The results showed that raising the pH increased the
rate of efflux of phosphate and pyruvate but not that of
citrate, malate or α-ketoglutarate, although at equilibrium
the gradients of citrate, malate and α-ketoglutarate are
greatly influenced by the pH of the suspending media. Likewise, other experiments (7) have shown that stimulation of
citrate or α-ketoglutarate efflux from mitochondria, observed in longer term incubations upon increasing the external
pH or decreasing the internal pH, is abolished by inhibiting
phosphate transport with mersalyl.

To permit this scheme for converting Δ pH gradients into
anion gradients to function well as a mechanism for the
accumulation of substrate anions in the face of an electrical
potential gradient (negative inside), all exchanges should
be electroneutral. Since citrate is a trivalent and malate
is a divalent anion at pH 7 the exchange of these two species
would seem to violate this rule.

The electrogenic exchange of divalent malate for trivalent
citrate in the presence of a large negative electrical potential gradient should cause an uneven distribution of citrate
(low inside, high outside), opposite to that actually measured in vivo or in vitro (18). It has been shown (14)
however by measurement of pH changes in the media during the
exchange of citrate for malate that an extra proton is transported with each molecule of citrate so that the net exchange
is electro-neutral.

The same situation exists for the exchange of phosphate
with malate. Measurements of pH changes during the exchange
have shown that proton movements compensate for the difference
of charge (12).

Formally, divalent citrate exchanges with divalent malate
and divalent malate exchanges with divalent phosphate. The
exchanging ionic species would thus appear to have little
relationship to the predominant ionic species existing in
solution at pH 7. It is the carrier itself which confers
electro-neutrality. Thus the carrier may be specific for the
unusual anionic form, or it may bind the predominant anion
species and have a separate binding site for charge compensating protons.

The problem of facilitated inward transport is solved
in a somewhat different manner for the transport of fatty
acids. It has been shown that long chain fatty acids must

be converted to carnitine derivatives in order to be transported into the mitochondria.

The scheme for transporting long chain acyl groups into the mitochondria, as originally proposed by Fritz and Yue (19), involved conversion of fatty acids to acyl CoA derivatives , transfer of the acyl group from CoA to carnitine and finally penetration of acyl carnitine derivatives into the membrane and transfer of the acyl unit to CoA within the matrix space. The observation that free carnitine would not penetrate the intact rat liver mitochondrial membrane prompted Yates and Garland (20) to modify the original model. The modification includes the proposal that two membrane bound carnitine acyl CoA transferases exist, one on the inner and one on the outer surface of the membrane. The enzyme catalyzes the transfer of the acyl unit without transferring carnitine across the membrane, directly to matrix CoA. However, in a recently published paper Pande (21) has described a new mitochondrial membrane transport system which catalyzes the exchange of carnitine for acyl carnitines in heart mitochondria. This observation has been confirmed by Tubbs (22). A measurable pool of endogenous carnitine exists in heart but not in liver mitochondria. Thus it is possible to demonstrate with radioactive isotopes a carnitine-carnitine exchange in heart but not in liver mitochondria. Carnitine and acyl carnitine are neutral species at pH 7, thus the transport of a negative anion against a potential gradient is avoided. However, transport in the physiological direction (acyl carnitine, in; carnitine, out) may be facilitated by the electrical potential gradient if long chain acyl carnitine is protonated upon binding to lipid bilayers, as described by Levitsky and Skulachev (23).

When an anionic metabolite is not a substrate, but is instead a product of mitochondrial metabolism needed in the cytoplasm, outward transport may be facilitated if the anion is transported in an electrical way and transport involves the net movement of electrical charges through the membrane. This has been demonstrated for the adenine nucleotide carrier which catalyzes the exchange of ATP for ADP (24,25). ADP is a trivalent and ATP a tetravalent anion at pH 7. The exchange has been shown to be at least partially electrogenic, and measurements of pH changes and cation movements during the exchange process indicate that efflux of ATP is not accompanied by a charge compensating equivalent movement of protons. No more than 40% of the ATP transported is charge compensated (26). The gradients of ATP and ADP across the mitochondrial membrane, which occur when the mitochondria are equilibrated with adenine nucleotides, reflect the electrical character of the carrier. Thus the ratio of ATP/ADP will under normal energized conditions be higher outside the mitochondria than inside (27,28). When the electrical potential gradient across the mitochondria is lowered by uncoupling agents (29) or by the use of valinomycin and high concentrations of K^+ (2), the difference in the internal and external ratios of ATP/ADP disappears.

Aspartate may also be considered a product of mitochondrial metabolism, since it is an end product for the oxidation of glutamate and cannot be further oxidized itself. It is needed in the cytosol as a source of carbon for gluconeogenesis (30,31), for the synthesis of urea, and as a substrate in the malate-aspartate cycle for transporting reducing equivalents from the cytosol into the mitochondria (32,33). Aspartate efflux from mitochondria is brought about by a specific carrier-mediated exchange of glutamate for aspartate. The glutamate-aspartate carrier originally identified by Azzi et al (34) has been shown recently to be electrogenic (35-37). The electrogenic character of this exchange is imposed by the carrier. The transported anions, glutamate and aspartate have very similar pK's and both exist almost exclusively as monovalent anions at neutral pH. Studies showed that in the presence of an electrical potential gradient, the exchange occurred almost exclusively in the direction glutamate in, aspartate out. Subsequent measurements of protons, accompanying the exchange of limited amounts of glutamate and aspartate, indicated the entry into the mitochondria of one proton with each molecule of glutamate. Efflux of aspartate occurred without an accompanying proton. Thus formally the process may be described as the exchange of neutral glutamic acid for the aspartate anion. Again it should be emphasized, however, that the electrical character of the exchange is imposed by the specificity of the carrier and not by the nature of the pK's of the transported species.

The studies described indicate the nature of the energy source used for the facilitated transport of anions into mitochondria and show how the specificity of the carrier within the membrane can make use of the proton motive force to drive anion transport in whichever direction the needs of the cell dictate. However, questions remain unanswered regarding the mechanism of action of the carriers. Various models have been proposed for the mitochondrial anion exchange carriers but little progress has been made in elucidating the nature of the translocating process itself. Figure 1 is an illustration of several possibilities.

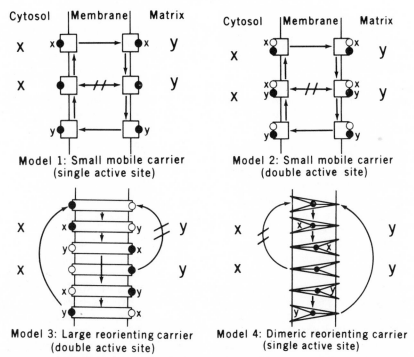

Fig. 1. *Several possibilities for translocation mechanisms of exchange carriers in the mitochondrial membrane.*

In each model we have depicted a mechanism for an exchange carrier which facilitates the 1:1 exchange of anion x in the cytosol for anion y in the matrix. In model 1, the carrier has a single binding site. X and y compete for this site, when they are on the same side of the membrane, and they are therefore competitive inhibitors of each other. When the level of y in the matrix is raised the apparent Vmax of the carrier should increase, and the apparent Km of x in the cytosol may also increase due to a greater availability of carrier sites. The carrier must move across the membrane and rotate 180° during the translocation step. This model has been criticized on thermodynamic grounds by S. J. Singer (38), who emphasizes the difficulty of moving the presumably hydrophylic binding site for the substrate through the hydrophobic core of the membrane.

Model 2 is similar in most respects to model 1 with the exception that two substrate binding sites are present on the same side of the membrane. The translocation step can occur only when a single carrier binding site is filled.

In model 3 the carrier spans the membrane and has binding

sites simultaneously exposed to both sides of the membrane. X and y should be competitive inhibitors when on the same side of the membrane, since there is only one substrate site exposed on each membrane surface. Increasing the level of y in the matrix should increase the apparent Vmax but may have no effect on the cytosolic Km of x since the availability of external binding sites may be unaffected by the presence of bound y on the inside. However in an ordered binding mechanism the possibility exists that binding y on the internal face could cause a conformational change that would lower the Km on the external face. The translocation step involves a $180°$ rotation which would occur only when both substrates were simultaneously bound.

Model 4 is a reorienting dimer of the type proposed by Singer in a recent review article (38). According to this model, the binding site would be located in a crevice or pore of the dimeric protein. The translocation step would involve no movement of the binding site relative to the membrane surfaces, but reorientation of the dimer would expose the binding site to the opposite surface of the membrane. This model is kinetically indistinguishable from model 1 with one possible exception. In the case of electrophoretic exchange where $x^{(-n)}$ exchanges with $y^{-(n+1)}$ by means of a mechanism similar to model 1, an increase of the membrane potential would be expected to cause an increase in the apparent Vmax of the carrier, and little or no change in the Km of $y^{-(n+1)}$, but possibly an increase in the Km of cytosolic $x^{(-n)}$ due to increased availability of carrier sites.

On the other hand it is possible that the influence of the electrical potential gradient on model 4 would be quite different. Since the reorientation step involves no movement of the charged species relative to the potential gradient, little or no change of Vmax may occur. The potential gradient might facilitate movement of the charged $y^{-(n+1)}$ down the pore of its binding site, causing a change in the concentration of y at the substrate site and thus a change in its apparent Km.

In discussing the above models, one must also consider that the binding sites may undergo conformational changes so that the substrate specificity of the binding site varies. Such a change could conceivably occur by protonation of the carrier. Should this occur, then two substrates for the exchange would be non-competitive inhibitors of each other if they are bound to different forms of the carrier.

Extensive studies of the kinetics of the adenine nucleotide carrier (2,26) have produced data in support of models 1 or 4. However studies of the influence of the potential

gradient on kinetic parameters have shown somewhat unexpectedly an influence of energy on the Km of ATP as well as the
Vmax of the carrier (39). Thus the data tend to support
model 4.

Kinetic analyses have been studied for the phosphate (40),
dicarboxylate (41,42), α-ketoglutarate (43) and tricarboxylate
carriers (44). Varying the pH of the media together with
phosphate levels has shown that high pH competitively inhibits
phosphate transport from outside to inside (8). This would
indicate that phosphate is transported as the neutral acid,
or that there is a separate binding site for a H^+ which must
be filled prior to the binding of the phosphate anion. The
pK of the binding site would have to be rather acidic to
account for the data. Other data (17) showing that initial
rates of efflux are faster at more alkaline pH's are difficult to reconcile with model 1 if one assumes either a
separate binding site for the proton or transport of phosphate as a neutral acid.

Kinetic analysis of the dicarboxylate carrier produced
data in support of model 2 (41). Dicarboxylic acid, with
carboxyl groups in the cis configuration, bind to one site
on the carrier; they can exchange with each other, and on a
single side of the membrane act as competitive inhibitors
for each other. However extramitochondrial phosphate, also
a substrate for the carrier, inhibits transport of extramitochondrial dicarboxylic acids in a non-competitive manner.
This result suggests that phosphate is either bound to a
separate site or binds to a different form of the carrier
resulting from a conformational change induced by carrier
protonation. In addition, there is data to support the conclusion that the dicarboxylate binding site includes a divalent metal ion since transport is inhibited competitively
with low concentrations of bathophenanthroline (42).

The α-ketoglutarate carrier has also been studied extensively (45,46) and in this case data are available concerning the effects of simultaneously varying substrate concentrations on either side of the membrane. When initial
rates of exchange of α-ketoglutarate and malate were measured
as a function of three different levels of external anion,
at each of three different levels of internal anion, it was
found that Vmax varied with the internal anion concentration,
but that the Km of the external anion was unaffected by the
level of internal anion.

It appeared therefore that the availability of external
binding sites bore no relationship to binding of substrate
on the internal surface. Thus the conclusion was drawn that
substrates may bind independently of each other (i.e. at

KATHRYN F. LaNOUE AND MARC E. TISCHLER

separate sites). Therefore the authors suggested a model
similar to that of model 3 in figure 1, where both substrates
bind to the carrier on either side of the membrane before
the initiation of the translocating step. These experiments,
however, have been criticized on the basis of the narrow
range over which the anion concentrations were varied and
more recent reports (47) indicate that the results are significantly different when measurements are made over a wider
concentration range.

The substrates of the tricarboxylate carrier when present
on the same side of the membrane compete for one binding
site (44). This binding site may also contain a divalent
metal ion since transport is competitively inhibited by low
concentrations of bathophenanthroline (42). The influence
of pH on the carrier is interesting since like the phosphate
and glutamate-aspartate carrier it transports protons as well
as anions.

The pH profile for transport indicates a pH optimum at 7.
Increasing the pH above 7 increases the Km for external citrate, without effecting the Vmax, while the inhibition
below 7 is non-competitive.

This is consistent with a model in which the proton binds
at a separate carrier site with a pK group in the region of
pH 6-7. Thus above pH 7, an increase in the media $[H^+]$ may
facilitate binding of trivalent citrate, but below pH 7 an
increase in media $[H^+]$, by also acidifying the matrix, may
retard the release of protons and trivalent citrate from the
carrier binding site into the matrix.

Kinetic experiments are currently under way on the glutamate-aspartate carrier and some preliminary results will
be presented here. Experimental details of the techniques
used are described fully in recent publications (35-37).

The pH profile of the exchange of glutamate for aspartate
was measured under energized conditions using oxalacetate
and glutamate to generate intramitochondrial aspartate
continuously. Extramitochondrial glutamate could then exchange with the aspartate formed inside the matrix. The rate
of appearance of extramitochondrial aspartate was measured
by centrifuging the mitochondria through a layer of silicone
oil (48) into perchloric acid at various time intervals.

Mitochondria were inhibited with rotenone, and energy was
provided by the oxidation of ascorbate plus TMPD (N,N,N',N'-
tetramethyl-p-phenylene diamine). The extramitochondrial pH
was varied with the use of a combination buffer containing
10 mM Tris and 10 mM PIPES (piperazine-N,N'-bis (2-ethane
sulfonic acid)). The intramitochoncrial pH was estimated
with the use of ^{14}C-DMO (5,5-dimethyloxazolidine-2,4-dione)

(49). The internal pH was varied with the use of phosphate and nigericin (50,51) at each of the different buffer pH's. The results are shown in Fig.2. Flux was stimulated at alkaline pH even though previous data had shown that a proton was transported into the mitochondria with glutamate. The experiment shows that flux is determined by the matrix pH and not the media pH. Thus when plotting flux versus external pH, three families of curves are obtained depending on the Δ pH across the membrane. However only one curve is obtained on plotting matrix pH versus flux. This is consistent with a model in which the proton binding site is separate from the amino acid binding site and has a somewhat alkaline pK (i.e., 7.93).

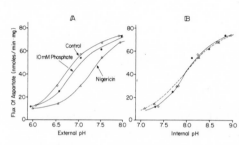

Fig.2. Effect of media and matrix pH on the glutamate-aspartate exchange. The buffer used in the control (o—o) and nigericin (Δ—Δ) incubations consisted of 100 mM KCL, 10 mM PIPES, 10 mM Tris-HCL, 6% dextran, 20 mM sucrose, 1 mM oxaloacetate, 5 mM Dl-cycloserine, 5 mM ascorbate, 0.2 mM TMPD. 0.01 mM rotenone, 1 μCi DMO [2-14C] and where added 20.6 ng nigericin per mg mitochondrial protein. Where phosphate (10 mM) was used (●—●) only the 20 mM sucrose was omitted. Glutamate was added 2 min after the incubation was begun with mitochondria (4.9 mg protein per ml), and samples were taken for mitochondrial separation. Flux of aspartate was determined from the linear rate of aspartate appearance in the media. Mitochondrial spaces were determined in parallel runs. The dashed line in B is a theoretical pH titration curve derived from the Henderson-Hasselbach equation. The temperature was 28°C.

Other data indicate that aspartate is a non-competitive inhibitor of glutamate entry, thus implying that aspartate and glutamate bind to different forms of the carrier, deprotonated and protonated, respectively. Figure 3 is a <u>Dixon</u> plot of the inverse of the rate of respiration of heart mitochondria versus the concentration of aspartate in the media, at different levels of external glutamate. A replot of the slopes of these lines versus the reciprocal glutamate concentration does not pass through the origin, thus showing that the inhibition is non-competitive (52,53). The Km for glutamate entry, determined in this way, was 7 mM and the Ki of aspartate was 4.3 mM.

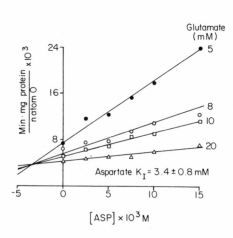

Fig. 3. *Inhibition of glutamate dependent respiration in rat heart mitochondria by external aspartate. Rat heart mitochondria (0.58 mg/ml) were incubated in a buffer (pH 7.2) containing 130 mM KCL, 0.05 mM EDTA, 20 mM KH_2PO_4, 20 mM PIPES, varied concentrations of glutamate and aspartate and 3 mM ADP. Respiration was initiated with malate (5 mM) and measured polarographically.*

The effect of varying the internal level of aspartate on the rate of glutamate entry has been measured using proton uptake as a measure of the rate of transport. Proton uptake has been shown previously to be stoichiometric with glutamate uptake. The internal aspartate concentration was varied by incubating glutamate-loaded, washed mitochondria with oxalacetate for varying periods of time (2 seconds - 1 minute) (35). This converts glutamate to aspartate via glutamate oxalacetate transaminase. Incubation was stopped with the addition of aminooxyacetate, an inhibitor of the transaminase (54). Subsequent addition of external glutamate initiates transport which can be measured as proton uptake with a very sensitive pH meter in a lightly buffered media. The results are shown in Fig. 4 and indicate that increasing internal aspartate increases the Km of external glutamate in a manner almost proportional to the accompanying increase of Vmax. These data are consistent with models 1 and 4. The degree to which the lines on the double reciprocal plot of flux vs. substrate concentration are parallel indicate the degree of irreversibility of some step between the carrier-glutamate complex facing the external media and the free carrier facing the matrix side of the membrane (55). The fact that the lines are parallel or nearly parallel indicate that the availability of the carrier on the external surface is a function of the forward reaction.

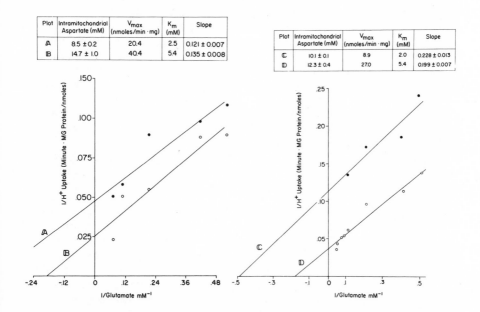

Plot	Intramitochondrial Aspartate (mM)	V_{max} (nmoles/min · mg)	K_m (mM)	Slope
A	8.5 ± 0.2	20.4	2.5	0.121 ± 0.007
B	14.7 ± 1.0	40.4	5.4	0.135 ± 0.008

Plot	Intramitochondrial Aspartate (mM)	V_{max} (nmoles/min · mg)	K_m (mM)	Slope
C	10.1 ± 0.1	8.9	2.0	0.228 ± 0.013
D	12.3 ± 0.4	27.0	5.4	0.199 ± 0.007

Fig. 4. Kinetics of the glutamate-aspartate exchange at different matrix aspartate concentrations. The kinetics were measured by following linear proton uptake. Glutamate-loaded mitochondria were incubated in 130 mM KCL, 6% dextran, 0.01 mM rotenone, sucrose [^{14}C], ^3H_2O, 1 mM oxaloacetate, and 1 mM PIPES, pH 7.0. Aminooxyacetate (5 mM) was added at different times after the incubation began to vary the aspartate loading. Prior to addition of varied glutamate concentrations, a sample was taken for analysis of the matrix aspartate concentrations.

Since aspartate acted as an inhibitor on the outer surface of the membrane it seemed plausible that glutamate should inhibit transport on the inner surface. Therefore, attempts were made to measure the Km of internal aspartate at high and low internal glutamate levels. Aspartate was varied as in the previous experiment with the use of aminooxyacetate acid and glutamate was varied by varying the time between the addition of aminooxyacetic acid and the initiation of transport by the addition of external glutamate to the media.

During the time period between the additions of aminooxyacetate and glutamate, internal glutamate is oxidized in the presence of oxalacetate to α-ketoglutarate and NH_3 by the combined action of malate dehydrogenase and glutamate dehydrogenase. In this experiment transport was supported by energy from the oxidation of ascorbate plus TMPD, and was measured directly as aspartate efflux by sampling the incubation media at 20 second intervals after the addition of glutamate. The mitochondria were centrifuged through silicone oil into perchloric acid. Aspartate was measured in the media and matrix spaces as a function of time to determine flux. The double reciprocal plot of flux vs. aspartate concentration at two different levels of internal glutamate (5 mM vs. 17 mM) show that glutamate is not an inhibitor and therefore has no effect on flux.

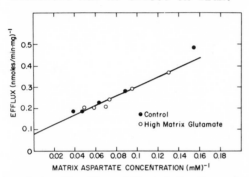

Fig. 5. Effect of matrix glutamate on the efflux of aspartate. Glutamate-loaded mitochondria (5.8 mg protein per ml) were incubated in a buffer containing 150 mM KCL, ^3H_2O (1 μCi), 6% dextran and 20 mM MOPS (pH 7.2 at 10°C. The experiment was begun with the addition of oxalacetate (1 mM). After 1.5 min aminooxyacetic acid (5 mM) was added. In two experiments (control) more oxalacetate (4 mM) was added at 3 min followed by ascorbate (5 mM) plus TMPD (0.2 mM) at 14 min. At 14.5 min ^{14}C-sucrose was added followed 0.5 min later by 2.5 mM glutamate to initiate aspartate efflux. Samples were taken for rapid separation by centrifugation immediately prior to glutamate addition and at 15, 30, 60, and 90 sec afterwards. In two other experiments ascorbate plus TMPD were added at 1.6 min after initiation of the incubation and ^{14}C-sucrose at 1.8 min. External glutamate (2.5 mM) was added at 2 min to initiate the efflux of aspartate. Samples were taken as in the control. Aspartate levels in the supernatant solution were measured with an amino acid analyzer and that in the mitochondrial fraction was measured enzymatically. Fluxes shown are the rates of aspartate appearance in the media at 0, 30 and 60 sec after glutamate addition. The aspartate concentrations shown are the amounts in the matrix divided by the matrix volume.

Other data not presented here have shown that the glutamate-glutamate exchange on the glutamate-aspartate carrier is very slow. Taken as a whole the kinetic data obtained thus far are consistent with both models 1 and 4. It appears likely that there is a separate proton binding site on the carrier which confers the electrogenic character to the exchange transport. This proton binding site has a relatively alkaline pK group of about 7.9 and is thus easily distinguishable from any of the glutamate or aspartate pK's. The lack of glutamate-glutamate exchange, the dependence of the external kM of glutamate on the amount of internal aspartate, as well as the inability of glutamate to inhibit aspartate transport at the inner side of the membrane, all indicate that the substrate binding site on the inner surface is relatively specific for aspartate. Thus the binding site outside is very different in character from the internal binding site and this type of asymmetry is more compatible with model 4 than with model 1.

It would appear from presently available data that model 4 may be a reasonable representation of the translocation step for adenine nucleotides and glutamate-aspartate exchange. The indications are that these two carriers at least do not make a 180° turn during translocation but present one surface to the cytoplasm and another surface to the matrix. This is shown by inhibitor specificity for sideness in the case of the adenine nucleotide carrier (56) and a similar substrate binding specificity in the case of the glutamate-aspartate carrier.

Changes in the Km of ATP translocation with membrane potential provide a further indication of the plausibility of model 4. The crevices in which substrate binding sites may be located represent potentially separate compartments of significant proportions. Substrate concentrations in these crevices may vary independently of concentrations in the matrix space or cytosol due to changes of membrane potential, and interactions with specific enzymes, e.g. glutamate oxalacetate transminase in the case of the glutamate-aspartate carrier (57) and ATPase in the case of the adenine nucleotide carrier (58). These possibilities should be taken into account in planning future experiments.

REFERENCES

1. Slater, E.C., Quagliariello, E., Papa, S. and Tager, J.M. (1969) *in* The Energy Level and Metabolic Control in Mitochondria, (Papa, S., Tager, J.M., Quagliariello, E. and Slater, E.C., eds.), pp. 1-12, Adriatica Editrice, Bari.
2. Klingenberg, M. (1970) Essays in Biochemistry 6, 119.
3. Harris, E.J., Höfer, M.P. and Pressman, B.C. (1967) Biochemistry 6, 1348.

4. Harris, E.J. and Pressman, B.C. (1969) Biochim. Biophys. Acta 172, 66.
5. Mitchell, P. (1968) *in* Chemiosmotic Coupling and Energy Transduction, Glynn Research, Bodmin, England.
6. Mitchell, P. and Moyle, J. (1969) Eur. J. Biochem. 9, 149.
7. Palmieri, F. and Quagliariello, E. (1969) Eur. J. Biochem. 8, 473.
8. Palmieri, F., Quagliariello, E. and Klingenberg, M. (1970) Eur. J. Biochem. 17, 230.
9. Quagliariello, E. and Palmieri, F. (1970) FEBS Letters 8, 105.
10. Palmieri, F., Genchi, G. and Quagliariello, E. (1971) *in* Biological Aspects of Electrochemistry, Experientia Suppl. 18, 505.
11. Meijer, A.J., Brouwer, A., Reijngoud, D.J., Hoek, J.B. and Tager, J.M. (1972) Biochim. Biophys. Acta 283, 421.
12. Quagliariello, E. and Papa, S. (1971) *in* Biological Aspects of Electrochemistry, Rome.
13. Hoek, J.B., Lofrumento, N.E., Meijer, A.J. and Tager, J.M. (1971) Biochim. Biophys. Acta 226, 297.
14. Papa, S., Lofrumento, N.E., Kanduc, D., Paradies, G. and Quagliariello, E. (1971) Eur. J. Biochem. 22, 134.
15. Chappell, J.B. (1969) Brit. Med. Bull. 24, 150.
16. Chappell, J.B. and Haarhoff, K.M. (1966) *in* Biochemistry of Mitochondria, (Slater, E.C., Kaniuga, Z. and Wojtzak, L., eds.) pp. 75-91, London and Warsaw.
17. Papa, S. and Lofrumento, N.E., Qualiariello, E., Meijer, A.J. and Tager, J.M. (1970) Bioenergetics 1, 287.
18. Williamson, J.R. (1975) *in* Gluconeogenesis: Its Regulation in Mammalian Metabolism, (Hanson, R.W. and Mehlman, M.A., eds.), John Wiley and Sons, New York, in press.
19. Fritz, I.B. and Yue, K.T.N. (1963) J. Lipid Res. 4, 279.
20. Yates, D.W. and Garland, P.B. (1970) Biochem. J. 119, 547.
21. Pande, S.V. (1975) Proc. Nat. Acad. Sci. U.S.A. 72, 883.
22. Ramsay, R.R. and Tubbs, P.K. (1975) FEBS Letters 54, 21.
23. Levitsky, D.O. and Skulachev, V.P. (1972) Biochim. Biophys. Acta 275, 33.
24. Pfaff, E. and Klingenberg, M. (1968) Eur. J. Biochem. 6, 66.
25. Klingenberg, M., Heldt, H.W. and Pfaff, E. (1969) *in* The Energy Level and Metabolic Control in Mitochondria, (Papa, S., Tager, J.M., Quagliariello, E. and Slater, E.C., Eds.) p. 236, Adriatica Editrice, Bari.
26. Klingenberg, M. (1972) *in* Mitochondria: Biogenesis and Bioenergetics/Biomembranes: Molecular Arrangements and Transport Mechanisms Vol. 28,

(Van den Bergh, S.G., Borst, P., Van Deenen, L.L.M., Riemersma, J.C., Slater, E.C. and Tager, J.M., eds.), pp. 147-162, North Holland Publishing Co., Amsterdam.

27. Elbers, R., Heldt, H.W., Schmucker, P., Soboll, S. and Wiese, H. (1974) Hoppe-Seylers Z. Physiol. Chem. 355, 378.

28. Heldt, H.W., Klingenberg, M. and Milovancev, M. (1972) Eur. J. Biochem. 30, 434.

29. Slater, E.C., Rosing, J. and Mol, A. (1973) Biochim. Biophys. Acta 292, 534.

30. Lardy, H.A., Paektau, V. and Walter, P. (1965) Proc. Nat. Acad. Sci. U.S.A. 53, 1410.

31. Anderson, J.H., Nicklas, W.J., Blank, B., Refino, C. and Williamson, J.R. (1971) in Regulation of Gluconeogenesis: 9th Conference of the Gesellshaft für Biologische Chemie (Söling, H.-D and Willms, B., eds). pp 293-315, G. Thieme-Verlag, Stuttgart.

32. LaNoue, K.F. and Williamson, J.R. (1971) Metabolism 20, 119.

33. Borst, P. (1963) in Funktionelle und Morphologische Organisation der Zelle (Karlson, P., ed.) pp. 137-158, Springer-Verlag, New York.

34. Azzi, A., Chappell, J.B. and Robinson, B.H. (1967) Biochem. Biophys. Res. Commun. 29, 148.

35. LaNoue, K.F., Meijer, A.J. and Brouwer, A. (1974) Arch. Biochem. Biophys. 161, 544.

36. LaNoue, K.F., Bryla, J. and Bassett, D.J.P. (1974) J. Biol. Chem. 249, 7514.

37. LaNoue, K.F. and Tischler, M.E. (1974) J. Biol. Chem. 249, 7522.

38. Singer, S.J. (1974) Ann. Rev. Biochem. 43, 805.

39. Souverijn, J.H.M., Huisman, L.A., Rosing, J. and Kemp, A.Jr. (1973) Biochim. Biophys. Acta 305, 185.

40. Coty, W.A. and Pedersen, P.L. (1974) J. Biol. Chem. 249, 2593.

41. Palmieri, F., Prezioso, G., Quagliariello, E. and Klingenberg, M. (1971) Eur. J. Biochem. 22, 66.

42. Meisner, H., Palmieri, F., and Quagliariello, E. (1972) Biochemistry 11, 949.

43. Palmieri, F., Quagliariello, E. and Klingenberg, M. (1972) Eur. J. Biochem. 29, 408.

44. Palmieri, F., Stipani, I., Quagliariello, E. and Klingenberg, M. (1972) Eur. J. Biochem. 26, 587.

45. Sluse, F.E., Ranson, M. and Liebécq, C. (1972) Eur. J. Biochem. 25, 207.

46. Sluse, F.E., Goffart, G. and Liebécq, C. (1973) Eur. J. Biochem. 32, 283.

47. Sluse, F.E. (1974) Post-FEBS Colloquium on Ion Transport and Mitochondrial Function, Budapest, Hungary, September, 1974.

48. LaNoue, K.F., Walajtys, E. and Williamson, J.R. (1973) J. Biol. Chem. 248, 7171.

49. Adanki, S., Cahill, F.D. and Sotos, J.F. (1968) Anal. Biochem. 25, 17.

50. Pressman, B.C. (1968) Fed. Proc. 27, 1283.

51. Cockrell, R.S., Harris, E.J. and Pressman, B.C. (1967) Nature 215, 1487.

52. Cleland, W.W. (1967) Ann. Rev. Biochem. 36, 77.

53. Cleland, W.W. (1963) Biochim. Biophys. Acta 67, 188.

54. Hopper, S. and Segal, H.L. (1962) J. Biol. Chem. 237, 3189.

55. Plowman, K.M. (1972) in Enzyme Kinetics, pp. 40-55, McGraw Hill, New York.

56. Klingenberg, M. and Bucholz, M. (1973) Eur. J. Biochem. 38, 346.

57. Tischler, M.E., Pachence, J., Blackwell, B., Williamson, J.R. and LaNoue, K.F. (1975) Arch. Biochem. Biophys., in press.

58. Vignais, P.V., Vignais, P.M. and Doussiere, J. (1975) Biochim. Biophys. Acta 376, 219.

MITOCHONDRIAL METABOLISM AND CELL REGULATION*

J.R. Williamson

Anion Transport, Distribution and Energetics

The citric acid cycle provides reducing equivalents to
the electron transport chain for ATP synthesis and also, via
ancilliary reactions, provides substrate for biosynthetic
reactions in the cytosol. The relative importance of these
roles in the overall cell metabolism of the tissue will thus
vary with the particular organ and the dietary status of the
animal. These multiple and changing demands necessitate a
rather complex detailed regulation of the individual enzymes
of the citric acid cycle in order to allow a nonuniform flux
through various segments. In principle, regulation may be
achieved by changes in any of three basic parameters: substrate availability, phosphate potential (ATP/ADP x P_i) and
pyridine nucleotide oxidation-reduction state. Apart from
obvious changes in the relative concentrations of primary
substrates supplied to an organ by the blood, substrate a-
vailability to specific enzymes of the citric acid cycle is
also regulated by the permeability of the mitochondrial membrane to particular intermediates. Transport of metabolic
anions across the mitochondrial membrane occurs by carrier-
mediated 1:1 exchange mechanisms which are specific for individual anions.

The major anion translocators so far described and studied in some detail are shown in Fig. 1 (see 1-4 for reviews). Basically there are three types of translocators:
electrogenic, electroneutral proton-compensated, and electroneutral. With electrogenic transport, there is a charge imbalance of the transported anions, and flux through the adenine nucleotide and glutamate-aspartate carriers is energe-
tically driven by maintenance of an electrical potential
(negative inside) generated by energy conservation reactions
of the electron transport chain. The phosphate, glutamate,
pyruvate and tricarboxylate translocators promote electroneutral exchanges of anions, but since transport of a proton is involved to achieve electrical balance, these exchanges are very sensitive to alterations of pH. Thus entry, of

* This work was supported by grants AM-15120 and HL-14461
from the National Institutes of Health and by contact
NIH-71-2494 under the Myocardial Infarction Program,
National Heart and Lung Institute.

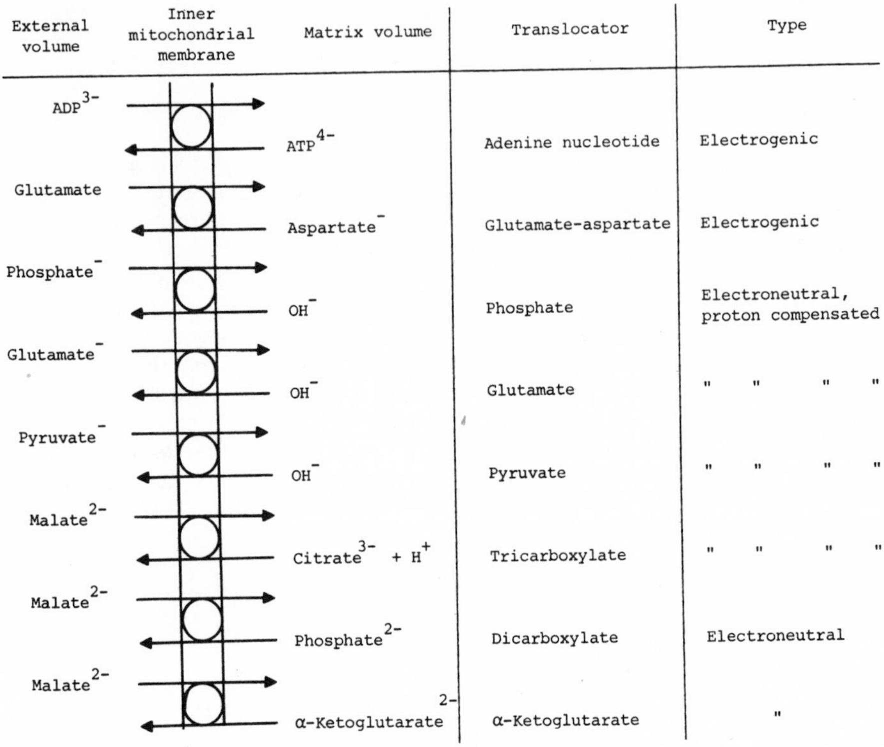

External volume	Inner mitochondrial membrane	Matrix volume	Translocator	Type
ADP^{3-}		ATP^{4-}	Adenine nucleotide	Electrogenic
Glutamate		$Aspartate^-$	Glutamate-aspartate	Electrogenic
$Phosphate^-$		OH^-	Phosphate	Electroneutral, proton compensated
$Glutamate^-$		OH^-	Glutamate	" " " "
$Pyruvate^-$		OH^-	Pyruvate	" " " "
$Malate^{2-}$		$Citrate^{3-} + H^+$	Tricarboxylate	" " " "
$Malate^{2-}$		$Phosphate^{2-}$	Dicarboxylate	Electroneutral
$Malate^{2-}$		$\alpha\text{-Ketoglutarate}^{2-}$	α-Ketoglutarate	"

Fig. 1. *Mitochondrial anion translocators.*

phosphate, glutamate, etc. is inhibited by a rise of pH of the medium. On the other hand, the electroneutral dicarboxylate and α-ketoglutarate translocators are relatively insensitive to pH changes since no ancilliary proton movements are involved. Only the phosphate and adenine nucleotide translocators are directly involved in respiration and are common to all tissues, while the presence of the other translocators (e.g. citrate-malate exchange) correlates with specialization of the organ for particular metabolic functions such as fatty acid synthesis. In tissues other than liver, which has a uniquely high capacity for ketogenesis, citric acid cycle flux depends on the rate of mitochondrial electron transport since acetyl-CoA generated from pyruvate or β-oxidation of fatty acids is removed by oxidation in the cycle. Thus, studies with isolated heart mitochondria oxidizing pyruvate have shown a direct correlation between in-

creased pyruvate uptake and oxygen consumption for the
change from resting (state 4) to active (state 3) respiration induced by addition of ADP and phosphate (5-7). Activation of electron transport is associated with an oxidation
of the mitochondrial NAD system (8) and a decrease (9-11) or
no change (12) of the intramitochondrial ATP/ADP ratio. Basic questions concerning regulation of the citric acid cycle, therefore, are (a) definition of the relationships between the NAD redox state and the adenine nucleotide phosphorylation state in the cytosol and mitochondria, (b) elucidation of the sensitivities of individual enzymes of the
cycle to feedback effects from the electron transport chain
caused by alterations of either the pyridine nucleotide oxidation-reduction state or the phosphorylation state of the
intramitochondrial adenine nucleotides, and (c) definition
of the interrelationships between transport of intermediates
across the mitochondrial membrane and their rate of formation and utilization in the citric acid cycle.

According to the proposals of Mitchell (13,14), oxidation-
reduction reactions at the three energy conservation sites
of the electron transport chain generate an electrochemical
potential difference between protons in the matrix and the
media of mitochondria, where the overall protonmotive force
(Δp) in mV is comprised of an electrical term $\Delta\Psi$ for the
membrane potential and a term for the $[H^+]$ difference:

$$\Delta p = \Delta\Psi - Z \Delta pH \qquad\qquad (Eq. 1)$$

where $Z = (2.3 RT/F)$. Under normal conditions of mitochondrial respiration, $\Delta\Psi$ is the main component of the total
protonmotive force. This force is used to drive the synthesis of ATP from ADP and P_i, and hence should be at least e-
qual to the free energy of synthesis of ATP for given conditions prevailing in the cell. Therefore,

$$\Delta p = \Delta G'/nF + \frac{1}{n} Z \log [ATP]/[ADP][P_i] \qquad (Eq. 2)$$

where $\Delta G'$ is the free energy of hydrolysis of ATP under appropriate experimental conditions, $[ATP],[ADP],[P_i]$ refer to
concentrations in the cytosol and n is the number of protons
transported across the mitochondrial membrane per molecule
of ATP synthesized.

Estimations of the size of the membrane potential have
been made in respiring mitochondria by measuring the potassium or rubidium concentration gradient across the mitochondrial membrane in the presence of valinomycin to increase
monovalent cation permeability, and applying the Nernst e-
quation:

$$\Delta E = RT/F \, \log \frac{[M^+]_i}{[M^+]_o} \tag{Eq. 3}$$

where $[M^+]_i$ and $[M^+]_o$ represent the activities of the cation in the internal (matrix) and external (medium) phases, respectively. The pH gradient is estimated separately from the pH change observed on lysing mitochondria with Triton X-100 (15) or more directly from the distribution of $[^{14}C]$ 5,5-dimethyl-2,4-oxazolidine dione using modifications of the method described by Addanki et al (16). The overall proton-motive force is then calculated from equation 1. Measured values of $\Delta\Psi$ for state 4 conditions range from 125 mV (17) to 150-168 mV (15,18) with -Z ΔpH = 23 to 77 mV, corresponding to a Δp of 148 to 228 mV. Both $\Delta\Psi$ and -Z ΔpH were found to decrease during active respiration in the presence of ADP, with Δp falling in the range from 143 to 204 mV (15, 17,18). However, even the highest measured value of Δp is insufficient by 50 to 80 mV to account for production of ATP against the phosphate potential maintained both by state 4 and intermediate states if an n value of 2 is used in equation 2 (18). The stoichiometry of the number of protons synthesized per mole of ATP synthesized is still a controversial issue, but a number of authors have suggested that it may be greater than 2 and even variable as the rate of electron transport flux increases (14,18-21).

Relative values of ΔpH and $\Delta\Psi$ have important consequences in relation to the distribution of anions at equilibrium across the mitochondrial membrane. Studies with isolated mitochondria (22,23) have ascertained that for those anions which are transported as the undissociated acid (e.g. acetate) or by electroneutral exchange, the distribution of anion A of charge n between the internal (i) and external (e) media is equal to the nth power of the $[H^+]$ distribution:

$$\frac{A_i^{n-}}{A_e^{n-}} = F \, \frac{[H^+]_e}{[H^+]_i} \tag{Eq. 4}$$

or $\log (A_i^{n-}/A_e^{n-}) = \log F + n \, \Delta$pH $\tag{Eq. 5}$

where F is the ratio of the extra and intramitochondrial activity coefficients of the anion. Thus, as a first principle, most of the metabolic anions with specificity for the translocators will be expected to have a higher concentration in the mitochondria than in the cytosol, so that calculation of large inverse concentration gradients must be viewed with suspicion (24). Since equilibrium conditions correspond to zero flux, it is obvious that net exchange between anions can occur only if their relative gradients deviate from the equilibrium values. Thus the driving force for transport of anions by electroneutral exchange is manifested in anion gradients, with the energy ultimately being derived from energy conservation in the electron transport chain which is responsible for maintenance of the pH gradient.

In contrast, the distribution and driving force of anions transported by electrogenic exchange is expected to be determined by the membrane potential. Thus for anion A_o^{n-} exchanging with anion $B_i^{(n+1)-}$, the distributions of A and B at equilibrium across the mitochondrial membrane is given by:

$$\Delta\Psi = Z \log \left[A_i/A_o \times B_o/B_i \right] \qquad \text{(Eq. 6)}$$

As discussed elsewhere (25), it is probable that anions transported by electrogenic exchange are not in equilibrium with the membrane potential. Nevertheless, the membrane potential accounts for a higher concentration of electrogenically transported anions in the cytosol than in the mitochondria. Thus unlike electroneutrally transported anions, the concentration gradient of ATP and aspartate across the mitochondrial membrane may be considerably less than 1. This has been confirmed directly for adenine nucleotide transport. Studies both with isolated rat liver mitochondria (26-28), and in intact liver (29,30) or isolated hepatocytes (31,32) after separation of the mitochondria from the cell by special techniques have shown that the cytosolic ATP concentration is greater than that in the mitochondria. Kinetic restrictions on the electrogenic translocators probably prevent equilibrium of anion distribution with the membrane potential under normal conditions. Thus, extramitochondrial ATP has been shown to inhibit entry of ADP into mitochondria (28,33,34) while extramitochondrial aspartate inhibits glutamate transport on the glutamate-aspartate translocator (35).

Maintenance of an electrical potential across the mitochondrial membrane imposes vectorial properties on the ADP: ATP and glutamate:aspartate exchange reactions. Thus, influx of glutamate coupled to aspartate efflux is much faster than transport in the reverse direction (35). The physiological effect of electrogenic aspartate transport is to make the malate-aspartate cycle (36) unidirectional for the transport of reducing equivalents into mitochondria, and it also permits maintenance of a lower NAD redox potential in

the mitochondria(approx. -300 mV) compared with the cytosol
(approx. -250 mV). The ratio of (NAD/NADH) cytosol to (NAD/
NADH) mitochondria may achieve values of about 100 (4,37).
Energy required to transport NADH into the more reduced environment of the mitochondrial matrix, therefore, is derived
directly from the respiration-driven proton pump. The protonmotive force provides a static head to maintain the membrane potential which otherwise would be dissipated by the
electrical imbalance associated with aspartate transport.
Energy is thus used which would otherwise be expended for
the synthesis of ATP. Preliminary experiments indicate that
the aspartate:ATP stoichiometry is at least 2.

The physiological effect of electrogenic adenine nucleotide transport is to make the extramitochondrial ATP/ADP ratio up to 100 times greater than that in the mitochondria
(2). This phenomenon has been observed with separation studies performed with isolated liver mitochondria (26-28) as
well as intact liver (29,31). In experiments with perfused
rat hearts under different conditions of left ventricular
work (38), the cytosolic and mitochondrial ATP/ADP ratios
were calculated from knowledge of the total tissue adenine
nucleotide, creatine phosphate and creatine contents, the mitochondrial adenine nucleotide content, the tissue mitochondrial content, and an assumed equilibrium of creatine phos-
phokinase with cytosolic ATP and ADP. These calculations
showed that the cytosolic ATP/ADP ratio was about 2 orders
of magnitude greater than that of the mitochondria and
changed by a proportionately much greater extent than the
intramitochondrial ATP/ADP ratio with alterations of heart-
work (38).

*Relation of Phosphate Potential to Pyridine
Nucleotide Redox State*

The difference between the ATP/ADP ratios in the cytosol
and mitochondria has important implications in relation to
the interactions between the adenine nucleotide and pyridine
nucleotide systems in the different cellular compartments.
The redox state of the NAD system in the cytosol is linked
to the cytosolic phosphate potential by the glycolytic enzymes 3-P-glycerate kinase and glyceraldehyde-3-P dehydrogenase according to the relationships:

$$GAP + P_i + NAD = 1,3\ PGA + NADH \qquad\qquad (Eq.\ 7)$$

$$1,3\ PGA + ADP = 3\ PGA + ATP \qquad\qquad (Eq.\ 8)$$

Thus, $\dfrac{[NAD]}{[NADH]} = \dfrac{[ATP]}{[ADP][P_i]} \times \dfrac{[3\ PGA]}{[GAP]} \times \dfrac{1}{K_{PGK} \cdot K_{GAPDH}}$

(Eq. 9)

where the equilibrium constants, K_{GAPDH} and K_{PGK} for the reactions in equations 7 and 8 refer to pH 7. The cytosolic NAD/NADH may be estimated from measurements of the lactate/ pyruvate ratio by assuming equilibrium of lactate dehydrogenase. When this was done, and values for metabolites measured in total liver extracts substituted in equation 9, the overall reaction could be demonstrated to be in near-equilibrium for a number of metabolic conditions in liver (39,40). On this basis, it was concluded that the components of lactate dehydrogenase, glyceraldehyde-3-P dehydrogenase and 3-phosphoglycerate kinase systems were in near equilibrium in vivo and that the redox state of the cytosolic NAD system was controlled by the phosphorylation state of the adenine nucleotide system. Thus, if the ratio of 3 PGA/GAP remains approximately constant, an oxidation of the cytosolic NAD system is expected to correlate with an increased cytosolic phosphate potential and vice versa. However, in view of the fact that the true cytosolic phosphate potential may be considerably greater than that estimated from total tissue measurements of ATP, ADP and P_i, it must be concluded that the reaction depicted by equation 9 is not as close to equilibrium as previously supposed. The deviation from true thermodynamic equilibrium for near-equilibrium systems is dependent on flux; the greater the flux the greater the deviation, and this fact must always be considered in the interpretation of metabolic data. Nevertheless, a high cytosolic phosphate potential is probably an important factor responsible for maintaining the cytosolic NAD redox potential in a relatively oxidized state under normoxic conditions.

The cytosolic phosphate potential also influences the mitochondrial NAD redox state through its effect on respiratory control of the electron transport chain. The rate-limiting step of oxidative phosphorylation is thought to be the adenine nucleotide translocation step, and Davis and Lumeng (28) have reported an operational $\underline{K_m}$ for cytosolic ADP of 13 µM for respiring rat liver mitochondria. Control of respiration in the intermediate range between state 4 and state 3 is considered to be exerted by the cytosolic phosphate potential (28,41,42). Under conditions of state 4 respiration, studies with isolated mitochondria have established the probability that the overall process of oxidative phosphorylation at the three energy conservation sites is near thermodynamic equilibrium (42,43). Since oxidative phosphorylation involves conversion of extramitochondrial

ADP to ATP, translocation of the adenine nucleotides across the mitochondrial membrane is an intermediary step, so that the equation relating the production of 3 moles of ATP from ADP during electron transport from NADH to cytochrome oxidase is expressed as:

$$NADH + 2 \text{ cyt } a_3^{3+} + 3 \text{ ADP} + 3 \text{ P}_i = NAD + 2 \text{ cyt } a_3^{2+} + 3 \text{ ATP}$$

(Eq. 10)

Thus,
$$\frac{[NAD]}{[NADH]} = K \cdot \frac{[a_3^{3+}]}{[a_3^{2+}]} \cdot \frac{[ADP]^3 [P_i]^3}{[ATP]^3}$$

(Eq. 11)

where concentration terms for ATP, ADP and P_i refer to those in the extramitochondrial space. Equation 11 illustrates that oxidation of the mitochondrial NAD system is reciprocally related to the cube power of the extramitochondrial phosphate potential. This expression explains why the mitochondrial NAD system in the presence of substrate is highly reduced under state 4 conditions and becomes more oxidized in state 3 when the cytosolic phosphate potential falls. An expression similar to that of equation 11 relating the square of the phosphate potential to the redox states of NAD and cytochrome c for the first two phosphorylation sites of the electron transport chain has been tested experimentally in isolated mitochondria, ascites tumor cells and perfused liver (44-46). Close approximation to equilibrium were obtained for mitochondria under state 4 conditions, but even in the intact tissues good agreement was obtained between measured and calculated equilibrium constants indicating apparent near-equilibrium even under conditions approaching state 3 respiratory rates. However, total tissue values for ATP, ADP and P_i were used for the cytosolic phosphate potential, which was therefore underestimated by about an order of magnitude. Contrary to the author's interpretation, therefore, the measurements in the intact cells illustrate a deviation of the phosphorylating electron transport chain from thermodynamic equilibrium with the phosphate potential when electron transport flux increases. Oshino et al (47) present other arguments suggesting that assumptions of equilibrium may be erroneous at high electron flow rates.

Anion Transport in Relation to
Gluconeogenesis and Ureogenesis

Different biosynthetic functions in the liver require net flux of certain anions across the mitochondrial membrane, depending on the nature of the substrate supplied. However, since the specificity of the anion translocators necessi-

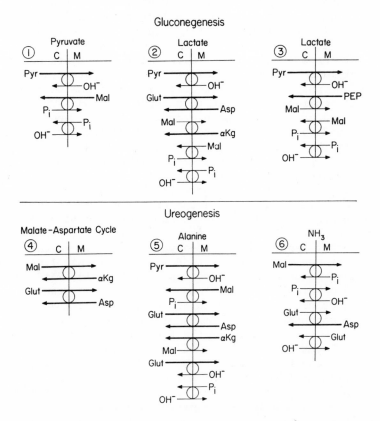

Fig. 2. Transport steps required for different metabolic functions in liver.

tates transport of the exchanging anion with a 1:1 stoichiometry, the second anion may be transported back in the opposite direction via another translocator, which theoretically could provide the limiting step for net transport of the first anion. This point is illustrated by the diagrams in Fig. 2. The net movement of anions is depicted by large heavy arrows, while ancilliary, obligatory transport of other anions is depicted by small arrows. The first two diagrams depict gluconeogenesis from pyruvate and lactate, respectively, in rat liver, where pyruvate carboxylase is a mi-

tochondrial enzyme and P-enolpyruvate carboxykinase is cytosolic. In species such as guinea pig, rabbit or birds
where P-enolpyruvate carboxykinase is partly or mainly also
a mitochondrial enzyme, P-enolpyruvate may be transported
directly out of the mitochondria instead of via malate or
aspartate (third diagram). Although not initially obvious
from the requirements for net anion movements, it is evident
that all three situations involve malate transport on the
dicarboxylate translocator as well as phosphate transport
on the phosphate-hydroxyl exchange in the opposite direction
to that required for oxidative phosphorylation. The involvement of the malate:phosphate exchange in gluconeogenes from
lactate and pyruvate accounts for its sensitivity to inhibition in the intact liver to 2-n-butylmalonate (48,49)
which is a specific inhibitor of malate transport on the
dicarboxylate translocator (50,51). A similar situation ap-
plies to gluconeogenesis from alanine (diagram 5), while
urea synthesis from both alanine or added ammonia (diagram
6) which requires net efflux of aspartate, involves also a
glutamate:hydroxyl exchange as an ancilliary transport step.
On the other hand, the malate-aspartate cycle, which is required for net transport of reducing equivalents from the
cytosol to mitochondria, involves only the glutamate:aspar-
tate and α-ketoglutarate:malate translocators (diagram 4).

Evidence presented in detail elsewhere (4,52) suggests
that the concentrations of most of the transported anions,
at least in the cytosol, are in the apparent K_m region for
the translocators. The rate of anion transport in vivo,
therefore, may be expected to be controlled both by the
concentration of the various anions on either side of the
mitochondrial membrane as well as by the apparent K_m and
Vmax values of the translocators. In addition, specific
anion interaction effects may be important, as exemplified
by aspartate inhibition for glutamate entry on the glutamate-aspartate translocator (35). It has also been suggested that the rate of urea synthesis in liver may be limited
by the glutamate-hydroxyl translocator because of its low
Vmax and relatively high apparent K_m for glutamate (53). In
general, however, there is as yet insufficient reliable kinetic information on the translocators to draw any firm
conclusions. Evidence pertaining to malate transport being
rate-limiting for ureogenesis in the intact liver cell, together with data on effects of glutamate and aspartate
transport kinetics on cellular metabolism is discussed
elsewhere (25,52).

Feedback Interactions between the Respiratory Chain and the Citric Acid Cycle

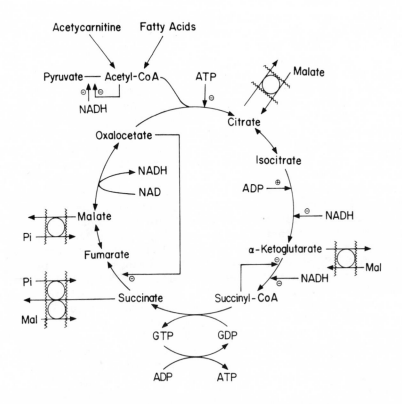

Fig. 3. Scheme showing possible feedback interactions from the electron transport chain to the citric acid cycle as mediated by adenine and pyridine nucleotides.

In Fig. 3 the reaction steps of the citric acid cycle are depicted together with interactions of the intermediates with the anion translocators and the interactions of individual enzyme steps with the intramitochondrial adenine and pyridine nucleotides. For simplicity, ancilliary reactions such as transamination of intramitochondrial oxalacetate with glutamate via spartate aminotransferase, conversion of pyruvate to oxalacetate via pyruvate carboxylase, and reaction of α-ketoglutarate, glutamate and ammonia via glutamate dehydro-

genase are omitted from this scheme. Multisite interaction
is essential to achieve uniform flux through the citric acid
cycle in mammalian mitochondria because of the interaction
of the substrates with the anion transport systems and the
fact that many of the enzyme steps are far displaced from
equilibrium and therefore not subject to normal mass action
control by product inhibition. Regulation is achieved by kinetic control of those enzymes which are associated with a
large negative free energy change, namely citrate synthase,
isocitrate dehydrogenase and α-ketoglutarate dehydrogenase.
In addition, an important regulatory function is achieved by
control of the rate of acetyl-CoA production from pyruvate
and fatty acids via pyruvate dehydrogenase, β-oxidation and
acylcarnitine transferase. As shown in Fig. 3 and discussed
by Atkinson (54), direct inhibition of citrate synthase by
ATP and stimulation of NAD-linked isocitrate dehydrogenase
by ADP has been observed with the isolated enzymes, and these
interactions were initially thought to provide the major sti-
mulus via changes of the intramitochondrial energy charge
for increased flux through the citric acid cycle upon stimulation of respiration by ADP (54,55). It is now recognized
that this concept is an oversimplification, and that the major feedback effects are exerted by changes of the pyridine
nucleotide-oxidation-reduction state, as described in detail
in later sections. Investigations of the control of the citric acid cycle in intact mitochondria have depended upon alterations of the NAD oxidation-reduction potential independently of the intramitochondrial ATP/ADP ratio by the appropriate use of oligomycin and uncoupling agents to inhibit
energy transfer between the electron transport chain and the
adenine nucleotides. However, a number of difficulties and
discrepancies in the experimental data indicate that it is
necessary to be cautious in the interpretation of results
and particularly in extrapolation of findings with isoalted
enzymes and mitochondria to the intact tissue. The greatest
problem arises in relation to knowledge of intramitochondrial concentration of metabolites both because of uncertainty of the matrix activity coefficients and binding to matrix
enzymes of those metabolites present in low concentrations.
Furthermore, it is a matter of dispute whether the intramitochondrial ATP/ADP ratio of liver mitochondria changes appreciably with changes of extramitochondrial ATP/ADP ratio
at physiological concentrations of extramitochondrial adenine nucleotides (c.f. 28). In any case, with mammalian
mitochondria the change of the intramitochondrial ATP/ADP
ratio upon increasing the respiratory rate from the resting
state 4 value appears to be small relative to the change of

the NADH/NAD ratio. Electron transport flux is controlled by extramitochondrial phosphate potential, as discussed earlier, and causes reduction of pyridine nucleotides as flux decreases. Thus, feedback inhibition to the mitochondrial dehydrogenases is secondary to phosphate acceptor control of electron transport. Blowfly flight muscle mitochondria offer an interesting contrast to mammalian mitochondria. First, the mitochondrial membrane is impermeable to intermediates of the citric acid cycle (56), presumably due to the absence of most of the anion translocators. Second, the transition from resting (state 4) to active (state 3) respiration under suitable conditions is associated with a reduction rather than an oxidation of the mitochondrial pyridine nucleotides, despite a higher adenine nucleotide phosphorylation level in the resting state (57,58). In conjunction with further studies on measurements of the mitochondrial content of citric acid cycle intermediates and CoA derivatives (59,60), it was concluded that the primary event is an activation of NAD-isocitrate dehydrogenase (with feedback activation of citrate synthase) as a result of a lowered ATP/ADP ratio during active respiration. Thus, direct activation of the dehydrogenases by the adenine nucleotide phosphorylation state is stronger than the effect of respiratory control on the NAD redox state, and changes of the NAD/NADH ratio, therefore, are considered to play a secondary role in regulation of citric acid flux in this species of mitochondria. Further discussions will be restricted to mammalian mitochondria, particularly from liver and heart, with emphasis in the regulation of pyruvate dehydrogenase, citrate synthase, isocitrate dehydrogenase and α-ketoglutarate dehydrogenase.

Pyruvate dehydrogenase: The pyruvate dehydrogenase complex contains three functional enzymes required for the conversion of pyruvate to acetyl-CoA; pyruvate dehydrogenase itself, dihydrolipoyl transacetylase and the flavoprotein dihydrolipoyl dehydrogenase (61). The activity of the enzyme complex is inhibited by the products of pyruvate oxidation, acetyl-CoA and NADH, and these inhibitions are reversed by CoA and NAD, respectively (62-65). In addition, the activity of the pyruvate dehydrogenase component is regulated by another mechanism (66,67) involving phosphorylation of PDH_a (active) to an inactive phosphorylated form (PDH_b) as depicted in Fig. 4. Phosphorylation and concomitant inactivation of pyruvate dehydrogenase is catalyzed by a Mg ATP^{2-} kinase (PDH_a kinase) having a $\underline{K_m}$ for ATP of 0.02 mM (68). The PDH_a kinase is competitively inhibited by ADP with $\underline{K_i}$ of 0.03 to 0.1 mM depending on the K^+ concentration (68,69). There is thus a potentiality for strong control of PDH_a kinase activity by the

Pyruvate Dehydrogenase Phosphorylation–Dephosphorylation

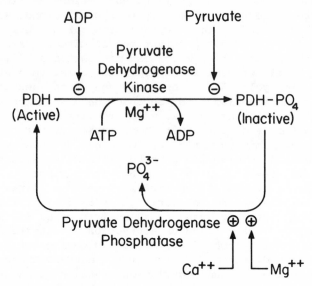

Fig. 4. Regulation of pyruvate dehydrogenase activity by the phosphorylation dephosphorylation cycle.

intramitochondrial ATP/ADP ratio, which has been confirmed experimentally in studies with isolated mitochondria (70-72). Pyruvate is also an inhibitor of PDH_a kinase, and $\underline{Ki}$ values for pyruvate of about 0.2 mM have been found for the heart complex, while higher values of 1 to 2 mM were reported for liver and kidney (68). The inhibitory effect of pyruvate on PDH_a kinase is apparently independent of that of ADP. Dephosphorylation of pyruvate dehydrogenase is catalyzed by a phosphatase (PDH_b phosphatase) which requires for its activity both free Mg^{2+} and Ca^{2+} (61,68,71,73-78). The apparent $\underline{Km}$ for Mg^{2+} is about 2 to 3 mM, while that for Ca^{2+} appears to be up to 2 orders of magnitude lower. Pettit $\underline{et}$ $\underline{al}$ (77) have suggested that Ca^{2+} acts by binding PDH_b phosphatase to dihydrolipoyl transacetylase and lowers the apparent $\underline{Km}$ of the PDH_b for the phosphatase. In addition, K^+ or NH_4^+ have been shown to be required for inhibition of PDH_a kinase by ADP, while these cations had no effect on PDH_b phosphatase (79).

Although control of pyruvate dehydrogenase phosphorylation-dephosphorylation cycle by PDH_b phosphatase can readily be demonstrated by altering the divalent cation content of mitochondria (71,78), it is probable that the major regulation under physiological conditions is achieved by control of PDH_a kinase activity rather than through control of PDH_b phosphatase by changes of the mitochondrial Mg^{2+} and Ca^{2+} concentrations (71,72). However, recent findings of Pettit, Pelley and Reed (80) indicate a separate type of control of both PDH_a kinase and PDH_b phosphatase activities. These authors observed an independent stimulation of PDH_a kinase by acetyl-CoA and NADH and an inhibition by CoA and NAD. Also, NADH inhibited PDH_b phosphatase with the inhibition being reversed by NAD, while effects of acetyl-CoA and CoA on the phosphatase were relatively small (80). Thus the activities of PDH_a kinase and PDH_b phosphatase appear to change in a concerted fashion with a change of the NADH/NAD ratio. These effects may explain earlier findings of inhibition of pyruvate dehydrogenase activity by high acetyl CoA/CoA and NADH/NAD ratios, and the well-demonstrated inhibitory effects of fatty acids on pyruvate oxidation and pyruvate dehydrogenase interconversion in various mammalian tissues (see 72,80). The multiplicity of factors regulating pyruvate dehydrogenase activity permit a versatility of metabolic functions and allow, for example, the rate of production of acetyl-CoA from pyruvate for fatty acid synthesis to be relatively independent of the requirements of the citric acid cycle for acetyl-CoA. In general, however, high NADH/NAD and ATP/ADP ratios associated with low rates of electron transport flux and a low citric acid cycle activity will result in an active PDH_a kinase and an inactive PDH_b phosphatase, so that most of the pyruvate dehydrogenase is in the inactive phosphorylated form. The poise of this equilibrium can be altered independently by changes of the pyruvate concentration or the acetyl-CoA/CoA ratio via their effects on PDH_a kinase. Thus, the relative rates of oxidation of glucose and fatty acids will be directly dependent on changes of these parameters.

Citrate synthase and isocitrate dehydrogenase: Citrate synthase activity is regulated directly by the relative availability of either of its two substrates oxalacetate and acetyl-CoA. However, ligand interactions with citrate synthase and separate factors controlling the intramitochondrial concentrations of oxalacetate and acetyl-CoA make the overall regulation rather complex. The apparent Km of citrate synthase for oxalacetate at saturating acetyl-CoA concentrations is in the range of 1 to 5 μM (81-83). The intramitochondrial oxalacetate concentration is expected to be determined primarily by the malate concentration and the NADH/NAD ratio via equilibrium with malate dehydrogenase. Unless the

intramitochondrial malate concentration rises appreciably
due to an increase of the cytosolic malate concentration or
an inhibition of malate efflux, oxalacetate availability in
state 3 will increase primarily as a result of the lowered
NADH/NAD ratio. Calculations of the mitochondrial oxalacetate concentration based on the malate content and an assumed equilibrium of malate dehydrogenase yielded values of
0.36 μM (84) or lower (81), while the calculated value from
measured α-ketoglutarate, glutamate, and aspartate contents
of mitochondria in state 4 assuming equilibrium of aspartate
aminotransferase was 10 μM (84). In these experiments the
measured oxalacetate concentration was 36 μM. The data indicate that although competition for oxalacetate between citrate synthase and aspartate aminotransferase can be demonstrated in mitochondria under certain conditions (85), oxalacetate must be compartmented due to protein binding or possibly to a concerted interaction between malate dehydrogenase and citrate synthase (86). However, since citrate synthase in intact mitochondria or tissue operates well below
its V$\underline{max}$, it is probable that the effective oxalacetate concentration available to citrate synthase is in the regulatory region of its K$\underline{m}$. In the transition from state 4 to
state 3 respiration in rat heart mitochondrial oxidizing pyruvate and malate or acetylcarnitine and malate, an 8-fold
increase of flux through citrate synthase was associated
with only a 3-fold fall of the NADH/NAD ratio (10), suggesting that the increase of oxalacetate concentration produced
as a result of the lowered NAD oxidation-reduction potential
was insufficient by itself to account for the increased flux
through citrate synthase. In these experiments the acetyl-
CoA levels decreased slightly and therefore could not account for the flux change. Although there is uncertainty
on the two points (a) that the mitochondrial NADH/NAD ratio
measured from the total contents of NADH and NAD is strictly
proportional to the ratio of free NADH/NAD, and (b) that the
deviation of malate dehydrogenase from equilibrium remains
constant, an additional interaction affecting oxalacetate
availability in heart via product inhibition by citrate appears to be of significance. Studies with isolated beef
heart synthase have shown that citrate is a competitive inhibitor of oxalacetate with a K$\underline{i}$ of 1.6 mM (82), while in
the above experiments (10) the intramitochondrial citrate
content fell from 2.6 to 1.6 mM between state 4 and state 3
thereby having the effect of decreasing the apparent K$\underline{m}$ of
citrate synthase for oxalacetate. The fall of the citrate
content is interpreted as due principally to an activational
interaction at isocitrate dehydrogenase caused by the lo-

wered NADH/NAD ratio. This follows from the finding that
NADH is an inhibitor of NAD-linked isocitrate dehydrogenase
(Ki of 20 µM) competitive with NAD (87). A decreased ATP/
ADP ratio in state 3 relative to state 4 may also contribute
towards the activation since ADP lowers the apparent Km for
isocitrate (87). It should be noted that the coupled regulation of citrate synthase and isocitrate dehydrogenase via
changes of the citrate content is particularly effective in
heart mitochondria because of the poor permeability of the
membrane to citrate (7,88).

Flux through citrate synthase in intact mitochondria is
also effectively regulated via changes of acetyl-CoA availability. Kinetic studies with isolated citrate synthase show
that the apprent Km for acetyl-CoA is 5÷10 µM (81-83), which
is more than an order of magnitude lower than the range of
acetyl-CoA contents found in isolated mitochondria (10). However, a number of nucleotides are inhibitory toward citrate
synthase, competitive with respect to acetyl-CoA, and non-
competitive with respect to oxalacetate (82,89). These include ATP (Ki = 0.5-1.0 mM), NADH (Ki = 1.5 mM), NADPH (Ki =
0.8 mM), and succinyl-CoA (Ki = 130 µM) all of which compete for an adenine binding site on the enzyme. Much interest has been generated by the possibility of a direct ATP
inhibition of citrate synthase since this provides a classical example of end-product feedback to the first step of a
sequence (54). However, only ATP and not MgATP is inhibitory (90). This finding is highly significant in view of
the probability that ATP in the mitochondria is fully chela-
ted with Mg^{2+} (71,84). Furthermore, in contrast to earlier
reports (55,91), more recent studies with heart (10,84,85)
and liver (92,93) mitochondria showed no correlation between
intramitochondrial ATP levels and flux through citrate synthase when other variables were maintained constant. Direct
inhibition of citrate synthase by NADH would tend to reinforce regulation via the NADH/NAD ratio on the oxalacetate
concentration, but the high inhibitory constant probably
makes it quantitatively rather unimportant. On the other
hand, considerable evidence has been accumulated from studies with isolated heart (10,84,85) and liver mitochondria
(94) in favor of an indirect energy-linked control mediated
by changes in the ratio of succinyl-CoA/acetyl-CoA. This
type of regulation appears to be most important under conditions of low intramitochondrial NADH/NAD ratio (i.e. with a
relative high oxalacetate availability) and low acetyl-CoA
concentrations. In addition, it provides feedback regulation between the activities of α-ketoglutarate dehydrogenase
and citrate synthase as discussed later. The rather high ra-

tio of the Michaelis constant for succinyl-CoA and acetyl-CoA observed with the isolated beef heart enzyme (K_i/K_m = 16) suggest that succinyl-CoA should be a relatively weak inhibitor (82). However, increasing ionic strength increased the apparent K_m for acetyl-CoA (95), which together with other factors (82), may account for the observations that in heart (10) as well as liver (55) mitochondria the concentration of acetyl-CoA giving half maximum stimulation of flux through citrate synthase was 0.1-0.2 mM, assuming an activity coefficient of unity. In experiments using rat heart mitochondria incubated with pyruvate or acetylcarnitine and malate (84), Dixon plots of the reciprocal of flux through citrate synthase against succinyl-CoA concentration at three different ranges of acetyl-CoA provided a value of 270 µM for the apparent K_i for succinyl-CoA, which is only 2-fold greater than the value measured with purified beef heart citrate synthase (82). Evidence that intramitochondrial concentrations of acetyl-CoA and succinyl-CoA may under some conditions be in a kinetically favorable range for regulation of citrate synthase has also been provided by studies with rat heart mitochondria incubated with a variety of substrates at approximately constant NADH/NAD ratios (85).

α-Ketoglutarate dehydrogenase and succinate thiokinase: As first noted by Garland (96), α-ketoglutarate dehydrogenase is inhibited by both its products, NADH and succinyl-CoA. Further studies with purified enzyme from pig heart (97) showed that inhibition by succinyl-CoA was competitive with respect to CoA (K_i of 6.9 µM compared with a K_m for CoA of 2.7 µM) and was independent of the NADH/NAD ratio. Inhibition by NADH was noncompetitive with respect to both NAD and α-ketoglutarate and was observed at both high and low succinyl-CoA levels. Independent or concurrent regulation of flux through α-ketoglutarate dehydrogenase by the succinyl-CoA/CoA and NADH/NAD ratios have been described in experiments with both heart (10,84,85) and liver (94,97) mitochondria. A kinetic analysis of flux through α-ketoglutarate dehydrogenase with varying succinyl-CoA/CoA ratios at an approximately constant NADH/NAD ratio has been made with guinea pig liver mitochondria incubated in different respiratory states with α-ketoglutarate as substrate (97). These studies provided a value of 1.6 for the ratio of the K_m for CoA to the K_i for succinyl-CoA in intact mitochondria compared with a mean value of 0.46 obtained with isolated α-ketoglutarate dehydrogenase, indicating that the kinetic behavior of the isolated enzyme is similar to that of the enzyme functioning in the mitochondria. Further studies with rat heart mitochondria (85) in which α-ketoglutarate was generated intra-mitochondrially showed that inhibitory effects of high suc-cinyl-CoA/CoA or NADH/NAD ratios were modulated by the intra-mitochondrial α-ketoglutarate concentration. Thus, addition of glutamate to mitochondria incubated with malate and ace-tylcarnitine or acetoacetate increased flux through α-keto-glutarate dehydrogenase by increasing the intramitochondrial α-ketoglutarate concentration independently of the degree of product inhibition. These experiments also illustrate that the intramitochondrial α-ketoglutarate concentration is in a favorable region for kinetic control of the enzyme.

Changes of the succinyl-CoA/CoA ratio in isolated mito-chondria showed a good correlation with changes of the ATP/ADP ratio (10,85,97). However, the nucleotides in direct equilibrium with the succinyl-CoA/CoA ratio via succinate thiokinase are GTP and GDP. Measurements of the GTP content of liver mitochondria showed that the level remained approx-imately constant despite 10-fold changes of the succinyl-CoA/CoA ratio (97). Calculations of the GTP/GDP ratio made by assuming equilibrium at succinate thiokinase indicated that although changes with different respiratory states of the mitochondria were in the same sense as those of the ATP/ADP ratio, the mass action ratio of reactants and products of nucleoside diphosphate kinase differed by up to three orders of magnitude from the equilibrium constant of 0.9 (97). It is apparent from these and other studies (98) that nucleo-side diphosphate kinase does not equilibrate the phosphate potentials of the intramitochondrial adenine and guanine nu-cleotides. This conclusion, although controversial (99), provides a rationality for the different nucleotide specifi-city of those mitochondrial enzymes involved in substrate interconversions, e.g., succinate thiokinase, P-enolpyruvate carboxykinase, and the ATPase responsible for energy conser-vation in the electron transport chain. Disequilibrium be-tween the adenine and guanine nucleotides allows changes of the GTP/GDP and succinyl-CoA/CoA ratios to be larger than those of the intramitochondrial ATP/ADP ratio, which as pre-viously discussed changes by a much smaller amount than the extramitochondrial ATP/ADP ratios because of the electroge-nic nature of adenine nucleotide transport.

Coordination between the Citric Acid Cycle and Malate-Aspartate Cycle

Since the heart has a limited ability to synthesize di-carboxylic acids from pyruvate, the citric acid cycle nor-mally operates as a coordinated unit. However, the permea-bility of the heart mitochondrial membrane to α-ketogluta-

rate, unlike that to citrate, is high. Hence there is a
competition for intramitochondrial α-ketoglutarate between
efflux from the mitochondria and further metabolism in the
citric acid cycle to succinyl-CoA via α-ketoglutarate dehydrogenase. It may be noted that net α-ketoglutarate efflux
from the mitochondria in heart will only occur on a continued basis during operation of the malate-aspartate cycle,
when an equivalent amount of α-ketoglutarate is formed by
transamination between oxalacetate and glutamate via aspartate aminotransferase (see Fig. 5). In liver, the intrami-

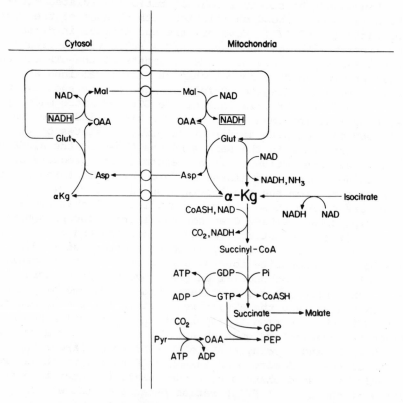

*Fig. 5. Reactions producing and utilizing α-ketoglutarate
in the mitochondrion.*

tochondrial reactions of α-ketoglutarate are more complicated since it may also be used or produced by glutamate dehydrogenase, depending on the ammonia availability (Fig. 5).
Furthermore, in those species containing mitochondrial P-en-

olpyruvate carboxykinase, GTP produced by the conversion of succinyl-CoA to succinate is used directly for P-enolpyru-vate synthesis. The convergence of the citric acid cycle, the malate-aspartate cycle, and the oxidative deamination of glutamate at the common intermediate, α-ketoglutarate makes control of its further oxidation an important feature of mitochondrial metabolism. Factors which regulate the distribution of α-ketoglutarate utilization between the citric acid cycle and the malate-aspartate cycle are the α-ketoglutarate and malate concentration gradients across the mitochondrial membrane and the mitochondrial energy state, as reflected in both the NADH/NAD and ATP/ADP ratios. The apparent K_m of the α-ketoglutarate-malate carrier for α-ketoglutarate on the matrix side of the mitochondrial membrane is 1.5 mM (100) compared with a value of 0.67 mM for the apparent K_m of α-ketoglutarate dehydrogenase for α-ketoglutarate in the intact heart mitochondria (85). Since both apparent K_ms are similar, transport and oxidation of α-ketoglutarate will both be regulated over the same range of intramitochondrial α-ketoglutarate concentrations. Interactions between transport of anions across the mitochondrial membrane and the enzyme reactions of the citric acid cycle appear to be mediated primarily by a kinetic control of the anion exchange translocators. Regulation of net transport flux is affected by the activities of enzymes in both the mitochondrial and cytosol spaces through their regulation of the steady-state concentrations of the particular transported intermediates in either compartment.

The interaction between the partial components of the malate-aspartate cycle with α-ketoglutarate dehydrogenase has been investigated with rat heart mitochondria metabolizing glutamate (85). These experiments showed that the initial rate of α-ketoglutarate efflux increased with increasing malate concentrations in the medium with a half-maximum effect at 0.36 mM. Increased transport of α-ketoglutarate from the mitochondria resulted in a fall of intramitochondrial α-ketoglutarate concentration and decreased flux through α-ketoglutarate dehydrogenase. Likewise, the concentration of extramitochondrial glutamate regulates the rate of aspartate efflux (101). Anionic interactions at the substrate binding sites of the carriers also affect their kinetics. Thus, extramitochondrial α-ketoglutarate is a competitive inhibitor of malate on the α-ketoglutarate carrier (100), while extramitochondrial aspartate is a non-competitive inhibitor of glutamate on the glutamate-aspartate carrier (35). Feedback interactions from the phosphorylating electron transport chain, mediated by changes of both the succinyl-CoA/CoA and

NADH/NAD ratios which regulate the activities of α-ketoglutarate dehydrogenase and citrate synthase, secondarily affect the rate of α-ketoglutarate efflux and hence the activity of the malate-aspartate cycle by altering the intramitochondrial α-ketoglutarate concentration (85). Thus the supply of reducing equivalents to the electron transport chain via the malate-aspartate cycle is coordinated with the total rate of generation of NADH via the citric acid cycle by interrelated feedback effects to both citrate synthase and α-ketoglutarate dehydrogenase.

Coordination of Citric Acid Cycle Flux

From studies with isolated mitochondria it is concluded that energy-linked control of the citric acid cycle is largely mediated by indirect effects via changes of the NADH/NAD ratio and succinyl-CoA levels rather than by a direct interaction of nucleotides with citrate synthase (102) or the lipoyl dehydrogenase moeity of α-ketoglutarate dehydrogenase (103). The importance of a direct interaction of ADP with isocitrate dehydrogenase in rat liver mitochondria during the state 4 to state 3 transition has also been questioned (104) since these studies indicated a better correlation of flux through isocitrate dehydrogenase with the NADH/NAD ratio than the intramitochondrial ATP/ADP ratio. Changes of the intramitochondrial NADH/NAD ratio appear to provide the primary signal for coordination of flux through the electron transport chain with the irreversible steps of the citric acid cycle. An additional negative feedback loop links the activities of α-ketoglutarate dehydrogenase and citrate synthase via inhibitory interactions of succinyl-CoA. However, this interaction may be neutralized by an increase of acetyl-CoA concentration at the level of citrate synthase and by an increase of α-ketoglutarate concentration at the level of α-ketoglutarate dehydrogenase. The net effect, which determines overall flux in the citric acid cycle,is strongly influenced in liver by the activities of ancilliary enzymes such as β-oxidation, pyruvate dehydrogenase, pyruvate carboxylase, aspartate aminotransferase, and glutamate dehydrogenase which, in turn, will depend on the relative availabilities of primary substrates.

The intramitochondrial ATP/ADP ratio is apparently effective in regulating pyruvate metabolism in rat liver mitochondria. At low pyruvate concentrations, decreases of the intramitochondrial ATP/ADP ratio induced by addition of creatine kinase and varying amounts of creatine correlated with a progressive decrease in pyruvate carboxylation to oxalacetate and increase of pyruvate oxidation to acetyl-CoA (27).

The intramitochondrial ATP/ADP ratios appear to be much lower in liver (27,97) than in heart (10,85) mitochondria under comparable respiratory state conditions and may be susceptible to larger changes due to the presence of intramitochondrial ATP-consuming reactions. Inhibition of citrate-synthase, induced either by decreased oxalacetate availability (93) or by an increase of the succinyl-CoA/acetyl-CoA ratio (94) increases the rate of ketogenesis in rat liver mitochondria supplied with palmityl-carnitine. It is also evident that the rate of β-oxidation is regulated at both the NAD and flavin-linked steps by feedback from the electron transport chain (105-107). Nevertheless, ketogenesis in vivo may still be regarded as an overflow response of the liver caused by the rate of generation of acetyl-CoA by β-oxidation being greater than the capacity of the citric acid cycle for its oxidation.

Finally, in concluding this review, the physiological significance of the concepts developed mainly from studies with isolated enzymes and mitochondria may be placed in perspective by mention of data pertaining to regulation of the citric acid cycle in the intact perfused rat heart. A temporary unspanning of the citric acid cycle in perfused rat heart has been observed during the transition from metabolism of endogenous fatty acids to metabolism of exogenous glucose in association with alterations of tissue metabolite levels (108). Cycle unspanning was the direct result of the relaxation time needed for the enzymes to readjust tissue metabolite levels to accomodate the switch from fatty acid to glucose metabolism. In further studies, the transition from low to high work loads has been investigated in rat hearts perfused with pyruvate, glucose or palmitate (109-113). Oxygen consumption increased 2-to 4-fold, while the pyridine nucleotides changed toward a more oxidized state corresponding to a state 4 to state 3 transition of the mitochondria. In one study, increased respiration correlated with a fall of the total tissue phosphate potential and a rise of the calculated free ADP concentration in the cytosol from 40 to 80 µM while the calculated intramitochondrial ATP/ADP ratio decreased only from 0.55 to 0.40 (109). Citrate and α-ketoglutarate contents of the heart fell rapidly while malate and oxalacetate contents increased (109). Analysis of the kinetic data indicated that coordination of flux in the citric acid cycle with increased flux through the electron transport chain was mediated primarily as a result of the decreased mitochondrial NADH/NAD ratio due to activational interactions at pyruvate dehydrogenase citrate synthase, isocitrate dehydrogenase and α-ketoglutarate de-

101

hydrogenase although minor imbalances of cycle flux were apparent within the first 60 sec after the work transition. Citrate synthase appeared to be regulated directly by an increase of mitochondrial oxalacetate concentration and indirectly by release of citrate inhibition resulting from the activation of isocitrate dehydrogenase. Although it is difficult to interpret data obtained in the intact tissue unambiguously with relation to compartmentation of metabolites, the control mechanisms appear to be strictly analogous to those obtained with isolated heart mitochondria during the state 4 to state 3 transition with pyruvate and malate as substrates (10).

So far no evidence has been obtained with the intact heart to indicate that succinyl-CoA plays an important role in the regulation of citrate synthase. In the few instances in which tissue succinyl-CoA levels have been measured or estimated (108,110,111), an increase of succinyl-CoA content was associated with an increase of flux through citrate synthase. It would appear that in the intact tissue, the mitochondrial acetyl-CoA level is sufficiently high to overcome the inhibitory effect of succinyl-CoA particularly when fatty acids are supplied as fuel (110). On the other hand, changes of succinyl-CoA levels do appear to be important in the regulation of α-ketoglutarate dehydrogenase, particularly for a coordination of citric acid cycle flux with that of the malate-aspartate cycle (108).

REFERENCES

1. Chappell, J.B. (1969) in Inhibitors: Tools in Cell Research, (Bücher, T., Sies, H., eds.)p. 335, Springer-Verlag, Heidelberg.
2. Klingenberg, M. (1970) Essays in Biochemistry 6, 119.
3. Meijer, A.J. and Van Dam, K. (1974) Biochim. Biophys. Acta 346, 213.
4. Williamson, J.R. (1976) in Gluconeogenesis: Its Regulation in Mammalian Metabolism, (Hanson, R.W., Mehlman, M.A., eds.) John Wiley and Sons, New York (in press).
5. Stuart, S.C. and Williams, G.R. (1968) Biochemistry 5, 3912.
6. McElroy, F.A. and Williams, G.R (1968) Arch. Biochem. Biophys. 126, 492.
7. LaNoue, K.F., Nicklas, W.J. and Williamson, J.R. (1970) J. Biol. Chem. 245, 102.
8. Chance, B. and Williams, G.R. (1956) Advan. Enzymol. 17, 63.
9. Stucki, J.W., Brawand, F. and Walter, P. (1972) Eur. J.

Biochem. $\underline{27}$, 181.

10. LaNoue, K.F., Bryla, J. and Williamson, J.R. (1972) J. Biol. Chem. $\underline{247}$, 667.
11. Klingenberg, M. (1972) *in* Mitochondria: Biogenesis and Bioenergetics/Biomembranes: Molecular Arrangement and Transport Mechanisms, Vol. 28, (Van Den Bergh, S.G., Borst, P., Van Deenen, L.L.M., Riemersma, J.C., Slater, E.C. and Tager, J.M., èds.) pp. 147-162, North-Holland, Amsterdam.
12. Davis, E.J., Lumeng, L. and Bottoms, D. (1974) Fed. Eur. Biochem. Soc. Lett. $\underline{39}$, 9.
13. Mitchell, P. (1968) Chemiosmotic Coupling and Energy Transduction, Glynn Research Ltd., Bodwin, England.
14. Greville, G.D. (1969) Curr. Top. Bioenerg. $\underline{3}$, 1.
15. Mitchell, P. and Moyle, J. (1969) Eur. J. Biochem. $\underline{7}$, 471.
16. Addanki, S., Cahill, R.O. and Sotos, J.F. (1968) J. Biol. Chem. $\underline{243}$, 2337.
17. Padan, E. and Rottenberg, H. (1973) Eur. J. Biochem. $\underline{40}$, 431.
18. Nicholls, D.G. (1974) Eur. J. Biochem. $\underline{50}$, 305.
19. Mitchell, P. (1969) *in* Mitochondria - Structure and Function (Ernster, L. and Drahota, Z., eds.) pp. 219-231, Academic Press, London.
20. Rottenberg, H. (1975) J. Bioenerg. (in press).
21. Papa, S., Guerrieri, F., Simone, S. and Lorusso, M. (1973) *in* Mechanisms in Bioenergetics, p. 451, Academic Press, New York.
22. Papa, S., Lofrumento, N.E., Quagliariello, E., Meijer, A.J. and Tager, J.M. (1970) Bioenergetics $\underline{1}$, 287.
23. Palmieri, F., Quagliariello, E. and Klingenberg, M. (1970) Eur. J. Biochem. $\underline{17}$, 230.
24. Greenbaum, A.L., Gumaa, K.A. and McLean, P. (1971) Arch. Biochem. Biophys. $\underline{143}$, 617.
25. Williamson, J.R. (1975) FEBS Meeting, Paris.
26. Heldt, H.W., Klingenberg, M. and Milovansev, M. (1972) Eur. J. Biochem. $\underline{30}$, 434.
27. Walter, P. Zwez, S.M. and Brawand, F. (1974) *in* Regulation of Hepatic Metabolism (Lundquist, F. and Tygstrup, N. eds.) pp. 79-89, Munksgaard, Copenhagen.
28. Davis, E.J. and Lumeng, L. (1975) J. Biol. Chem. $\underline{250}$, 2275.
29. Elbers, R., Heldt, H.W., Schmucker, P., Soboll, S. and Wiese, H. (1974) Hoppe-Seylers Z. Physiol. Chem. $\underline{355}$, 378.
30. Heldt, H.W. (1975) FEBS Meeting, Paris.
31. Zuurendonk, P.F. and Tager, J.M. (1974) Biochim. Bio-

phys. Acta <u>333</u>, 393.

32. Zuurendonk, P.F., Akerboom, T.P.M. and Tager, J.M. (1975) FEBS Meeting, Paris.

33. Klingenberg, M. (1970) FEBS Lett. <u>6</u>, 145.

34. Souverijn, J.H.M., Huisman, L.A., Rosing, J. and Kemp, A., Jr. (1973) Biochim. Biophys. Acta <u>305</u>, 185-198.

35. Tischler, M.E., Pachence, J., Williamson, J.R. and LaNoue, K.F. (1975) Arch. Biochem. Biophys. (in press).

36. Borst, P. (1963) *in* Funktionelle und Morphologische Organisation der Zelle (Karlson, L. ed.) pp. 137-162.

37. Krebs, H.A. and Veech, R.L. (1969) *in* The Energy Level and Metabolic Control in Mitochondria (Papa, S., Tager, J.M., Qualiariello, E. and Slater, E.C. eds.) p. 329, Adriatic Ed., Bari, Italy.

38. Illingworth, J.A., Ford, W.C.L., Kobayashi, K. and Williamson, J.R. *in* Recent Advances in Studies on Cardiac Structure and Metabolism, Vol. 8, (Roy, P.-E. and Harris, P. eds.) University Park Press, Baltimore (in press).

39. Veech, R.L., Raijman, L. and Krebs, H.A. (1970) Biochem J. <u>117</u>, 499.

40. Stubbs, M., Veech, R.L. and Krebs, H.A. (1972) Biochem. J. <u>126</u>, 59.

41. Owen, C. and Wilson, D.F. (1974) Arch. Biochem. Biophys. <u>161</u>, 581.

42. Slater, E.C., Rosing, J. and Mol, A. (1973) Biochim. Biophys. Acta <u>292</u>, 534.

43. Wilson, D.F. and Erecinska, M. (1972) *in* Biomembranes: Molecular Arrangements and Transport Mechanisms (van Deenan, L.L.M., Riemersma, J.C. and Tager, J.M. eds.) pp. 119-132, North Holland, Amsterdam.

44. Erecinska, M., Veech, R.L. and Wilson, D.F. (1974) Arch. Biochem. Biophys. <u>160</u>, 412.

45. Wilson, D.F., Stubbs, M., Veech, R.L., Erecinska, M. and Krebs, H.A. (1974) Biochem. J. <u>140</u>, 57.

46. Wilson, D.F., Stubbs, M., Oshino, N. and Erecinska, M. (1974) Biochemistry <u>13</u>, 5305.

47. Oshino, N., Sugano, T., Oshino, R. and Chance, B. (1974) Biochem. Biophys. Acta <u>368</u>, 298-310.

48. Williamson, J.R., Anderson, J. and Browning, E.T. (1970) J. Biol. Chem. <u>246</u>, 7632.

49. Söling, H.D., Kleineke, J., Willms, B., Janson, G. and Kuhn, A. (1973) Eur. J. Biochem. <u>37</u>, 233.

50. Robinson, B.H. and Chappell, J.B. (1967) Biochem. Biophys. Res. Commun. <u>28</u>, 249.

51. Meijer, A.J. and Tager, J.M. (1969) Biochim. Biophys. Acta <u>189</u>, 136.

52. Meijer, A.J., Gimpel, J.A., DeLeeuw, G.A., Tager, J.M. and Williamson, J.R. (1975) J. Biol. Chem. (in press).
53. McGivan, J.D. and Chappell, J.B. (1975) FEBS Lett. 52, 1.
54. Atkinson, D.E. (1969) *in* Citric Acid Cycle Control and Compartmentation (Lowenstein, J.M. ed.) pp. 137-161, Marcel Dekker, New York.
55. Garland, P.B., Shepherd, D., Nicholls, D.G., Yates, D. W.and Light, P.A. (1969) *in* Citric Acid Cycle Control and Compartmentation (Lowenstein, J.M. ed.) pp. 163-212, Marcel Dekker, New York.
56. Van Den Bergh, S.G. and Slater, E.C. (1962) Biochem. J. 82, 362.
57. Hansford, R.G. (1972) Biochem. J. 127, 271.
58. Hansford, R.G. (1975) Biochem. J. 146, 537.
59. Hansford, R.G. (1974) Biochem. J. 142, 509.
60. Johnson, R.N. and Hansford, R.G. (1975) Biochem. J. 146, 527.
61. Reed, L.J., Linn, T.C., Hucho, F., Namihira, G., Barrera, C.R., Roche, T.E., Pelley, J.W., and Randall, D.D. (1972) *in* Metabolic Interconversion of Enzymes (Wieland, O., Helmreich, E. and Holzen, H. eds.) pp. 281-293, Springer-Verlag, Berlin.
62. Garland, P.B. and Randle, P.J. (1964) Biochem. J. 91, 6c.
63. Bremer, J. (1969) Eur. J. Biochem. 8, 535.
64. Wieland, O., von Jagow-Westerman, B. and Stukowski, B. (1969) Hoppe-Seyler's Z. Physiol. Chem. 350, 329.
65. Tsai, C.S., Burgett, M.W. and Reed, L.J. (1973) J. Biol. Chem. 248, 8348-8352.
66. Linn, T.C., Pettit, F.H., and Reed, L.J., (1969) Proc. Nat. Acad. Sci. U.S.A. 62, 234.
67. Linn, T.C., Pettit, F.H. Hucho, F. and Reed, L.J. (1969) Proc. Nat. Acad Sci. U.S.A. 64, 227.
68. Hucho, F., Randall, D.D., Roche, T.E., Burgett, M.W., Pelley, J.W. and Reed, L.J. (1972) Arch. Biochem. Biophys. 151, 328.
69. Roche, T.E. and Reed, L.J. (1974) Federation Proc. 33 (abstr.) 1427.
70. Martin, B.R., Denton, R.M., Pask, H.T. and Randle, P.J. (1972) Biochem. J. 129, 763-773.
71. Walajtys, E.I., Gottesman, D.P. and Williamson, J.R. (1974) J. Biol. Chem. 249 1857.
72. Taylor, S.I., Mukherjee, C. and Jungas, R.L. (1975) J. Biol. Chem. 250, 2028.
73. Reed, L.J., Pettit, F.H., Roche, T.E. and Butterworth, P.J. (1973) *in* Protein Phosphorylation in Control Mechanisms (F. Huijing and Lee, E.Y.C. eds.) pp. 83-

95, Academic Press, New York.

74. Severson, D.L., Denton, R.M., Pask, M.T. and Randle, P.J. (1974) Biochem. J. <u>140</u>, 225.

75. Denton, R.M., Randle, P.J. and Martin, B.R. (1972) Biochem. J. <u>128</u>, 161.

76. Sies, E.A. and Wieland, O. (1972) Eur. J. Biochem. <u>26</u>, 96.

77. Pettit, F.H., Roche, T.E. and Reed, L.J. (1972) Biochem. Biophys. Res. Commun. <u>49</u>, 563-571.

78. Schuster, S.M. and Olson, M.S. (1974) J. Biol. Chem. <u>249</u>, 7159.

79. Roche, T.E. and Reed, L.J. (1974) Biochim. Biophys. Res. Commun. <u>59</u>, 1341-1348.

80. Pettit, F.A., Pelley, J.W. and Reed, L.J. (1975) Biochem. Biophys. Res. Commun. <u>65</u>, 575-582.

81. Srere, P. (1972) *in* Current Topics in Cellular Regulation (Horecker, B.L. and Stadtman, E.R. eds.) Vol. 2, pp. 1-27, Academic Press, New York, 1970.

82. Smith, C.M. and Williamson, J.R. (1971) Fed. Eur. Biochem. Soc. Lett. <u>18</u>, 35.

83. Matsuoka, Y. and Srere, P.A. (1973) J. Biol. Chem. <u>248</u>, 8022.

84. Williamson, J.R., Smith, C.M., LaNoue, K.F. and Bryla, J. (1972) *in* Energy Metabolism and the Regulation of Metabolic Process in Mitochondria (Mehlman, M.A. and Hanson, R.W. eds.) pp. 185-210, Academic Press, New York.

85. LaNoue, K.F., Walajtys, E.I. and Williamson, J.R. (1973) J. Biol. Chem. <u>248</u>, 7171.

86. Srere, P.A. (1972) *in* Energy Metabolism and the Regulation of Metabolic Processes in Mitochondria (Mehlman, M.A. and Hanson, R.W. eds.) pp. 79-91, Academic Press, New York.

89. Srere, P.A. Matsuoka, Y. and Mukherjee, A.J. (1973) J. Biol. Chem. <u>248</u>, 8031.

90. Kosicki, G.W. and Lee, L.P.K. (1966) J. Biol. Chem. 241, 3571.

91. Garland, P.B. (1968) *in* Metabolic Roles of Citrate (Goodwin, T.W. ed.) pp. 41-60, Academic Press.

92. Olson, M.S. and Williamson, J.R. (1971) J. Biol. Chem. <u>246</u>, 7794.

93. Lopes-Cardozo and Van Den Bergh, S.G. (1972) Biochim. Biophys. Acta <u>283</u>, 1.

94. Williamson, J.R., Smith, C.M. and Bryla, J. (1974) *in* Regulation of Hepatic Metabolism (Lundquist, F. and Tygstrup, N. eds.) pp. 620-683, Munksgaard, Copenhagen.

95. Wu, J.-Y. and Yang, J.T. (1970) J. Biol. Chem. 245, 3561.
96. Garland, P.B. (1964) Biochem. J. 92, 10C.
97. Smith, C.M., Bryla, J. and Williamson, J.R.(1974) 249, 1497.
98. Bryla, J., Smith, C.M. and Williamson, J.R (1973) J. Biol. Chem. 248, 4003.
99. Garber, A.J. and Ballard, F.J. (1973) J. Biol. Chem. 245, 2229.
100. Sluse,F.E., Goffart, G. and Liebecq, C. (1973) Eur. J. Biochem. 32, 283.
101. Williamson, J.R., Safer, B., LaNoue, K.F., Smith, C.M. and Walajtys, F. (1973) in Rate Control of Biological Process (Davies, D.D. ed.) pp. 241-281, Cambridge University Press, Cambridge.
102. Jangaard, N.O., Unkeless, J. and Atkinson, D.E. (1968) Biochim. Biophys. Acta 151, 225-235.
103. Olson, M.S. and Allgyer, T.T. (1973) J. Biol. Chem. 248, 1590.
104. Wojtczak, A.B. (1974) Poster presentation, 9th FEBS Meeting, Budapest, Hungary.
105. Bremer, J. and Wojtczak, A.B. (1972) Biochim. Biophys. Acta 280, 515.
106. Lopez-Cardozo, M. and Van Den Bergh, S.G. (1974) Biochim. Biophys. Acta 357, 53.
107. Pande, S.V. (1971) J. Biol.Chem. 246, 5384.
108. Safer, B. and Williamson, J.R. (1973) J. Biol. Chem. 248, 2570.
109. Illingworth, J.A., Ford, W.C.L., Kobayashi, K. and Williamson, J.R (1974) in Recent Advances in Studies on Cardiac Structure and Metabolism (Roy, P.-E. et al eds.) University Park Press, Baltimore (in press).
110. Oram, J.F., Bennetch S.L. and Neely, J.R. (1973) J. Biol. Chem. 248, 5299.
111. Neely, J.R., Rovetto, M.J. and Oram, J.F. (1972) Progr. Cardiovas. Dis. 15, 289.
112. Neely, J.R. (1975) in Regulation of Cardiac Metabolism (Morgan, H., Opie, L. and Wildenthal K. eds.)(in press).
113. Williamson, J.R., Ford, W.C.L.,Kobayashi, K., Illingworth, J.A. and Safter. B. (1975) in Regulation of Cardiac Metabolism (Morgan, H., Opie, L. and Wildenthal, K. eds.) (in press).

KINETIC AND BINDING PROPERTIES OF THE ADP/ATP CARRIER AS A FUNCTION OF THE CARRIER ENVIRONMENT

P.V. Vignais, G.J.M. Lauquin and P.M. Vignais

INTRODUCTION

The inner membrane of mitochondria contains a number of highly specific carriers. The transport processes catalyzed by these carriers represent a means for the cell to control both the redox potential and the phosphate potential of the cytosol. In actively respiring mitochondria, the ADP/ATP transport system functions to import ADP and to export ATP. It thereby contributes to maintain a phosphate potential higher in the cytosol than in the mitochondria (1-3).

The kinetic properties of the ADP transport system have been studied extensively during the past ten years and some aspects of the transport mechanism itself have been revealed through the use of specific inhibitors (cf. 3).

In this chapter, after a short summary of the general features of the ADP/ATP transport, we shall examine in more detail the response of kinetic parameters of the ADP/ATP carrier to the environmental conditions in the mitochondrial membrane and we shall discuss in the light of the experimental results some possible mechanisms for the ADP/ATP transport.

Summary of the General Features of the ADP/ATP Transport

The ADP/ATP carrier catalyzes a one to one transmembrane exchange of intramitochondrial ADP or ATP with ADP or ATP added to mitochondria. Among natural nucleotides, only ADP and ATP are recognized and rapidly transported (4-9). However, besides ADP and ATP, synthetic adenine-nucleotide analogues can also be transported by the ADP/ATP carrier (8,10,11). Some of them, for example the methylene phosphoric analogues of ADP and ATP, are metabolically inert and their use has been decisive to verify that all the adenine nucleotides of the matrix space are exchangeable (8).

The ADP/ATP transmembrane exchange obeys saturation kinetics of Michaelis-Menten type (8,12). However, the K_m values for external ADP and ATP are dependent on the energy conditions of the mitochondria; this will be discussed in the next section.

The ADP/ATP transport is highly temperature-dependent; for example, a well-defined transition was found at 8° in the Arrhenius plot of the rate of ADP transport in rat liver mitochondria (8,12). This is possibly related to a physical phase transition of the membrane phospholipids surrounding the ADP carrier.

Two specific inhibitors of the ADP/ATP transport have been identified: atractyloside (13-15) and its derivatives including carboxyatractyloside (16,17) and bongkrekic acid (18-21). These inhibitors differ by their structures and the dependency of their inhibitory effect on pH. Atractyloside and its bio-synthetic percursor, carboxyatractyloside, which is a more powerful inhibitor, are amphiphilic molecules. Their diterpene moiety constitutes the hydrophobic portion, the glucose disulphate the hydrophilic portion (Fig. 1). Atractyloside and carboxyatractyloside can therefore be expected to be retained on the outer face of the inner mitochondrial membrane; they bind to it specifically and do not traverse it. Experiments carried out with radioactively-labeled inhibitors have proved this to be the case (17). Atractyloside and carboxyatractyloside will be, therefore, referred to as non-penetrant inhibitors. They are probably inserted by their hydrophobic diterpene moiety into the phospholipid bilayer in direct contact with the ADP carrier; whereas the polar glucose disulphate end protrudes from the phospholipid bilayer and interacts with the polar part of the ADP carrier.

In contrast, bongkrekic acid (Fig. 2) which is a tricarboxylic acid with pK values of the order of 5 (22) is inhibitory only at pH values below 7 (20,21), i.e., when a fraction of the carboxylic groups are protonated. This may be related to the observation that high affinity binding sites for bongkrekic acid are displayed only at pH below 7 (3). Such a pH dependence for inhibitory activity and binding affinity was not found for atractyloside and carboxyatractyloside. Therefore, to be inhibitory, bongkrekic acid must penetrate into the inner mitochondrial membrane.

Atractyloside, carboxyatractyloside and bongkrekic acid differ not only by their localization in the inner mitochondrial membrane, but also by the type of inhibition that they determine. Atractyloside is a competitive inhibitor of the ADP/ATP transport (17), carboxyatractyloside a non-competitive inhibitor (17) and bongkrekic acid an uncompetitive inhibitor (23).

Beside these specific inhibitors of the ADP/ATP transport, another class of non-specific, competitive inhibitors is represented by long-chain acyl-CoAs (24-27). Their role in the physiological regulation of the oxidative phosphorylation of the cytosolic ADP by mitochondria is plausible but not yet proven.

Thanks to the use of atractyloside as a competitive inhibitor capable of removing the ADP specifically bound to the ADP-carrier sites, it has been possible to discriminate specific and unspecific ADP-binding sites and to evaluate the

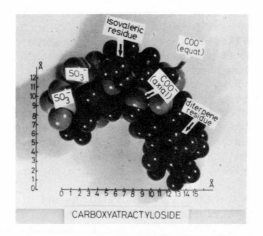

Fig. 1. *Space filling model of carboxyatractyloside. Note the bent shape of the molecule with two distinct portions, a hydrophilic one corresponding to the diterpene moiety and a hydrophilic one corresponding to glucose disulphate.*

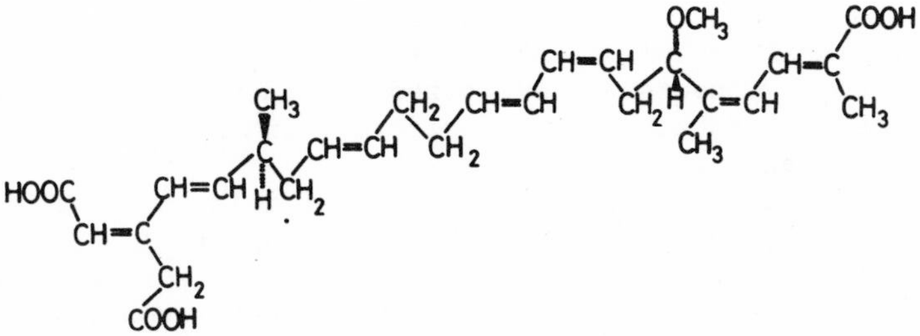

Fig. 2. *Structure of bongkrekic acid. Protonation of the carboxylic groups allows the penetration of bongkrekic acid into the mitochondrial membrane.*

111

number of ADP carriers per mitochondrion assuming one ADP binding site per carrier (3,17,28). The number of ADP carriers is one to two times higher than the number of cytochrome a molecules in rat liver mitochondria, and two to three times higher in rat heart mitochondria (3,28,29). Based on the number of carriers and on the rate of transport, a turnover number ranging from 1600 to 2000 per min at 20° for rat liver mitochondria has been calculated (3,30).

A decisive proof that ADP/ATP transport system is made of proteins has been provided by the genetic approach. Nuclear mutants specifically altered in the kinetic and binding parameters of the ADP/ATP transport system have been isolated (31-33). Investigations with such mutants may bring interesting clues as to the means whereby the ADP/ATP carrier synthesized on cytoribosomes is inserted into the inner mitochondrial membrane.

More recent investigations, based on the high affinity binding of atractyloside, have allowed us to isolate an atractyloside-binding protein which may be part or the whole of the ADP/ATP transport system (34). A specific ADP-binding protein (35) and a carboxyatractyloside-binding protein (36) have been isolated in other laboratories.

Now we shall see how the membrane environment modifies the kinetic and binding properties of the ADP/ATP transport system and conversely how the functioning of the ADP/ATP carrier influences the molecular properties of its environment.

Effect of Energization on the Kinetic Parameters of the ADP/ATP Transport and on the Binding Constants of Bongkrekic Acid

The so-called mitochondrial energization corresponds to a state of mitochondria where the redox energy is stored and not used for the synthesis of ATP. Energization can be readily obtained by addition to mitochondria of an oxidizable substrate plus an inhibitor of the phosphorylation itself, for example, oligomycin. Conversely, energized mitochondria can be de-energized by addition of a respiratory inhibitor or of an uncoupling agent.

The effect of energization on the affinity of mitochondria for external ATP was reported by Souverijn et al. (37), who showed that in rat liver mitochondria the K_m for ATP could be shifted from a value as low as 1 μM in the presence of uncoupler to a value of more than 100 μM after preincubation of mitochondria with an oxidizable substrate. The K_i for ATP in the ADP uptake is also dependent on the energy state of the mitochondria and, roughly the same as the K_m for ATP (37) under both high and low energy conditions. At the same time, we found that the K_m for external ADP in rat liver mitochondria depends on the intramitochondrial species of adenine nucleotides (3). The K_m for external ADP was lower than 3 µM when ATP was the predominant intramitochondrial nucleotide. It was two to three times higher when ADP was the predominant intramitochondrial nucleotide. More recent experiments performed with rat heart mitochondria and illustrated in Fig. 3A and 3B confirmed these results which can be summarized as follows. The affinity for external ATP is lower in energy rich conditions than in energy poor conditions; the converse is true for external ADP. In other words, the ADP transport system is able to respond, by a change of affinity for its substrates, to membrane modifications induced by the energy state of the mitochondria. This change of affinity may be due to changes in the geometry of the binding site of the carrier. It is interesting to note that there is a parallelism between the changes in K_m and V_{max} values depending on the energy state of mitochondria when ADP is the external substrate (Fig. 3A) (cf 3).

These results may be compared to those obtained from binding experiments carried out with radioactively-labelled bongkrekic acid and rat heart mitochondria (Fig. 4). Energization changed the hyperbolic shape of the binding curve into a sigmoidal shape. Furthermore, it increased by 10-20% the number of high affinity sites for bongkrekic acid. The effects of energization on the V_{max} value for ADP import and on the number of high affinity sites for bongkrekic acid are opposite, as if the import of ADP and the binding of bongkrekic acid were two mirror images of a same process. To illustrate this statement and to retain the hypothesis of the ADP/ATP transport system being a mobile carrier, it may be recalled that to be an inhibitor, bongkrekic acid has to penetrate the inner mitochondrial membrane. It may be therefore inferred that bongkrekic acid interacts with the ADP/ATP carrier when this latter is facing the matrix space of mitochondria. On the other hand, it may be assumed that the V_{max} value for external ADP (or ATP) increases with the number of carrier units which are available on the outer face of the inner mitochondrial membrane. Under these conditions, our results could be interpreted to mean that energization leads to the masking of a fraction of carriers or to their inaccessibility from the outer side of the inner membrane, perhaps due to the fact that they are retained with a modified conformation on the inner face of the mitochondrial membrane or in the lipid core of the membrane. The masked carriers can be unmasked by de-energization.

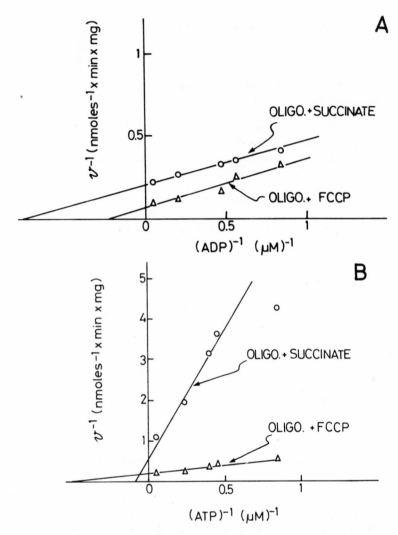

Fig. 3. *Effect of energization of mitochondria on the K_m and V_{max}
of the transport of external ADP (Fig. 3A) and ATP (Fig. 3B).
The incubation medium contained 110 mM KCl, 0.1 mM EDTA, 10 mM
morpholinoethane sulfonic acid (MES), pH 6.5, and either oligomycin 3 µg/ml plus 1 mM succinate (energized state) or oligomycin
3 µg/ml plus 0.5 µM FCCP (de-energized state). Rat heart mitochondria (1 mg protein) were added to 3 ml of medium in a
series of tubes and allowed to stand for 3 min at 20° and then
for 5 min at 0°. The incubation started with the addition of
[^{14}C]ADP or [^{14}C]ATP; it lasted for 20 sec at 0° and was ended
by addition of 5 µM carboxyatractyloside followed by rapid
centrifugation. The rate of ADP or ATP transport was calculated
as described in (8).*

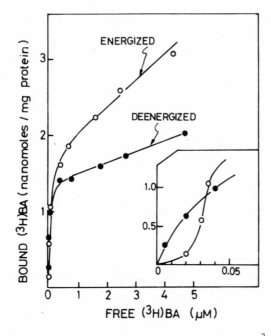

Fig. 4. Effect of energization on the binding of [³H]bongkrekic
acid to rat heart mitochondria. Medium for energized and de-
energized mitochondria were the same as in Fig. 3; 0.1 ml of
a suspension of rat heart mitochondria (1 mg protein) was added
to 5 ml of medium in a series of tubes and allowed to stand
for 3 min at 20°, before addition of increasing concentrations
of [³H]bongkrekic acid. To allow complete equilibration be-
tween bound and free bongkrekic acid incubation with [³H]
bongkrekic acid was carried out for 3 min at 20° and then for
40 min at 0°. The incubation was ended by centrifugation and
the radioactivity of the pellet measured by scintillation.

Other types of masking-unmasking cycles of binding sites
related to the ADP/ATP transport system will be described now.

Unmasking of ADP/ATP Carrier Sites Depending on the Membrane Environment

Unmasking of ADP and Bongkrekic Acid Sites in Rat Heart
Mitochondria: When rat heart mitochondria are incubated with
[¹⁴C]ADP nearly to equilibrium, the labelled ADP does not bind
to the totality of the carriers and a small fraction remains
apparently unreactive. The subsequent addition of bongkrekic
acid leads to a substantial increase of ADP binding, corresponding to the binding of ADP by the initially unreactive
carriers. On this basis, Klingenberg and Buchholz (38) have

postulated that the carrier units not loaded by ADP are accumulated and immobilized on the outer face of the inner mitochondrial membrane and that, upon addition of ADP, they become
loaded with ADP and move to the inner face of the membrane.
In recent experiments carried out with labelled bongkrekic acid,
we have found that the converse process holds as well. In the
absense of added ADP, only a fraction of high affinity sites
binds bongkrekic acid; a marked increase (15-20%) of bongkrekic
acid sites is induced by addition of ADP (Fig. 5). Most

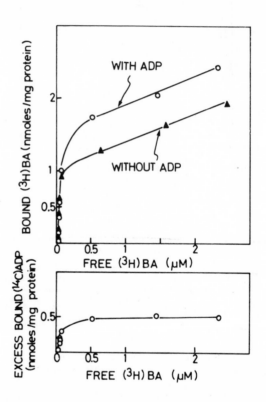

Fig. 5. *Effect of ADP on the binding affinity and capacity of rat
heart mitochondria for bongkrekic acid. Reciprocal effect of
bongkrekic acid on the binding of ADP. Rat heart mitochondria
(1.4 mg protein) were preincubated for 3 min at 20° with or
without 5 μM [^{14}C]ADP, in a series of tubes containing 110 mM
KCl, 0.1 mM EDTA and 10 mM MES, pH 6.5. Then [^{3}H]bongkrekic
acid was added at increasing concentrations and the tubes were
let to stand for 3 min at 20° and for 40 min at 0°. The incubation was ended by centrifugation and the radioactivity of the
pellet measured by scintillation.*

interesting, the number of binding sites for bongkrekic acid unmasked by ADP is roughly equal to the number of ADP binding sites unmasked by bongkrekic acid. It must be added that the enhancing effect of ADP is shared by ATP, but not by other nucleotides. The increased ADP binding has, therefore, its counterpart in the increased binding of bongkrekic acid. Whatever the mechanism of the ADP/ATP transport is, the above-mentioned experiments clearly point to the unmasking of specific carrier sites when the functioning of the carrier is induced by ADP.

Unmasking of Atractyloside and Carboxyatractyloside Binding Sites by Added ADP: When labeled carboxyatractyloside is added in the presence of EDTA to inner membrane vesicles prepared from rat liver mitochondria, a saturating titration curve is obtained with a K_d value of about 10 nM and a number of binding sites corresponding to 1-2 moles of bound carboxyatractyloside per mole of cytochrome a (3,17). When EDTA is replaced by $MgCl_2$, the carboxyatractyloside titration curve remains saturating but the total number of sites is decreased by a factor of two to three. Addition of a minute amount of ADP restores the totality of carboxyatractyloside binding sites. The half maximum effect is given by 0.1 µM ADP and the effect is shared by ATP, but not by other natural nucleotides. One is therefore led to conclude that in the presence of $MgCl_2$, a substantial fraction of carboxyatractyloside sites is not available for carboxyatractyloside binding. In keeping with the fact that carboxyatractyloside is a non-penetrant inhibitor, the above results can be interpreted to mean that in the presence of $MgCl_2$, a fraction of the carriers is masked to the binding of carboxyatractyloside because it is located on the inner face of the inner mitochondrial membrane and that external ADP mobilizes these masked carriers which then can interact with carboxyatractyloside. Similar results have been obtained with atractyloside and may be interpreted in the same way (39). This explanation, which appears straightforward in the concept of a mobile carrier, is, however, not as simple in view of the observation that sonication of the Mg particles in the presence of carboxyatractyloside does not increase the number of carboxyatractyloside binding sites. As an alternative, one may imagine that a certain number of unoccupied ADP carriers have become inefficient in specific binding due to conformational changes and that they can recover the ability to bind specific ligands after reaction with ADP or ATP.

Sensitization of the ADP/ATP Transport to -SH Reagents

When rat liver mitochondria are preincubated with N-ethyl-maleimide (NEM) and then ADP or atractyloside is added, a moderate inhibition of the ADP transport or of atractyloside binding is observed. However, when a trace of ADP is added together with NEM to the preincubation medium, the ADP transport and the atractyloside binding are dramatically inhibited (40,41). NEM is a lipophilic reagent which can cross the inner mitochondrial membrane. Its effect is shared by fuscin, another lipophilic -SH reagent, but not by mersalyl or p-CMB which are polar and non-penetrant reagents (42,43).

The addition of ADP to the preincubation medium which is decisive in the further inibition by NEM obviously induces the functioning of the ADP transport system. It also induces very significant morphological (44,45) and chemical (40-43) changes. In this chapter we shall examine only the chemical changes related to NEM binding. These changes essentially consist of an unmasking of -SH groups (from one to two moles per mole of cytochrome a).

The kinetics, the temperature and pH dependence of the -SH unmasking and the specific inducing effect of ADP or ATP have been reported in detail (43). The unmasked -SH groups could belong to the ADP/ATP carrier. Assuming a mobile carrier, one is led to assume an asymmetry of this carrier with respect to the accessibility of its -SH groups (from the external face or the internal face of the inner mitochondrial membrane). The alternative hypothesis is that the unmasked -SH groups belong to membrane proteins in the close proximity of the ADP/ATP carrier. This would be in line with the gross morphological changes brought about by addition of traces of ADP to mito-chondria.

A complicating factor in the evaluation of these results is the effect of energization on the -SH unmasking and on the inhibition of the ADP transport or of atractyloside binding by NEM. As shown in Table I, energization per se has the same effect as added ADP. It strikingly enhances the inhibition of atractyloside binding by NEM. It also contributes to the unmasking of -SH groups (Table II). Furthermore, the sensiti-zation to NEM of the ADP transport in mitochondria pretreated by ADP is higher in energy-rich conditions (Table III).

The above results show that the ADP/ATP transport system can be rendered sensitive to NEM not only by addition of specific substrates, but also by energization of the mitochon-dria. It is tempting to relate these observations to those reported above concerning the effect of energization on the binding of bongkrekic acid. Energization of the mitochondrial

TABLE I

Effect of the Energy State of Mitochondria on the Sensitization of [^{3}H]atractyloside (ATR) binding to N-ethylamleimide

State of Mitochondria	Additions	Bound [^{3}H] ATR (mole/mole cyto. a)	Inhibition of Binding - %
Energized	none	0.94	-
	NEM	0.51	46
	ADP	0.46	48
	NEM + ADP	0.10	90
De-energized	none	0.91	-
	NEM	0.91	0
	ADP	0.79	12
	NEM + ADP	0.49	46

For energization, rat liver mitochondria (5 mg) were preincubated at 4° in a series of tubes containing 5 ml of 110 mM KCl, 0.2 mM EDTA, 10 mM Tris-sulphate, pH 7.3, oligomycin 10 µg and 5 mM succinate. In the preincubation medium for de-energization, succinate was replaced by 1 µM FCCP. After 5 min, 100 µM NEM or 20 µM ADP, or 100 µM NEM plus 20 µM ADP were added and preincubation lasted for an additional period of 5 min at 4°. Then [^{3}H]atractyloside was added to each tube to a final concentration of 0.36 µM and incubation was carried out for 45 min at 4°. After centrifugation, the radioactivity of the pellet was measured.

TABLE II

Effect of the Energy State of Mitochondria on the Binding of [^{14}C]N-ethylmaleimide

State of Mitochondria	Bound [^{14}C]NEM (1)*	(moles/mole cyt. a) (2)*	(1-2)*
Energized	15.1	13.8	1.3
De-energized	5.1	4.9	0.2

* (1) without ATR, (2) with ATR

Prior to preincubation, rat liver mitochondria in 0.25 M sucrose (10 mg/ml) were treated with 100 µM unlabeled NEM for 15 min at 4°. After washing by 0.25 M sucrose, mitochondria were preincubated for 5 min at 4° under energized or de-energized conditions, as described in Table I. Then the incubation was started by the addition of 10 µM [^{14}C]NEM plus 20 µM ADP and as indicated 10 µM atractyloside. It was carried out at 4° for 5 min. The difference between bound NEM in the absence and in the presence of atractyloside corresponds to the amount of SH groups unmasked by the binding of ADP to the ADP carrier.

TABLE III

Effect of the Energy State of Mitochondria on the Sensitization of the ADP Transport to N-ethylmaleimide

State of Mitochondria	Addition to pre-incubation medium	Rate of ADP transport nmoles/min/mg protein	Inhibition of the rate of ADP transport (%)
Energized	ADP	0.8	
	NEM + ADP	0.3	63
De-energized	ADP	2.0	
	NEM + ADP	1.7	15

Energized and de-energized rat liver mitochondria were obtained as described in Table I, and preincubated for 5 min at 4° with 5 μM ADP and when indicated 100 μM NEM. Then the incubation was started by the addition of [^{14}C]ADP (50 μM, final concentration); it lasted for 20 sec at 0° and was ended by addition of 5 μM carboxyatractyloside.

membrane and ADP binding could in fact induce similar modifications in the kinetic and binding parameters of the ADP/ATP transport system. Other examples of sensitization to inhibitors of transport processes have been reported; this is the case of the transport of choline in erythrocytes (46,47) and of sugars in bacteria (48,49) where the sensitization phenomenon was taken as indicative of two conformational states of the carrier, corresponding to the inward-facing and to the outward-facing carrier.

Since the energization in mitochondria depends on the functioning of the respiratory chain in the oxidative phosphorylation complex, an obvious question is the following: is there a steric and functional relationship between the ADP/ATP transport system and the oxidative phosphorylation complex?

Is the ADP/ATP Transport Directly Connected to the Oxidative Phosphorylation System?

There are a number of observations which strongly suggest that some connections exist between the ADP/ATP transport system and the F_1-ATPase component of the oxidative phosphorylation complex: a) external ADP is phosphorylated by F_1-ATPase without prior dilution in the matrix pool of adenine nucleotides (50); b) ATP synthesized from external ADP does not enter readily in the matrix space, most of it is directly exported outside mitochondria (50-52); c) the F_1-ATPase, the respiratory chain and the ADP/ATP transport system behave towards inhibitors as components of a multi-enzyme complex would do (50,51); d) the number of respiratory chains and F_1-ATPase molecules is similar to the number of ADP/ATP carrier units (50,51).

Taken together, these observations suggest that the F_1-ATPase and the ADP/ATP transport system are situated in close proximity in the inner mitochondrial membrane. Such a close proximity would explain why changes of conformation of the oxidative phosphorylation complex may influence the conformation and thereby the binding properties of the ADP/ATP transport system.

General Comments on Carrier Models

There are a number of facts which suggest that the ADP/ATP transport is mediated by a mobile carrier moving back and forth across the inner mitochondrial membrane and having its substrate site alternatively oriented towards either the outer or the inner side of the membrane. To be mobile in the lipid core of the membrane, the substrate carrier complex must be hydrophobic. The mobile carrier hypothesis fits well the following observations: 1) the Arrhenius plot of the rate of ADP transport is characterized by a break which may correspond to the transition temperature of the lipid phase surrounding the carrier; 2) an exchange-diffusion transport can be readily explained on the basis of a mobile carrier; 3) some experimental results presented in this paper and others (cf. 38) have been interpreted in terms of carrier units distributed either on the inner or on the outer face of the mitochondrial membrane. The mobile carrier hypothesis, however, does not fit with the observation that the ADP transport depends on the functioning of the oxidative phosphorylation complex. This latter type of interaction would fit better with a fixed pore model of transport than with a mobile carrier model. Also, by comparison with the slow flip-flop of phospholipids in artificial membranes (53) and in biological membranes, in particular in the inner mitochondrial membrane (54), a transversal movement of carrier is not very likely to occur (55). A mixed model involving a mobile rotating carrier and a fixed channel could actually account for the number of observations made on the ADP transport. In this mixed model, the mobile rotating carrier would be located in the outer half of the phospholipid bilayer whereas the fixed channel made of an aggregate of proteins would span the inner half of the phospholipid bilayer at the immediate proximity of F_1-ATPase. The opening and closing movements of the channel would be coordinated to the rotation of the carrier. By rotation, the mobile carrier would deliver external ADP to the fixed channel or reject in the external medium the mitochondrial ATP. In such a mixed model, the movements of the channel are expected to be under the control of the closely positioned oxidative phosphorylation complex and, therefore, to respond to the energy state of the mito-chondria. Conversely, external ADP or ATP would directly effect the kinetics of the rotating carrier. These comments are illustrated in Fig. 6.

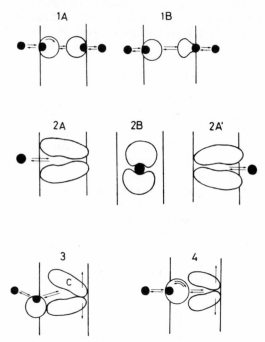

Fig. 6. *Models of Ad translocator. 1 - mobile carrier, symmetric (A), asymmetric (B). 2 - Channel (pore). States A, B, C represent steps for transport from one side of the membrane to the other. 3 - Association of a binding protein and a channel. 4 - Association of a mobile rotary carrier and a channel.*

REFERENCES

1. Heldt, H.W., Klingenberg, M. and Milovancev, M. (1972) Eur. J. Biochem. 30, 434.
2. Slater, E.C., Rosing, J. and Mol, A. (1973) Biochim. Biophys. Acta 292, 534.
3. Vignais, P.V., Vignais, P.M., Lauquin, G. and Morel, F. (1973) Biochimie 55, 763.
4. Brierley, G. and O'Brien, R.L. (1965) J. Biol. Chem. 240, 4532.
5. Kemp, A. and Groot, G.S.P. (1967) Biochim. Biophys. Acta 143, 628.
6. Winkler, H.H., Bygrave, F.L. and Lehninger, A.L. (1968) J. Biol. Chem. 243, 20.
7. Pfaff, E. and Klingenberg, M. (1968) Eur. J. Biochem. 6, 66.
8. Duée E.D. and Vignais, P.V. (1969) J. Biol. Chem. 244, 3920.
9. Souverijn, J.H.M., Weijers, P.J., Groot G.S.P. and Kemp, A. (1970) Biochim. Biophys. Acta 223, 31.
10. Duée, E.D. and Vignais, P.V. (1968) Biochem. Biophys. Res. Commun. 30, 420.
11. Vignais, P.V., Setondji, J. and Ebel, J.P. (1971) Biochimie 53 127.
12. Pfaff, E. Heldt, H.W. and Klingenberg, M. (1969) Eur. J. Biochem. 10, 484.
13. Kemp, A., Jr. and Slater, E.C. (1964) Biochim. Biophys. Acta 92, 178.
14. Chappell, J.B. and Crofts, A.R. (1965) Biochem. J. 95, 707.
15. Heldt, H.W., Jacobs, H. and Klingenberg, M. (1965) Biochem. Biophys. Res. Commun. 18, 174.
16. Vignais, P.V., Vignais, P.M. and Defaye, G. (1971) FEBS Letters 17, 281.
17. Vignais, P.V., Vignais, P.M. and Defaye, G. (1973) Biochemistry 12, 1508.
18. Henderson, P.J.F. and Lardy, H.A. (1970) J. Biol. Chem. 245, 1319.
19. Henderson, P.J.F., Lardy, H.A. and Dorschner, E. (1970) Biochemistry 9, 3453.
20. Kemp, A., Jr., Out, T.A., Guiot, H.F.L. and Souverijn, J.H.M. (1970) Biochim. Biophys. Acta 223, 460.
21. Kemp, A., Souverijn, J.H.M. and Out, T.A. (1970) in Energy Transduction in Respiration and Photosynthesis, (Quagliariello, E., Papa, S. and Rossi, C.S., eds.) pp. 959-969, Adriatica Editrice, Bari, Italy.
22. Lijmbach, G.W.M. (1969) Thesis, Delft.

23. Lauquin, G., Duplaa, A.M., Rousseau, A. and Vignais, P.V. (1975) 10th FEBS Meeting, Paris, Abstract # 1160.
24. Pande, S.V. and Blanchaer, M.C. (1971) J. Biol. Chem. 246, 402.
25. Vaartjes, W.J., Kemp, A., Souverijn, J.M.H. and Van den Bergh, S.G. (1972) FEBS Letters 23, 303.
26. Halperin, M.L., Robinson, B.H. and Fritz, I.B. (1972) Proc. Natl. Acad. Sci. (US) 69, 1003.
27. Morel, F., Lauquin, G., Lunardi, J., Duszyński, J. and Vignais, P.V. (1974) FEBS Letters 39 133.
28. Weideman, M.J., Erdelt, H. and Klingenberg, M. (1970) Eur. J. Biochem. 16, 313.
29. Winkler, H.H. and Lehninger, A.L. (1968) J. Biol. Chem. 243 3000.
30. Klingenberg, M. (1970) *in* Essays in Biochemistry, Vol. 6, (Campbell, S.N. and Dickens, F., eds.) pp. 120-159, Academic Press, New York.
31. Beck, J.C., Mattoon, J.R., Hawthorne, D.C. and Sherman, F. (1968) Proc. Nat. Acad. Sci. (US) 60, 186.
32. Kováč, J.C. and Hrušovská, E. (1968) Biochim. Biophys. Acta 153, 43.
33. Kolarov, J. and Klingenberg, M. (1974) FEBS Letters 45, 320.
34. Brandolin, G., Meyer, C., Defaye, G., Vignais, P.M. and Vignais, P.V. (1974) FEBS Letters 46, 149.
35. Egan, R.W. and Lehninger, A.L. (1974) Biochem. Biophys. Res. Commun. 59, 195.
36. Klingenberg, M., Riccio, P., Aquila, H., Schmiedt, B., Grebe, K. and Topitsch, P. (1974) *in* Membrane Proteins in Transport and Phosphorylation (Azzone, G.F., Klingenberg, M., Quagliariello, E. and Siliprandi, S., eds.) pp. 229-243, North-Holland Publishing Co., Amsterdam.
37. Souverijn, J.M.H., Huisman, L.A., Rosing, J. and Kemp, A. (1973) Biochim. Biophys. Acta 305, 185.
38. Klingenberg, M. and Buchholz, M. (1973) Eur. J. Biochem. 38, 346.
39. Vignais, P.V. and Vignais, P.M. (1971) FEBS Letters 13, 28.
40. Leblanc, P. and Clauser, H. (1972) FEBS Letters 23, 107.
41. Vignais, P.V. and Vignais, P.M. (1972) FEBS Letters 26, 27.
42. Vignais, P.M. and Vignais, P.V. (1973) Biochim. Biophys. Acta 325, 357.
43. Vignais, P.M., Chabert, J. and Vignais, P.V. (1974) *in* Symposium on Biomembranes, Structure and Function, pp. 307-313, 9th FEBS Meeting, Budapest.

44. Weber, N.E. and Blair, P.V. (1970) Biochem. Biophys. Res. Commun. 41, 821.

45. Stoner, C.D. and Sirak, H.D. (1973) J. Cell Biol. 56, 51.

46. Edwards, P.A. (1973) Biochim. Biophys. Acta 307, 415.

47. Edwards, P.A. (1973) Biochim. Biophys. Acta 311, 123.

48. Haguenauer-Tsapis, R. and Kepes, A. (1973) Biochem. Biophys. Res. Commun. 54, 1335.

49. Jimeno Abendano, J. and Kepes, A. (1973) Biochem. Biophys. Res. Commun. 54, 1342.

50. Vignais, P.V., Vignais, P.M. and Doussière, J. (1975) Biochim. Biophys. Acta 376, 219.

51. Bertina, R.M. and Out, T.A. (1973) 9th Intl. Congress Biochem., Stockholm Abstract Book, p. 236.

52. Bertagnolli, B.L. and Hanson, J.B. (1973) Plant Physiol. 52, 431.

53. Kornberg, R.D. and McConnell, H.M. (1971) Biochemistry 10, 1111.

54. Rousselet, A., Colbeau, A., Zakowski, A., Vignais, P.M. and Devaux, P.F. (1975) FEBS Meeting, Paris, Abstract # 1043.

55. Singer, S.J. (1974) Ann. Rev. Biochem. 43, 805.

THE ADENINE NUCLEOTIDE TRANSPORT OF MITOCHONDRIA

M. Klingenberg

Introduction

In 1964 it was first reported (1) that the inner mitochondrial membrane facilitates the exchange of ADP and ATP. Since then a number of other metabolites which function as intermediates in metabolite pathways of mitochondria were also found to be transported through the inner mitochondrial membrane. These include the substrates of the tricarboxylic acid cycle, of transamination reactions and of phosphate transfer reactions in oxidative and substrate level phosphorylation (for review see (2-4)).

It is interesting to note that unlike certain other transport systems most mitochondrial transport reactions involve a counter exchange. For dehydrogenation and phosphorylation inside the mitochondria substrates are taken up and products released in the cytosol.

The transport system most extensively investigated in the last ten years is the exchange between extra- and intra-mitochondrial ADP and ATP, a reaction which is essential to the transfer of energy from oxidative phosphorylation to extramitochondrial processes. This transport has been discovered on the basis of a systematic study of the permeation of nucleotides to the two mitochondrial membranes (1,4). The subsequent finding (4,5) that atractylate (ATR) is a specific inhibitor of the transport promoted considerably further research.

An outline of the various research stages on this transport is summarized in Table 1.

Abbreviations: ATR, atractylate;
                    BKA, bongkrekate;
                    CAT, carboxy atractylate;
                    SMP, sonic mitochondrial particles;
BHM, RLM, beef heart -, rat liver mitochondria

Table I
Sequence of Research on the Mitochondrial ADP,ATP
Carrier

1. Establishment of endogenous ANP-pool, role in phosphorylation (8), transport of ADP,ATP through inner mitochondrial membranes as exchange against endogenous pool (4,5).

2. Specificity (5,9,6,7,11), kinetics of transport, K_m, temperature dependence (10,12), specific inhibitors.

3. Regulation ("active" transport) by energization of membrane (4,9). Mechanism as electrogenic transport (13). Differentiation between homo- and hetero-exchange (14). Stoichiometry of H^+/ATP, ADP movements (14).

4. The ATP/ADP gradient, its dependence on membrane energization (15,16), correlation to membrane potential (17). Relation between ATP potential gradient and P/O.

5. Definition of carrier sites by the effect of ATR on the ADP binding. In- and outside localization of sites (18). The opposite effects of ATR and BKA on binding of ADP (19). Carrier site fixation exclusively in- or outside by ATR or BKA. Demonstration of a carrier site translocation (20).

6. Definition of carrier site with ^{35}S-ATR, ^{35}S-CAT, removal by BKA (21, 22).

7. Membrane conformation changes as a result of carrier orientation to membrane sides, kinetics of carrier translocation (23).

8. Definition of two carrier states according to membrane sidedness. C_m-state (active side) with high affinity for BKA, none for ATR, CAT and C_c-state (cytosol side) with high affinity for ATR, CAT and low for BKA (24).

9. The single site reorienting carrier mechanism (rotational or gated pore). Mobilizing (activating) step and translocating step. Contribution of protein-lipid interaction(25).

10. Carrier isolation. Solubilization of CAT-binding protein and NEM labelled protein. Definition of a hydrophobic binding protein with a mass of 29,000 dalton (24,26).

Metabolic significance of ADP,ATP transport.

The transport of ADP and ATP together with P_i is the most powerful transport system in eukaryotic cells which rely mainly on respiratory energy. The main function is the exchange of ADP, originating from energy consuming reactions in the cytosol, against ATP, generated in the mitochondria (cf. Fig.1). The ADP,ATP carrier is the exclusive link between inner- and extramitochondrial P_i-transfer reactions. This is based on the specificity for ADP and ATP which excludes AMP and all other nucleotides. It is located in the inner mito- chondrial membrane, whereas the outer mitochondrial membrane is largely unspecifically permeant to molecules up to M.W. 4000 (27).

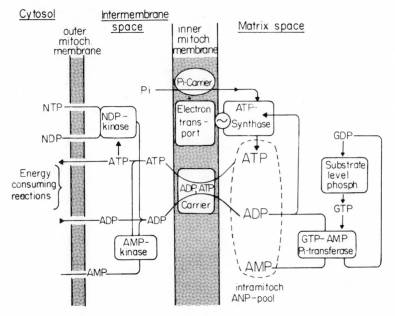

Fig.1 The role of the ADP,ATP carrier in mito- chondrial phosphate transfer reactions.
      The function of the intramitochondrial ANP pool as an intermediate in the synthesis of the extramitochondrial ATP: Localization of transport systems on the impermeable inner mitochondrial membrane and diffusion through the permeant outer mitochondrial

membrane. Exclusion of ANP by the inner
mitochondrial membrane and localization of
AMP re-utilizing phosphate transferases in
the perimitochondrial and intramitochondrial
space.

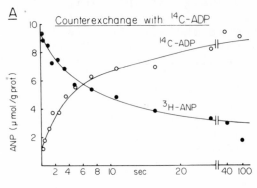

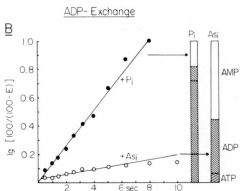

Fig.2
A. Counter-
exchange between
intra- and extra-
mitochondrial ANP.
Time dependence
measured through
"rapid automated
mixing and
sampling apparatus"
"R.A.M.S.A.")
(Klingenberg, un-
published) using
the CAT stop of
the exchange. Ex-
change started by
addition of ^{14}C-
ADP (200 µM) to
rat liver mito-
chondria in which
the endogenous
nucleotides had
been prelabelled
with ^{3}H-ADP.
B. Exchange of ANP
by varying the
endogenous ratio
of ADP+ATP. Pre-
incubation of the mitochondria with P_i or A_i,
respectively, in order to change the AMP
content as indicated in the columns. Endogenous ANP have been prelabelled by incubation
with ^{14}C-ADP.

The ADP,ATP transport represents an exchange of
ADP or ATP against the intramitochondrial nucleotide pool such that there is no net accumulation
in the mitochondria (4). The exchange kinetics
have been followed with the inhibitor-stop-method
(10) and, more recently, using an advanced rapid
mixing and sampling apparatus (12). Only evaluating

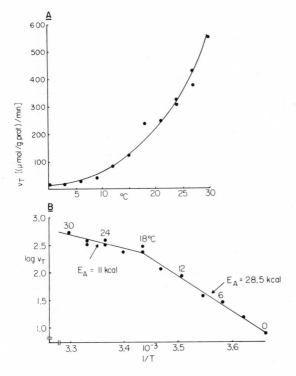

Fig.3
Temperature
dependence of
ADP-exchange.
Measurements on
rat liver mito-
chondria pre-
labelled with
^{14}C-ANP. The
exchange started
by the addition
of 200 μM ADP.
Each point re-
presents the
first order
kinetic measure-
ments obtained
by "R.A.M.S.A.".

the initial time course of the process, appropriate
exchange rates can be obtained. Since many reports
giving exchange rates in the literature are not
based on appropriate kinetics, the rate values have
to be revised. An example for the kinetics of the
exchange, as shown in Fig.2, illustrates two essential features: Firstly, simultaneous to the uptake of ^{14}C-ADP there is a release of endogenous
^{3}H-ANP from the endogenous pool. Secondly, the
exchange is limited by the size of the endogenous
pool ADP+ATP which can change on account of the
endogenous AMP. AMP cannot be released unless it
is interconverted to ADP.

The transport rates evaluated with this method
have been evaluated as a function of various parameters, such as temperature, concentration, energization of the membrane, etc. The temperature
dependence of the exchange rate (Fig.3) illustrates
a very steep increase of the exchange rate between
0° and 15°C with the activation energy of nearly

131

Table II

Kinetic Data

	RLM	BHM
Translocation activity (18°C) Maximum with ADP [μmol/min/g protein]	150-200	600
Turnover number (18°C) per CAT-binding site [min^{-1}]	500	500
Activation energy E_A [Kcal/Mol] ($0-18^{\circ}$C) ($18^{\circ}-30^{\circ}$C)	29 11	32 –

RLM: K_m (μM)

	energized	+FCCP
ADP	3.5(0.3) and 0.8(0.7)	2.0(0.3) and 0.5(0.7)
ATP	2.7(0.8) and 3 (0.2)	11.0(0.4) and 0.8(0.6)

( ) = portion of the two overlapping K_m.

30 Kcal and above this temperature with only $E_A =$
11 Kcal. As a result at low temperature the
exchange becomes a rate limiting step for the production of ATP by the mitochondria and the phosphorylation of endogenous nucleotides is several
times faster than that of the exogenous nucleotides.
The kinetic data are summarized in Table II.

Specificity

The most important aspect of the specificity to
the nucleotides is the virtual exclusion of AMP
from transport (Table III). Therefore, it can be
understood that on both sides of the inner mitochondrial membrane phosphate transfer reactions
exist which recover AMP produced for example by
substrate activation (cf. Fig.1). In the extramitochondrial space there is adenylate kinase, in
the intramitochondrial space are monophosphate
kinases, such as GTP-AMP transferase, etc.
The ADP/ATP exchange is very sensitive to variation in the base moiety of the molecules. Only
formycine phosphate has some activity. The deoxy-

Table III

Specificity of transport for
exogenous nucleotides

Nucleotides*	uptake activity (%)
ADP	Ξ,100
ATP	≃ 70
AMP	< 2
dADP	15
dATP	15
AMP:PNP	19
AMP·PCP	10
Formycin DP	15
G-.C-,U-IDP	≃ 0
3,5-ADP	0

*$18°C$, 200 μM nucleotide, in un-
coupled state (+FCCP); rat liver
mitochondria.

adenine nucleotides are also accepted and the changes in
the phosphate moiety are permitted such as in ANP-PNP and
ANP-PCP. Any other changes in the phosphate-group arrangements are inactive. In general, specificity of the transport is higher than for many kinases and for the ATP
synthesis in mitochondria. This explains the high specificity of oxidative phosphorylation for added ADP by intact
mitochondria as compared to that of sonic particles.

"Active" energy dependent exchange

The exchange between endogenous and exogenous ADP or ATP
can be considered as a carrier catalyzed process which is
basically non-energy dependent. A rapid exchange occurs
in uncoupled, de-energized mitochondria. Similar processes
have been formerly described as "facilitated diffusion."
The "energization" of the mitochondria influences the
exchange by superimposing an energetic distortion of the
transport rates of ATP versus ADP. In the energized state
ATP is taken up at a considerable lower rate than in the
uncoupled state (Fig.4). At the same time K_m is decreased.
With ADP this difference is much smaller (10,12,28).
Under competitive conditions, which correspond to those
in the cell, were both ADP and ATP are added outside, the

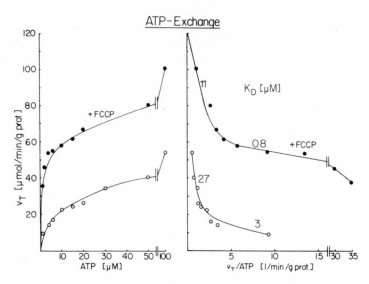

Fig.4   Concentration dependence of ADP exchange. Influence
of uncoupling.
Each point corresponds to a first order rate
evaluation of the kinetics obtained with "R.A.M.S.A."
at 18°C with rat liver mitochondria.

ability of mitochondria is striking to select about ten times
ADP in preference to ATP (16). This explains the finding
that with mitochondria added ADP is nearly completely converted to ATP with a relatively sharp break in the respiratory rate. In other experiments the efflux of ADP versus
ATP is shown to be changed by energization in the opposite
direction such that the efflux of ATP is strongly preferred
to that of ADP (29).

Interpretation of this energy-driven exchange comes from
considering that ATP has one negative charge in excess to
that of ADP. In the exchange of ADP against ATP one negative charge is released from the mitochondria. A membrane
potential positive outside would electrophoretically pull
this charge outside under the expense of energy. The energy
is taken from the pool of energy provided by the respiratory
chain which provides positive charges outside (Fig.5).

As a result of the asymmetric specificity in uptake and
release, the external ratio ATP/ADP should be higher than
the internal ratio ATP/ADP. This prediction from the
kinetic studies has been demonstrated with mitochondria such
that a ratio ATP/ADP ≃ 30 can be reached in the energized

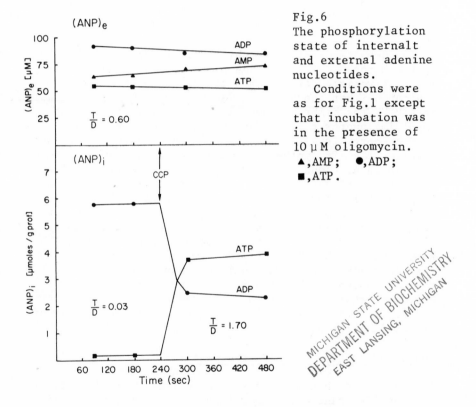

Fig.5
Scheme for electrogenic or electroneutral transport of ANP and ADP
through the mitochondrial membrane.

The scheme assumes that 3(+)
charges on the binding site are
compensated by 3(−) charges from
ADP or ATP.

Fig.6
The phosphorylation
state of internalt
and external adenine
nucleotides.

Conditions were
as for Fig.1 except
that incubation was
in the presence of
10 µM oligomycin.
▲,AMP; ●,ADP;
■,ATP.

state and below <1 under uncoupled state (see Fig.6) (15).
This fully supports the deductions from the kinetic studies.
A difference in the ratio ATP/ADP was also observed to exist
between the cytosol and the mitochondrial space and thus
proves its high physiological significance in the regulation
of the free energy of the ATP in the cytosol as compared to
the mitochondria (30).

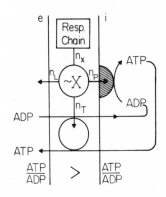

Fig.7
The division of the respiratory energy, n_x, into the part requires for ADP synthesis (n_p), for transport (n_T) and for leakage (n_L) :
$$n_x = n_p + n_T + n_L.$$

The dependence of the double ratio $(ATP/ADP)_e/(ATP/ADP)_i$ on the mitochondrial membrane potential could be experimentally verified (17). This can be taken as a strong indication that the preferentially electrophoretic extrusion of ATP by the membrane potential is the main reason for supplying energy to this "active" transport process.

The energy is extracted from the same energy source which supplies the ATP synthesis (Fig.7). Since it can be assumed that an equal amount of energy intermediates (in the form of charges released through the membrane) is generated per coupling site, the amount of energy left for the ATP synthesis is diminished when energy is deducted also for transport processes such as in the ATP extrusion. As a result the ratio $ATP/2e^-$ should be decreased (16,29). The total energy available for ATP formation is divided up such that less ATP is found as would correspond to maximum ATP/O ratios, however, at a higher energy potential. It can be estimated that about 2 to 3 Kcal free energy are added by the active transport resulting in the total free energy of external ATP of about 15 Kcal (13,14).

Inhibitors of the ADP,ATP transport.

There exist highly specific and effective inhibitors of the ADP,ATP transport. Most important are the antibiotics, ATR, CAT and BKA. The inhibitory effect of ATR on oxidative phosphorylation of mitochondria has been known for a considerable time, however, its site of action has been first misinterpreted. Only after the specific ADP/ATP exchange was discovered, ATR was recognized as an inhibitor of the transport. These results have been confirmed by several research groups.

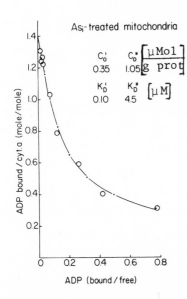

Fig. 8
Mass action plot of the ATR removable portion for evaluation of the binding parameters. The curve corresponds to a function computed with the parameters given in the figure for best fit with the measured values (cf. ref. 18).

Whereas inhibition by ATR is marginally competitive to the exchange, that of the CAT and BKA are uncompetitive(31,32). It is considered that all these ligands bind to the substrate site of the carrier (see below). The chemical structure of the inhibitors differs widely from that of the substrates. It is striking that all ligands have a minimum of three negative charges. For example, AMP^{-2} is not transported and also the desulfato-form of ATR and CAT largely ineffective (33). The common requirement of three negative charges might be caused by three positive charges at the binding site. This conclusion is also supported by the electrophoretic effect of the membrane potential on the ADP/ATP exchange as discussed above.

Definition of carrier sites.

The high specificity of the transport for ADP and ATP and the existence of very effective specific inhibitors is a strong indication that the transport is catalyzed by a specific carrier in the membrane. The carrier was postulated to be a protein because only a high molecular weight peptide can be envisaged to provide the high specificity for molecules such as ADP,ATP. It was therefore important to determine carrier sites on the membrane and possibly to define the catalytic mechanism.

The binding sites of the carrier on the mitochondrial membranes were first defined by binding with ADP and

Table IV

Comparison of binding data for ADP and inhibitors

	RLM		BHM	
	Number of binding sites (maximum)			
	μ Mol/g protein	Mol/Mol cyt.aa$_3$	μ Mol/g protein	Mol/Mol cyt.aa$_3$
ADP,ATP	0.1–0.3	1	0.6–1.0	1–1.3
ATR,CAT	0.15–0.35	1.2	1.2–1.8	2–2.5
BKA	–	–	1.2	2
	Dissociation constant K_d (μM)			
ADP,ATP	2 (0.9)		10, 0.9	(0.3:0.7)
ATR	0.2		0.1	
CAT	0.02		0.03	

ATP (18). It was important to discriminate this binding from ADP,ATP taken up in the endogenous pool and binding to other sites at the membrane. This was possible by appropriate use of ATR such that only that portion of ADP was attributed to the carrier which can be removed by ATR. The mass action plot of a concentration dependence revealed the existence of both high and low affinity binding sites at the carrier (Fig.8). The different affinities were interpreted to reflect that a portion of the binding sites with the high affinity is directed to the inner phase of the membrane. The maximum number of carrier sites for ADP as determined in this manner is given in Table IV.

These results indicate first that binding studies may give quite unexpectedly an insight into the dynamics of the carrier function. Further a number of surprising results arose which could not be explained by a free interaction of ligands with protein. The results could be more elegantly explained by taking into account that the binding involved a switching of carrier sites across the membrane. It can be expected that binding sites located inside should behave anomalously since they involve the translocation step before they can interact with the outside ligand. In interpreting these results simply in a more classical manner as

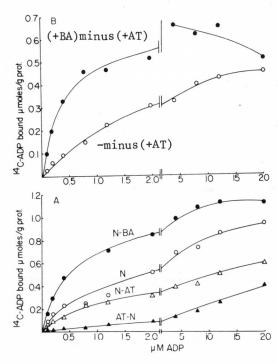

Fig.9
Influence of BKA on the ADP binding.

Concentration dependence under conditions where BKA or ATR are added either before or after ADP. The difference of the binding N–BKA minus N–ATR equals the specific binding portions to the carrier either under the influence of BKA (a') or without BKA (a). The plot demonstrates the increase of affinity for ADP under the influence of BKA (cf.ref.19).

"allosteric" or "conformational" effect, one would deprive themselves from exploiting these results for the carrier mechanism.

This problem became more prominent on the application of BKA to the binding. It was quite surprising to find that BKA, in contrast to ATR, was not removed in ADP but rather increases the binding (19,20). BKA in effect does not create more maximum binding sites but rather enhances the affinity for ADP or ATP of the membrane (Fig.9). Also in this case the specific binding is distinguished from unspecific binding by subtracting the level of ADP which cannot be removed by ATR. There are interesting differences whether BKA is added before or after ADP which have been investigated in great detail and which indicate that BKA in contrast to ADP can bind to carrier sites when it faces the inside. This is feasible because BKA as a lipophilic substance can be expected to be permeant to the membrane in contrast to the more hydrophilic ATR.

At first the afffinity increase for ADP by BKA would indicate the existence of a regulatory binding site for BKA from which BKA increases the binding of ADP. From the same

BA: increases (apparent) affinity

ATR: decreases affinity

Substrate and regulatory site:

Substrate site only:

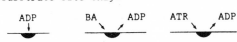

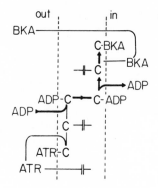

Fig.10 Scheme for the existence of two or one site on the carrier.

In the first case the inhibitors would bind to identical or different regulatory sites, separate from the substrate site. In the second case, there is only one type of binding site for substrates and inhibitory ligands.

Fig.11 Scheme based on the circulating (mobile) carrier model.

Reorientating carrier mechanism assuming a binary complex of BKA and carrier where BKA replaces ADP similarly to ATR. It is assumed that BKA can penetrate the membrane and ATR is impermeant.

site ATR should decrease the affinity of ADP (Fig.10). This interpretation would be in line with the classical interpretation of protein ligand interaction. The inhibition of transport would result from ATR removing the substrate or from BKA fixing the substrate so tight that it cannot dissociate.

There are, however, a number of peculiar effects of BKA on the binding which lead to a different model: The effect of BKA on the binding and on inhibition is relatively slow. It has a very strong dependence on the temperature and on pH. These and further data lead us to postulate that BKA binds the carrier mainly from the inside. Here it also removes ADP from the carrier. No ternary but only binary carrier-ligand complexes are formed where possibly inhibitors and substrates bind to the same side.

The apparent increase of ADP binding by BKA, as shown in Fig.9, can be explained on the basis of a simple single site reorienting carrier mechanism (Fig.11). Carrier sites located at first on the outside become activated by forming the ADP carrier complex to the inside. Here BKA displaces ADP and forms an impermeant tight BKA carrier complex. Eventually all outer carrier sites will bind ADP and then become trapped inside. Thus so much ADP is trapped inside as there have been originally carrier sites outside. This would correspond exactly to the apparent increase of ADP binding to the carrier.

In this picture (Fig.11) the opposite effects of ATR and BKA on the ADP binding demonstrate that all carrier sites can be either fixed in the outer or the inner position of the membrane. This corresponds to a single site carrier which can reorient the binding site essentially according to the formulation given by the mobile carrier mechanism. These effects are the first demonstration of a reorienting carrier mechanism on a molecular level. After presenting further data, we shall return to this discussion.

Quite unexpectedly certain configurational changes of the inner mitochondrial membrane in beef heart mitochondria can be related to the reorientation of the carrier sites (23). Mitochondria have a decontracted matrix when no ADP or ATR are added, i.e., when the carrier sites face the outer surface. They become partially contracted with ADP when carrier sites are in a steady-state turned both inside and outside. They become more contracted under the influence of BKA when all carrier sites face the inside. These conformation changes can be followed as turbidity changes in a mitochondrial suspension and therefore easily be used to record the kinetics of binding or reorientation since these processes could be shown to be closely parallel to the binding of the various ligands (Fig.12).

With these experiments and also direct measurements of the BKA uptake it became clear that only in the undissociated form ($BKA^{3-} + 3H^+ \rightleftharpoons BKAH_3$) BKA is made available for binding to the carrier. This is exactly what would be expected for the permeation of BKA through the membrane which requires the electroneutral form and would represent the step preceding the binding of BKA to the carrier on the inside. It explains the unusually high pH dependence of the rate for the BKA induced contraction which is found to follow $v = k (H^+)^{2.7}$. The rate increases approximately one hundred times from pH 7.2 to 6.5 (23).

Binding of ATR and CAT

The most direct measurement of the carrier sites on the membrane is possible by using radio-actively labelled

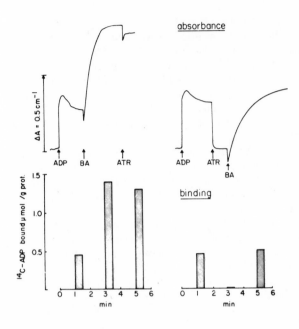

Fig.12
Comparison of ^{14}C-ADP binding with the absorption change of mitochondrial suspension as induced by addition of ligands of ADP, ATP carrier.

Photometric recording at 546 nm of suspension of beef heart mitochondria. In parallel experiments under the same conditions, binding of ^{14}C-ADP was measured. For more details see ref.23.

^{35}S-ATR or ^{35}S-CAT. With these compounds the maximum number of sites can be defined with more precision since the binding of ATR and CAT to the mitochondrial membrane is restricted exclusively to the carrier sites (21,22). The data as summarized in Table IV show that in the mitochondria there are about 1.2 to 2 binding sites for CAT per cytochrome a. With these data a relatively high density of occupation of carrier sites per surface can be calculated at about 10,000 $\mathring{A}^2$ per carrier.

The affinity for CAT is much higher than that for ATR such that CAT can replace ATR easily whereas CAT itself cannot be removed by unlabelled CAT. The difference in the affinity is also reflected by sensitivity to BKA. BKA can only remove ATR but not CAT. As shown in Fig.13, at the same time when BKA increasingly replaces ^{35}S-ATR, ^{3}H-ADP becomes "bound." This is exactly what would be expected from the reorientating carrier mechanism as illustrated in Fig.11.

The asymmetry of carrier sites

With the reorientating carrier mechanism established by these results the question arises to what extent the

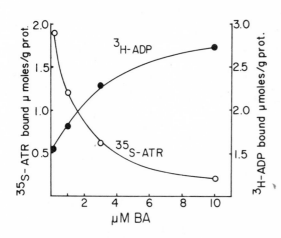

Fig.13
Removal of ^{35}S-ATR by BKA and con- comitant binding of ^{3}H-ADP.

Depleted BHM (1mg/ protein/ml) are in- cubated with 2 μM ^{35}S-ATR at 25°. BKA as indicated and 10 μM ^{3}H-ADP are added.

affinity to the various ligands of the carrier changes when it faces the out- or inside. The inversion of the mito- chondrial membrane by sonication exposes the inner face of the mitochondrial membrane to the outside and therefore makes it accessible to impermeant ligands. At this stage, it is useful to introduce the designation m-side for the matric face of the membrane and c-side for the cytosol face. in sonic particles the m-side is directly accessible to the various ligands.

The transport in these particles reveals a difference to mitochondria (24): exchange is largely insensitive to ATR but fully inhibited by BKA. The inhibition by BKA is not accompanied by the effects diagnostic for a preceding permeation through the membrane since it is independent of pH and without time delay. In sonic particles BKA appears to bind to the carrier directly from the outside. ATR inhibits the transport in the particles only when it is bound before sonication to the mitochondria. Obviously ATR can inhibit and bind to the carrier site only from the c-side, whereas BKA binds primarily from the m-side (Fig.14).

Binding studies with ^{35}S-CAT and BKA also support these conclusions on the different binding properties when the carrier site is in the c- or in the m-state. For example, in contrast to mitochondria ADP is removed by BKA from the particles and not by ATR.

Another indication of the asymmetry of carrier comes from the sensitivity to alkylating reagents, such as NEM. Trans- port is inhibited by NEM only when it is activated by ADP. Correspondingly the NEM alkylation is fully inhibited by ATR, however, it is not inhibited by BKA. The analogy of

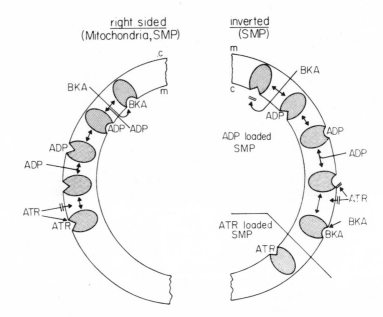

Fig.14   The asymmetry of the carrier specificity towards ATR
and BKA when the carrier is facing the m- or c-side
of the membranes (cf. ref. 24).
      The scheme interpretes the different binding
phenomena of ^{35}S-ATR, BKA and ADP on SMP and mitochondria. It attempts to explain that in ADP,ATP
loaded particles exchange can be inhibited by BKA
but not by ATR unless preloaded with ATR before
sonication.

the influence of NEM and BKA is further stressed by the
analogous effect on the membrane configuration. Obviously
for the alkylation by NEM the carrier must be brought in the
m-state, either by ADP or by BKA whereas in the c-state the
alkylation is inhibited.
      All these results indicate that the binding site of the
carrier as well as concomitant carrier conformation differ
when in the c-state and m-state. The different reactivities
in the two states are summarized in terms for the affinities
as follows:
affinity in c-state: medium for ADP and ATP, high for ATR
                     and CAT, low for BKA.
affinity in m-state: medium for ADP and ATP, very low for
                     ATP and CAT, high for BKA.

144

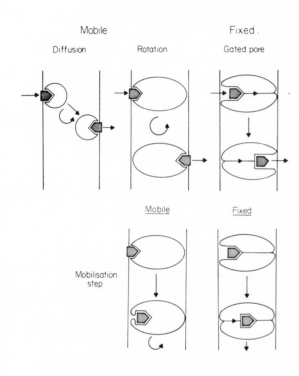

Fig.15
Single site re-
orientating carrier
models: mobile
carrier acting by
diffusion and
rotation, fixed
carrier trans-
locating the sub-
strate through
gated pore. The
binding site
switches from one
to the other side
of the membrane.
Conformation
change of the
carrier induced by
the substitute,
during the acti-
vation or mobili-
zation step, is
visualized as im-
bedding the sub-
strate into the
carrier.

It may be added that also towards the substrates ADP and
ATP a different affinity can be expected in the m- and c-
state. However, the affinity ratio ATP to ADP should be the
same for both states in the non-energized state in order to
fulfill thermodynamics. In the energized state even a
difference in the affinity ratios ATP to ADP in both states
may develop. This difference can create a gradient in the
ratio ATP/ADP, such that this ratio is high outside through
inside. It is maintained on the expense of energy by the
"active" exchange process, discussed above.

Mechanism of carrier transport.

The most useful description of carrier transport has been
in terms of the mobile reorientating carrier (12,29). This
description (as shown in Fig.5) must be regarded in analogy
to an enzyme mechanism for deriving laws of carrier kinetics.
This "mobile" carrier model has been useful for describing
transport kinetics. However, there has been criticism of
this simple formulation on the basis of kinetic studies, for

example in glucose transport through erythrocytes and more complicated dimeric or tetrameric models have been proposed.

The present studies on the ADP,ATP transport system showed that it is possible to fix all carrier sites in either one or the other position by inhibitors and thus provide at first experimental demonstration on the molecular level of a single site reorienting carrier mechanism. With substrate alone this demonstration would be more equivocal since substrate binding brings the carrier into a steady state distribution at both sides. The original mobile carrier mechanism does not specify whether the carrier makes rotational or translational movements through the membrane or whether it is fixed (Fig.15). In both cases carrier sites reorient from one to the other face of the membrane. The latter case may be also called fixed, gated pore mechanism since the reorientation requires that the binding site is either opened to one or the other face of the membrane for accepting or releasing the hydrophilic ligand.

There is an interesting difference concerning the symmetry of the binding site for mobile or fixed carrier (Fig.15). In a rotational model the substrate can be visualized to bind from both sides in the same orientation, in the gated pore the substrate with the opposite group to the carrier from one or the other side. When channeled through the gated pore, it leaves head first and reversely is bound tail first. In fact, the drastic asymmetry of affinity of the carrier sites observed for the binding of ATR and BKA favors the conclusion that the carrier follows a gated pore mechanism.

Isolation of the carrier protein

For the isolation of the carrier protein, two assay systems have been utilized which identify the protein after its solubilization. One relies on the high affinity binding of the carrier to CAT and the other on the covalent labelling with NEM. After many futile efforts the isolation and purification of the protein turned out at least for the CAT loaded carrier to be rather simple (24,26).

The crucial step is the saturation of the carrier with the ^{35}S-CAT in the mitochondria. Then on solubilization with Triton X-100 the carrier CAT binding is retained whereas on first solubilization with Triton X-100 the subsequent binding of CAT is not anymore possible. Obviously CAT helps to retain the carrier in a conformation against unfolding by the detergent. All other low molecular detergents readily destroy the CAT-carrier complex either in the intact

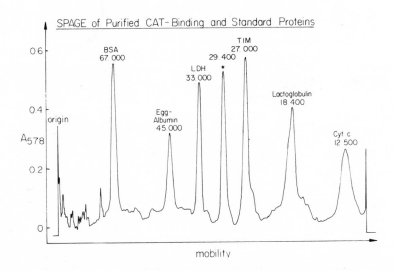

SPAGE of Purified CAT-Binding and Standard Proteins

Fig.16   Determination of molecular weight of purified CAT-
protein by SDS-PAGE. 5x240 mm gel (cf. ref.26).

mitochondria or in the Triton X-100 solubilized preparations.
    For the solubilization unusually high concentration of
Triton X-100 (about 5%) and high salt concentrations are
necessary indicating the high lipophilic character of the
protein. It belongs to one of the least extractable proteins
of the total mitochondria. The Triton X-100 extract is then
passed through hydroxylapatite and sepharose or agarose
columns. On polyacrylamide gel-electrophoresis a single
peak is obtained with an approximate M.W. of 29,000 (Fig.16).
    This hydrophobic protein can be estimated to amount to
6% of the total protein of BHM. It appears well possible,
that the solubilized CAT binding protein is actually a dimer
of two 29,000 M.W. units. The CAT is removed also by ADP
and BKA indicating that the protein has probably also the
binding sites which recognize ADP and BKA. This would
support the earlier concept of a single binding site for all
these three ligands.
    Several lines of evidence indicate the binding with CAT
induces a particular conformation to the carrier. Firstly,
CAT binding obviously protects the carrier from the
denaturation by Triton X-100 since binding of CAT is retained only when performed before the extraction (26).
Secondly, the separation of the CAT binding protein from
other proteins of the membrane and its adsorption properties
to columns differ strongly in the CAT complex as compared

to the free protein (P. Riccio and H. Aquila, unpublished).
Thirdly, antibodies have been produced against the isolated
CAT binding protein which reacts only against the protein
loaded with CAT. The antibody detect obviously a particular
conformation maintained in the CAT-carrier complex
(B.Buchanan, unpublished).

References
1. Klingenberg, M., Pfaff, E., and Kröger, A. (1964) in
   Rapid Mixing and Sampling Techniques in Biochemistry,
   pp. 333-337, Academic Press, New York.
2. Klingenberg, M. (1970) in Essays in Biochemistry Vol.6
   (Campbell, P.N., et al.,eds.) pp. 119-159, Academic
   Press, London/New York.
3. Meijer, A.J., and van Dam,K. (1974) Biochim.Biophys.
   Acta, Review Bioenergetics, Vol. 346, 213-244.
4. Klingenberg, M., and Pfaff, E. (1965) in Regulation of
   Metabolic Processes in Mitochondria (Tager,J.M.,et al.,
   eds.) pp. 180-201, Elsevier, Amsterdam.
5. Pfaff, E., Klingenberg, M., and Heldt, H.W. (1965)
   Biochim. Biophys. Acta 104, 312-315.
6. Duèe, E.D., and Vignais, P.V. (1969) J. Biol. Chem. 14,
   3920-3931.
7. Duèe, E.D., and Vignais, P.V. (1969) J. Biol. Chem. 14,
   3932-3940.
8. Heldt, H.W., and Klingenberg, M. (1965) Biochem. Z.343,
   433-451.
9. Pfaff, E., and Klingenberg, M. (1968) Eur. J. Biochem.6,
   66-79.
10. Pfaff, E., Heldt, H.W., and Klingenberg, M. (1969) Eur.
    J. Biochem. 10, 484-493.
11. Winkler, H.H., Bygrave, F.L., and Lehninger, A.L. (1968)
    J. Biol. Chem. 243, 20-29.
12. Klingenberg, M. (1975) in Membrane-bound Enzymes,
    Plenum Publishing Corp., New York, in press.
13. Klingenberg, M., Heldt, H.W., and Pfaff, E. (1969) in
    The Energy Level and Metabolic Control in Mitochondria
    (Papa, S., et al., eds.) pp. 237-253, Adriatica Editrice
    Bari.
14. Klingenberg, M., Wulf, R., Heldt, H.W., and Pfaff, E.
    (1969) in Mitochondria: Structure and Function, Vol.17
    (Ernster, L., and Drahota, Z.,eds.) pp. 59-77, Academic
    Press, London/New York.
15. Heldt, H.W., Klingenberg, M., and Milovancev, M. (1972)
    Eur. J. Biochem. 30, 434-440.
16. Klingenberg, M. (1972) in Mitochondria/Biomembranes,
    Proc.of the 8th FEBS Meeting,pp.147-162,Elsevier.

17. Klingenberg, M., and Rottenberg, H., unpublished.
18. Weidemann, M.J., Erdelt, H., and Klingenberg,M. (1970) Eur. J. Biochem. 16, 313-335.
19. Erdelt, H., Weidemann, M.J., Buchholz, M., and Klingenberg, M. (1972) Eur. J. Biochem. 30, 107-122.
20. Klingenberg, M., and Buchholz, M. (1973) Eur. J. Biochem 38, 346-358.
21. Klingenberg, M., Buchholz, M., Erdelt, H., Falkner, G., Grebe, K., Kadner, H., Scherer, B., Stengel-Rutkowski, L., and Weidemann, M.J. (1971) in Biochemistry and Biophysics of Mitochondrial Membranes (Azzone, G.F., et al.,eds.) pp. 465-486, Academic Press,New York/London.
22. Vignais, P.V., Vignais, P.M., Defaye, G., Chabert,J., Doussière, J., and Brandolin, G. (1972) in Biochemistry and Biophysics of Mitochondrial Membranes (Azzone,G.F., et al., eds.) pp. 447-464, Academic Press, New York.
23. Scherer, B., and Klingenberg, M. (1974) Biochemistry 13, 161-170.
24. Klingenberg, M., Riccio, P., Aquila, H., Schmiedt, B., Grebe, K., and Topitsch, P. (1974) in Membrane Proteins in Transport and Phosphorylation (Azzone, G.F., et al., eds.) pp. 229-243, North-Holland Publishing Co., Amsterdam.
25. Klingenberg, M. (1974) in Dynamics of Energy-Transducing Membranes (Ernster,L., et al., eds.) pp. 511-528, Elsevier Scientific Publishing Co., Amsterdam.
26. Riccio, P., Aquila, H., and Klingenberg, M. (1975) FEBS Lett., in press.
27. Pfaff, E., Klingenberg,M., Ritt, E., and Vogell, W. (1968) Eur. J. Biochem. 5, 222-232.
28. Souverijn, J.H.M., Huisman, L.A., Rosing, J., and Kemp, A.,Jr., (1973) Biochim.Biophys.Acta 305, 185-198.
29. Klingenberg, M.(1975) in Energy Transformation in Biological System, Ciba Foundation Symposium 31, pp. 105-124, Associated Scientific Publishers, Amsterdam.
30. Elbers, R., Heldt, H.W., Schmucker, P., Soboll, S., and Wiese, H. (1974) Hoppe-Seyler's Z.Physiol.Chem.Bd.355, 378-393.
31. Vignais, P.V., Vignais,P., and Defaye, G. (1973) Biochemistry 12, 1508-1518.
32. Henderson, P., Lardy, H., and Dorschner, E. (1970) Biochemistry 9, 3453-3457.
33. Vignais, P.V., Vignais, P.M., Defaye, G.,Lauquin,G., Doussière, J., Chabert, J., and Brandolin, G. (1973) in Mechanisms in Bioenergetics (Azzone,G.F.,et al.,eds.) pp. 323-333, Academic Press, New York/London.

CORRELATION OF MITOCHONDRIAL SWELLING AND LOCALIZATION OF MALATE DEHYDROGENASE ACTIVITY

A. Rendon and L. Packer

INTRODUCTION

Methods developed in recent years for the separation of mitochondrial outer and inner membranes have made it possible to localize mitochondrial proteins in different membranes and spaces (1,2). Factors modifying the localization of some mitochondrial proteins were investigated by Waksman and Rendon (3) who demonstrated that the malate dehydrogenase that is known to be located in the mitochondrial matrix in a sucrose medium could be released into the intermembranal space by succinate or other mitochondrial substrates.

The present work further explores the localization of malate dehydrogenase by examining the relation between mitochondrial swelling and release of the malate dehyrogenase activity.

METHODS

Rat liver mitochondria were prepared as described in reference 3, except a 0.25 M sucrose plus 1.0 mM Tris-EDTA, pH 7.5, medium was used. Solubilized malate dehydrogenase was obtained in the supernatant after sonicating mitochondria and centrifuging at 100,000 g for 1 hr at 4°C. The enzyme was assayed at 25°C both in intact mitochondria and in the supernatant fraction by monitoring in a Cary 14 spectrophotometer the decrease in absorbance at 340 nm due to the NADH oxidation.

The assay medium contained 0.25 M sucrose, 10 mM Na-phosphate buffer (pH 7.0), 0.2 mM NADH, 0.08 mM Na-oxaloacetate and 0.1 mg of mitochondrial protein in a volume of 3.0 ml. The method of Lowry et al. (5) was used for protein determination with serum albumin as standard. Respiration was measured using a Clark type oxygen electrode.

Mitochondrial preparations with respiratory control less than 3 were discarded.

RESULTS

The classical observation (cf. 6) that liver mitochondria are impermeable to the pyridine nucleotides has been amply confirmed (7,8). Thus, we can investigate malate dehydrogenase activity by measuring the rate of oxidation of external added

NADH upon addition of oxaloacetate to intact mitochondria as shown in Fig. 1.

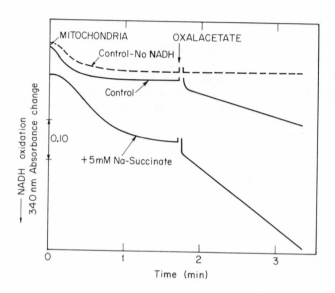

Fig. 1. Malate dehydrogenase activity of mitochondria. The reaction was assayed by NADH oxidation as in methods.

Since the inner mitochondrial membrane is impermeable to NADH, then the results of Fig. 1 show that Na-succinate treatment increases the malate dehydrogenase activity suggesting that this component induced the release of the enzyme from the matrix into the intermembranal space.

To further characterize this effect, experiments were performed using increasing concentrations of Na-succinate. In Fig. 2, we show that malate dehydrogenase activity follows a bell shaped curve with maximum activity at 5 mM succinate.

Control experiments with sonicated mitochondrial fractions show that succinate per se had no effect in enzyme activity. On the other hand, we have observed that antimycin A (1 µg/mg of protein) inhibits the effect of succinate in intact mitochondria, but has no effect on the solubilized enzyme activity.

In order to better understand the effect of succinate on the release of the malate dehydrogenase, light scattering studies were conducted using the same experimental conditions. It was found (Fig. 3) that both in controls and in the presence of succinate, significant mitochondrial swelling occurs. However when antimycin A was added, mitochondrial swelling was completely prevented.

152

MALATE DEHYDROGENASE ACTIVITY

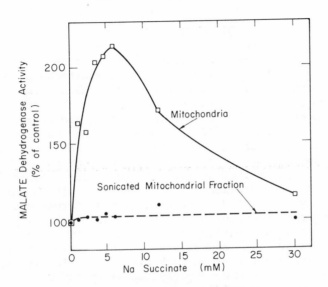

Fig. 2. *Influence of Na-succinate concentration on the malate dehydrogenase activity.*

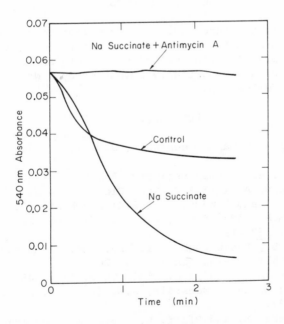

Fig. 3. *Effect of succinate and antimycin A on mitochondrial swelling. Conditions as in Fig. 1; Na-succinate (5 mM), antimycin A (1 μg/mg of protein).*

DISCUSSION

The release of proteins generally believed to be localized in the matrix into the intermembranal or extramitochondrial spaces poses many questions about the nature, mechanisms, properties and function of this phenomenon. In the present study it has been clearly demonstrated that an aid in answering some of these questions is afforded by utilizing the property of impermeablility of the inner membrane to pyridine nucleotides to develop a means to estimate this phenomenon.

Our results suggest that the ability to measure malate dehydrogenase in the intermembranal or the extramitochondrial compartments may be related to structural transformations of the inner membrane accompanying mitochondrial swelling.

In summary, our observations indicate a parallel occurrence of mitochondrial swelling and release of the enzyme. On the basis of the data given here, a mechanism regulating release of mitochondrial malate dehydrogenase in vitro may be suggested which is dependent upon oxidizable substrates and is regulated by the energized state of the mitochondria.

ACKNOWLEDGMENTS

We would like to acknowledge the excellent technical assistance of Mrs. R.L. Oropeza-Rendon and the helpful discussion of Drs. A. Quintanilha, H. Tinberg and C. Mehard. A. Rendon was partially supported by a NATO post-doctoral fellowship.

REFERENCES

1. Schnaitmann, C. and Greenawalt, J.W. (1968) J. Cell Biol. 38, 158.
2. Sottocasa, G.L., Kulyenstierna, B., Ernster, L. and Bergstrand, A. (1967) J. Cell Biol. 32, 415.
3. Waksman, A. and Rendon, A. (1974) Biochimie 50, 907.
4. Matlib, M.A. and O'Brien, P.J. (1975) Arch. Biochem. Biophys. 167, 193.
5. Lowry, O.H., Rosebrough, N.Y., Farr, A.L. and Randall, R.J. (1951) J. Biol. Chem. 193, 265.
6. Lehninger, A.L. (1951) J. Biol. Chem. 190, 345.
7. Purvis, J.L. and Lowenstein, J.M. (1961) J. Biol. Chem. 236, 2794.
8. Klingenberg, M. and Pfaff, E. (1966) in Regulation of Metabolic Processes in Mitohcohdria (Tager, J.M., Papa, S., Quagliariello, E. and Slater, E.C., eds.) BBA Library, Vol. 7, pp. 180-201, Elsevier, Amsterdam.

ON THE PROBLEM OF SITE SPECIFIC AGENTS IN OXIDATIVE PHOSPHORY-
LATION. THE ACTION OF OCTYLGUANIDINE AND K$^+$

M. Tuena de Gómez-Puyou, M. Beigel and A. Gómez-Puyou

In mitochondria, energy derived from electron transport
between NADH and oxygen may be used to drive the synthesis of
ATP in three sites. This energy may also be utilized to
support cation transport and transhydrogenation. As ATP can
induce reversed electron transport (transhydrogenation and
cation accumulation), the process is accepted to be fully
reversible. These observations, therefore, suggest that all
these processes share a common link. The identity of this
link has raised much controversy. At this time, the problem
is not completely settled and thus it is customary to explain
the mechanism of oxidative phosphorylation in terms of two
hypotheses: The chemical hypothesis and the chemiosmotic
hypothesis.

According to the chemical hypothesis (Fig. 1), energy
generated during electron transfer is directly transformed
into a chemical high energy intermediate of some ATP precursor
that is subsequently utilized for the synthesis of ATP (1).
A more recent variation of the chemical coupling mechanism
indicates that oxidation energy is primarily utilized for
creating an unstable, energy-rich conformation of a protein
which is also directly coupled to the enzymatic processes
that lead to ATP formation (2). So far no experiments have
unequivocally proved or disproved the validity of this concept.

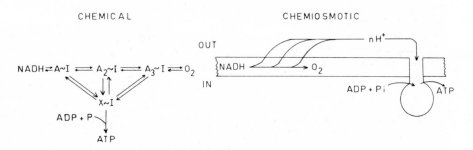

Fig. 1. Scheme of the mechanism of chemical and chemiosmotic
coupling. Note that in the former, the respiratory carriers
have to interact with the components of the ATP synthesizing
machinery. In chemiosmotic coupling, the interaction is not
necessary.

In chemiosmotic coupling (3,4), it is proposed that electron transfer in the respiratory chain directly induces the deposition of H^+ on the outer phase of the mitochondrial membrane, thus creating an electrochemical gradient (Fig. 1). The energy of the gradient is thereafter utilized for ATP formation through the ATPase complex. Although the identity of the proton carriers has not been established, recent experimentation has provided strong support in favor of Mitchell's concept (cf. 3,4).

Both versions of chemical coupling require a molecular connection between components of the respiratory chain and the ATPase complex of mitochondria. However, in chemiosmotic coupling the ATPase complex, at least in principle, could function as an entity molecularly independent of the respiratory chain whose function would be to transform the energy of an electrochemical gradient into the chemical energy of ATP.

In chemiosmotic coupling, the mechanism of action of certain agents that have been reported to possess a "site-specific" action may be of value in the understanding of the molecular events involved in oxidative phosphorylation. That is, there are some agents that are much more active at one particular phosphorylation site. Although two of the most widely used inhibitors of oxidative phosphorylation, oligomycin and DCCD, affect to a very similar extent the phosphorylation of ADP; Table I lists some of these agents as well as their reported site of action.

It is to be noted that all agents listed in Table I show a positive change at pH around 7, that is, their pKa is high enough so that at pH 7 they exist with a positive charge. Another common characteristic of these agents is that all those that inhibit oxidative phosphorylation show a certain lipophilicity. K^+, on the contrary, is the only cation whose action is to favor and not to inhibit oxidative phosphorylation.

TABLE I

Action and site of some agents that affect oxidative phosphorylation

	Site of Action	Action
Octylguanidine (5)	I	Inhibition
Tetrabutylammonium (6)	I	Inhibition
Ethidium bromide (7)	I	Inhibition
K^+ (8)	I	Stimulation
Phenethylbiguanide (9,10)	II	Inhibition
Decamethylene diguanidine (11)	III	Inhibition

The numbers in parentheses indicate the reference.

It is worth noting that not all agents inhibit the same phosphorylation site: octylguanidine (5), tetraalkylammonium salts (6), ethidium bromide (7) and K^+ modify the first site (8); phenethylbiguanide (9,10) and decamethylene diguanidine (11) affect the second and third phosphorylation sites, respectively. It is not clear whether the site of action of these agents is at the level of the respiratory chain or in the ATPase complex. This question is of particular importance because if these agents act exclusively on the ATPase complex, it would indicate that there might be several and distinct ATPase systems at various levels of the respiratory chain that could be distinguished on the basis of their sensitivity to either of these "site specific" agents.

The purpose of this work is to evaluate, with respect to the mechanism of oxidative phosphorylation, the action of two of these "site specific" agents, octylguanidine and potassium ions. Also, experiments that describe how the interaction of these agents, with certain components of the membrane, modifies the molecular mechanisms involved in oxidative phosphorylation will be shown.

On the Site of Action of Octylguanidine and K^+ on Oxidative Phosphorylation: We have been investigating the action of K^+ and octylguanidine on the first phosphorylation site because the effects of octylguanidine on some energy transfer reactions of mitochondria have been widely studied (12-14), and because this is the only site in which opposite effects of lipophilic cations and K^+ may be observed (15,16). Nevertheless, in the initial studies it was not clear whether these two cations act on the respiratory chain or in the ATPase complex.

Our preliminary experiments showed that in submitochondrial particles the aerobic oxidation of NADH, but not of succinate, was stimulated more than two fold by K^+ and strongly inhibited by octylguanidine; of interest was the finding that there was a competitive action between the two cations (15). These results suggest that K^+ and octylguanidine act on the NADH-ubiquinone segment, but do not eliminate the possibility that this was the consequence of their action in the ATPase complex.

Although Pressman (5) originally reported that octylguanidine did not affect uncoupler stimulated ATPase activity in whole mitochondria, more recent work has shown that the ATPase activity of submitochondrial particles is significantly inhibited by octylguanidine (14,17). It was also found that K^+ induces a strong stimulation of ATPase activity in submitochondrial particles; it should be acknowledged, however, that this stimulation only occurs at low levels of ATP (18).

In any case, ATPase activity and the oxidation of NADH through the respiratory chain showed parallel responses to

octylguanidine and K^+, i.e., K^+ stimulated ATPase and
respiration while octylguanidine inhibited both activities.
Obviously the effects of octylguanidine and K^+ suggest that
the action of cations at the NADH-ubiquinone span and site I
phosphorylation could be a reflection of a single effect on
the ATPase complex. Thus, the locus and mechanism of action
of K^+ and octylguanidine on the ATPase complex of the mitochondria was studied in more detail.

The ATPase complex of the mitochondria consists of various
components: a hydrophobic or membrane factor that directs H^+
to the catalytic site of ATPase, according to the predictions
of Mitchell (19,20); a protein that confers oligomycin sensitivity to ATPase activity (21); factor 1 or F_1 which possess
the catalytic site for ATP synthesis (F_1 is a soluble protein
that is characterized by an active oligomycin insensitive
ATPase) and an ATPase inhibitor (24) whose function is still
unknown. Theoretically, the cationic interference of function
of any of these components might lead to modification in the
activity of the whole complex.

When the effect of octylguanidine and K^+ on the ATPase
activity of soluble F_1 was studied, it was found that K^+
stimulated while octylguanidine inhibited the hydrolytic
activity of F_1. Under various experimental conditions, ATPase
activity of soluble and particulate F_1 showed that the binding
of the soluble enzyme to the membrane did not modify the
response of F_1 to the cations. Thus, the interactions of
octylguanidine and K^+ with F_1 are responsible for the modification of the activity of the whole ATPase complex.

A titration with octylguanidine of the coupled oxidation of
glutamate malate and succinate (14) and of the ATPase activity
of F_1 was carried out to ascertain whether octylguanidine
exerts independent actions in the respiratory chain and in the
ATPase complex, or whether its "site-specific" action could be
ascribed solely to its effect in F_1 (Fig. 2).

It was found that the octylguanidine inhibition curve of
the coupled oxidation of glutamate plus malate is hyperbolic
and apparently biphasic. On the other hand, the octylguanidine
inhibition curve of the ADP stimulated oxidation of succinate
is sigmoidal. While the octylguanidine concentration that
induces half-maximal inhibition of the State 3 oxidation of
succinate is quite close to that causing half-maximal inhibition of ATPase (240 and 330 µM, respectively), the titre for
octylguanidine inhibition of State 3 oxidation of glutamate
plus malate is significantly lower (half-maximal inhibition at
30 µM). The similarity in the concentration of octylguanidine
required to induce half-maximal inhibition of ATPase activity
and of the coupled oxidation of succinate is highly suggestive
that the interaction of ocytlguanidine with F_1 is responsible

for the inhibition of oxidative phosphorylation with succinate as substrate. On the other hand, the inhibition of the coupled oxidation of NAD-dependent substrates is probably due to the inhibition of the activity of ATP synthetase activity plus inhibition of electron transfer through the NADH-ubiquinone segment, since the concentration of octylguanidine required to inhibit the process is approximately five times lower with NAD-dependent substrates than with succinate.

In agreement with this conclusion, the data of Table II show that although the aerobic oxidation of NADH by submitochondrial particles is enhanced by Li^+, Na^+, K^+, Rb^+ and Cs^+, the ATPase activity is stimulated only by K^+, Rb^+ and Cs^+. That is, the cation requirements are much more strict in the ATPase system than in the respiratory chain.

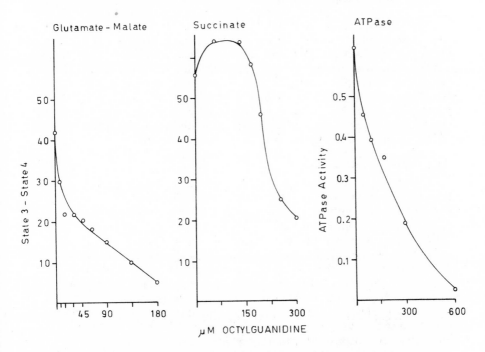

Fig. 2. *Effect of octylguanidine on the coupled oxidation of glutamate-malate and succinate and on the ATPase activity of F_1. Note that the concentration of octylguanidine required to induce half-maximal inhibition of the coupled oxidation of glutamate-malate is much lower.*

TABLE II

Effect of various cations on NADH oxidation and ATPase activity of submitochondrial particles

	Percent Stimulation	
	NADH Oxidation	ATPase Activity
Li^+	80	6
Na^+	115	11
K^+	120	44
Rb^+	120	44
Cs^+	–	23

NADH oxidation and ATPase activity were measured as described elsewhere (14,15), the chloride salts of the indicated cations were assayed, their concentration was 20 mM.

The differences in sensitivity to various monovalent cations and different concentrations of octylguanidine of the electron transfer and ATPase systems suggest that cations exert separate actions on the two systems. In the case of octylguanidine, the data indicate that the site-specific action of octylguanidine is merely apparent, the higher sensitivity being the consequence of its action on some component of the NADH-ubiquinone span. Nevertheless, it should be recalled that other guanidine derivatives (cf. Table I) apparently affect the second and third phosphorylation site. Once their action on ATPase is thoroughly studied, more insight will be gained on the problem of electron transport and phosphorylation. Nevertheless, the data that have been obtained with octyl-guanidine are entirely consistent with the mechanism of chemiosmotic coupling.

With respect to the action of K^+, the evidence strongly suggests that K^+ acts on the respiratory chain and on the ATPase complex. However, it is not clear at the moment why K^+, in K^+-depleted mitochondria, increases the P:O ratios with NAD-dependent substrates but not with succinate (8).

Cation Induced Conformational Changes in F_1: The following experiments explore the action of cations on F_1 and its relation to oxidative phosphorylation.

As ammonium ions were found to stimulate (similar to K^+) the ATPase activity of F_1 (18), the effect of amines substituted with alkyl chains of various lengths was studied to explore whether the lipophilicity of the amine influenced the response of F_1 to the cations. The results of Fig. 3 show that ammonium stimulates the activity, while ethylammonium does not induce appreciable modification. On the contrary butylammonium and hexylammonium inhibited ATPase activity, the latter being much more effective. Apparently those cations that are capable of undergoing hydrophobic interactions with the enzyme inhibit the activity. Therefore, we think that stimulation of ATPase activity by cations requires not only binding, but also a subsequent release. Maximal rates of ATPase activity would seem to depend on a cyclic movement of K^+.

Gitler and Montal (25) showed that the interaction of cations with some proteins drastically changes the solubility properties of the proteins. On the basis of this observation, we have proposed (16) that cations, upon interacting with a protein of the inner membrane, could induce changes in the position of the protein within the membrane; i.e., the inter-action of the cation with the protein would render the protein more hydrophobic and this could induce the protein to take a different location within the structure of the membrane. If this hypothesis is correct, the enzyme would present different characteristics in the presence of a cation, not only in the rate of its hydrolytic activity but also in its structural properties.

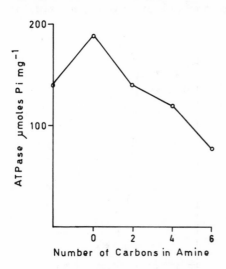

Fig. 3.   Effect of various alkyl amines on the ATPase activity of F_1. The conditions were described elsewhere (18), the concentration of the amines was 0.5 mM.

The sensitivity to cold of soluble F_1 has been known for some time (22,23), i.e., the activity of F_1 declines very rapidly upon exposure to temperatures of 4°, while at room temperature the activity is stable. On the other hand, the

ATPase activity of particulate F_1 is very stable at temperatures below 4°. These observations suggest that the attachment of F_1 to the membrane confers different properties to the enzyme.

The data of Fig. 4 illustrates that the cold-induced decay of the ATPase activity of F_1 is strikingly diminished by introducing octylguanidine into the system. K^+ do not exhibit this effect, on the contrary they induce acceleration in the rate of decay. These findings, therefore, indicate that octylguanidine confers to F_1 some of the properties characteristic of the membrane-bound enzyme which may be the consequence of a conformational change in the enzyme that is mediated by cations.

The fact that F_1 is capable of undergoing conformational changes does not invalidate the mechanism of chemiosmotic coupling. Racker (26) mentions that in 1965 Mitchell proposed that a conformational change may take place in ATPase in response to a membrane potential; his idea of a conformational change being entirely different from the concept of a conformational change as the primary event in oxidative phosphorylation.

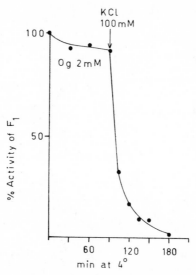

Fig. 4. *Protection by octylguanidine of the cold-induced inactivation of F_1. F_1 was incubated at 4° at the indicated times, its activity was measured, at the arrow 100 mM KCl was added and the decay in activity was followed. Note that in the absence of KCl, the activity remained constant. In the absence of octylguanidine the ATPase activity of F_1 declined by about 60% in a period of two hr.*

According to the mechanism of chemiosmotic coupling, protons on the outer phase of the membrane are captured by the
membrane factor of the ATPase complex, deposited on the catalytic site of F_1 and induce ATP synthesis (19,20). This concept necessarily implies that the site of catalysis has to be
in exact relation to the site of proton deposition, otherwise
dissipation of protons might be expected to take place.

In this respect, the action of octylguanidine in conferring
to soluble F_1 some of the properties of the membrane-bound
enzyme may be of importance in the understanding of the
molecular events that occur in oxidative phosphorylation.
Indeed, some recent findings of Dr. Sergio Papa at the
University of Bari are of particular interest. Papa found
that octylguanidine at certain concentation induced a tighter
coupling of oxidative phosphorylation in submitochondrial
particles (personal communication). In a sense, these experiments recall the coupling action of oligomycin described by
Lee and Ernster (13), except that the two compounds interact
with different subunits of the ATPase complex.

Although there is not satisfactory explanation for this
action of oligomycin, the action of octylguanidine observed
by Papa may be explained by assuming that F_1, upon interaction
with octylguanidine, adopts a more favorable position within
the membrane in relation to the site of H^+ deposition. In
this form, the efficiency of ATP formation may be increased.

Nevertheless, it is necessary to answer several questions
before accepting this hypothesis. First, have conformational
changes been observed in F_1 during oxidative phosphorylation.
At least in chloroplasts the evidence is quite clear. F_1
bound to chloroplast membranes take up tritium in a light-
dependent reaction (27) and binding of N-ethylmaleimide to
F_1 occurs upon illumination (28). In mitochondria the
evidence is controversial; structures that resemble F_1 have
appeared on the outside of the inner membrane (29,30). Also,
changes in fluorescence of aurovertin bound to F_1 that depend
on phosphate ADP and ATP have been reported (31,32). Although
all these finding are highly suggestive that conformational
changes do occur in F_1, it is not clear whether cations are
involved in these alleged conformational changes or what
factors control them.

Second, one might ask whether in the mitochondria there are
compounds that exert an action that is mimicked by octylguanidine. It should be recalled that in some types of mitochondria,
their F_1 component contains a peptide that inhibits ATPase
activity (24,33). Its amino acid composition has been studied;
it is a basic peptide with a molecular weight of approximately

11,000 that contains 8 lysine and 3 arginine residues. More-over, according to the method of purification of Pullman and Monroy (24), the peptide is soluble in ethanol which suggests that it possesses a significant degree of lipophilicity. Thus, certain resemblances exist between octylguanidine and the ATPase inhibitor; i.e., both are cations at pH around 7.0, both show a certain degree of hydrophobicity, both inhibit ATPase activity and both prevent cold inactivation of soluble F_1. Finally it is important to note that the action of the inhibitor (similar to that of octylguanidine) is reversed by K^+ (17,33). The function of the ATPase inhibitor in mitochon-dria is still unknown, but whether the peptide induces coupling as observed by Papa with octylguanidine is worth exploring in more detail.

REFERENCES

1. Slater, E.C. (1953) Nature 172, 975.
2. Boyer, P.D. (1964) in Oxidases and Related Redox Systems (King, T.E., Mason, H.S. and Morrison, M., eds.) p. 994, Interscience Publishers, New York.
3. Mitchell, P. (1966) Chemiosmotic Coupling in Oxidative and Photosynthetic Phosphorylation, Glynn Research Ltd., Bodmin, Cornwall, England.
4. Mitchell, P. (1966) in Regulation of Metabolic Processes in Mitochondria (Tager, J.M., Papa, S. Quagliariello, E. and Slater, E.C, eds.) p 65, Elsevier, Amsterdam.
5. Pressman, B.C. (1963) J. Biol. Chem. 238, 401.
6. Rogers, K.S. and Higgins, E.S. (1973) J. Biol. Chem. 248, 1742.
7. Peña, A., Chávez, E., Cárabez, A. and Tuena de Gómez-Puyou, M. (to be submitted).
8. Gómez-Puyou, A., Sandoval, F., Tuena de Gómez-Puyou, M., Peña, A. and Chávez, E. (1972) Biochemistry 11, 97.
9. Pressman, B.C. (1963) in Energy Linked Function of Mito-chondria (Chance, B., ed.) p. 181, Academic Press, New York.
10. Haas, D.W. (1964) Biochim. Biophys. Acta 92, 433.
11. Guillory, R.J. and Slater, E.C. (1965) Biochim. Biophys. Acta 105, 221.
12. Schatz, G. and Racker, E. (1966) J. Biol. Chem. 241, 1429.
13. Lee, C.P. and Ernster, L. (1966) in Regulation of Metabolic Processes in Mitochondria (Tager, J.M., Papa, S., Quaglia-riello, E. and Slater, E.C., eds.) p. 218, Elsevier, Amsterdam.

14. Papa, S., Tuena de Gómez-Puyou, M. and Gómez-Puyou, A. (1975) Eur. J. Biochem. $\underline{55}$, 1.

15. Lotina, B., Tuena de Gómez-Puyou, M. and Gómez-Puyou, A. (1973) Arch. Biochem. Biophys. $\underline{158}$, 520.

16. Gómez-Puyou, A. and Tuena de Gómez-Puyou, M. (1974) *in* Perspectives in Membrane Biology (Estrada-O, S. and Gitler, C., eds.) p. 303, Academic Press, New York.

17. Tuena de Gómez-Puyou, M., González, S. and Gómez-Puyou, A. (1974) Fed. Proc. $\underline{33}$, 158.

18. Tuena de Gómez-Puyou, M. Beigel, M. and Gómez-Puyou, A. To be submitted.

19. Mitchell, P. (1973) FEBS Letters $\underline{33}$, 267.

20. Mitchell, P. (1974) FEBS Letters $\underline{43}$, 189.

21. McLennan, D.H. and Tzagoloff, A. (1968) Biochemistry $\underline{7}$, 1603.

22. Pullman, M.E., Penefsky, H.S., Datta, A. and Racker, E. (1960) J. Biol. Chem. $\underline{235}$, 3322.

23. Penefsky, H.S., Pullman, M.E., Datta, A. and Racker, E. (1960) J. Biol. Chem. $\underline{235}$, 3330.

24. Pullman, M.E. and Monroy, G.C. (1963) J. Biol. Chem. $\underline{238}$, 3762.

25. Gitler, C. and Montal, M. (1972) Biochem. Biophys. Res. Commun. $\underline{47}$, 1486.

26. Racker, E. (1974) *in* Perspectives in Membrane Biology (Estrada-O, S. and Gitler, C., eds.) p. 623, Academic Press, New York.

27. Ryrie, I. and Jagendorf, A.T. (1971) J. Biol. Chem. $\underline{246}$, 3771.

28. McCarthy, R.E. and Fagen, J. (1973) Biochemistry $\underline{12}$, 1503.

29. Wrigglesworth, J.M., Packer, L. and Branton, D. (1970) Biochim. Biophys. Acta $\underline{205}$, 125.

30. Telford, J.N. and Racker, E. (1973) J. Cell Biol. $\underline{57}$, 580.

31. Chang, T.M. and Penefsky, H.S. (1973) J. Biol. Chem. $\underline{248}$, 2746.

32. Van de Stadt, R.J., Van Dam, K. and Slater, E.C. (1974) Biochim. Biophys. Acta $\underline{347}$, 224.

33. Hortsman, L.L. and Racker, E. (1970) J. Biol. Chem. $\underline{245}$, 1336.

ELECTRON TRANSFER AND ENERGY COUPLING AT THE NADH-UBIQUINONE
SEGMENT OF MITOCHONDRIA

Sergio Estrada-O and Carlos Gómez-Lojero

INTRODUCTION

The elucidation of the mechanism by which electrons are
transferred and coupled to ATP synthesis in the NADH-ubiquinone
reductase region, has admittedly been considered as one of the
most difficult and contentious problems to solve in the study
of the mitochondrial oxidation chain. Apparently contributing
to this difficulty are: a) the labile properties and unknown
topological distribution of a rich population of iron-sulfur
centers existing in this segment, which have hindered its
complete structural resolution; b) the lack of readily measurable absorption bands of the segment in the visible spectrum,
which has limited the possibility of using sensitive optical
techniques already found useful for other regions of the chain,
such as the cytochrome segment; c) an apparently confusing
literature which indicate the phenomenological inhibitory
action exerted on this region by a vast group of substances
apparently unrelated in their physical and chemical properties.
On the other hand, recent experimental developments have
clearly made emerge a useful account of the NADH-ubiquinone
reductase segment of the chain. The main experimental support
for this recent progress can be recognized in: a) the
resolution of multiple electron paramagnetic resonance (EPR)
signals which, at temperatures between that of liquid nitrogen
(77°K) and liquid helium (4.2°K), allowed the recognition and
titration of the relative oxidation-reduction potentials of
several independent iron-sulfur centers of the segment (1-4);
b) the comparative biological studies of electron transport
and energy coupling at the NADH-ubiquinone region utilizing
phenotypic variants of different yeast strains grown under
deprived conditions (5,6); c) the rational approach to identify the site of interaction of inhibitors with respect to the
oxygen or substrate side of the iron-sulfur centers of the
segment (7,8); d) the availablility of reconstitution models
of this membrane region in completely functional vesicular
lipid systems (9).
The present volume bieng mainly oriented for the benefit
of the beginning student, this chapter will deal with three
sections: first, a succinct review of the criteria existent
to evaluate the properties of the NADH-ubiquinone segment of
the mitochondrial oxidation chain; second, an analysis of the
approach followed to monitor the functioning of the NADH-Q

region through specific inhibitors, particularly as exemplified by the macrocyclic polyethers studied at our laboratory; thirdly, a summarized prospective view of the future directions of research expected in this area.

Electron Flow, H^+ Translocation and Energy Coupling at the NADH-ubiquinone Segment of Mitochondria

It is important to depart from the view that the transfer of electrons which occurs within the oxidation chain of mitochondria, chloroplasts or photosynthetic bacteria has been found to be associated to the transmembrane translocation of protons apparently catalyzed by the respiratory carriers located in such membranes (10,11). These observations have been foremost in supporting the notion that electron flow drives a H^+ pump in such membranes (12,13). In this respect, the NADH-ubiquinone region of the mitochondrial oxidation chain seems to be fully capable of implementing the variety of mechanisms suggested by which electron transport could drive a H^+ pump (10-13). Figure 1 diagramatically indicates the possible path for H^+ translocation as well as the arrangement of electron carriers at the NADH-ubiquinone region. Yet, it should be mentioned that unequivocal data which demonstrate the identification of the primary H^+ carrying species as well as the actual topological distribution of iron-sulfur centers at such segments are still missing. Experimental difficulty also exists for distinguishing between a primary coupling role for the H^+ gradient as envisaged by Mitchell (10,11) or a secondary role in which the transmembrane gradient is considered to be in equilibrium with a high energy state which is chemical (14) or conformational (15) in nature, or due to changes in proton activity in a localized environment within the membrane (16) which represent the primary event for energy coupling (17).

Identification and Possible Role of the Iron-sulfur Centers from the NADH-Q Region of Mitochondria

Recent work carried out with low temperature (4-20°K) electron paramagnetic resonance spectroscopy (EPR) has revealed that the span of the respiratory chain from NADH to ubiquinone contains at least four iron-sulfur centers in approximately equal concentration and individually equimolar to flavin mononucleotide content. These centers have been designated centers 1 to 4 (1,4). The standard oxidation-reduction potentials of the individual centers were found (6) to be -0.20 volts for center 2, -0.245 volts for centers 3 + 4, and -0.305 volts

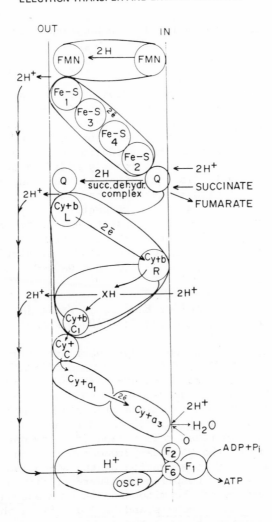

Fig. 1. Arrangement of the components of the oxidation chain and coupling device in the mitochondrial cristae membrane. Abbreviations: FMN, flavin mononucleotide; Q, Coenzyme Q; $Cy + b_L$ and $Cy + b_R$ designate two different cytochrome b components (left and right) (55); $C_y + c_1$, c, a_1 and a_3 designate different cytochrome species; $Fe-S_{1-4}$ denote the iron sulphur center of the NADH-dehydrogenase. Their distribution within the membrane is speculative.

for center 1; their rates of reduction being in the order 2>3+4>1. Thus, it appears that the possible reaction sequence of iron-sulfur components in the intramolecular mechanism of electron transport occuring from NADH to ubiquinone is:

The iron-sulfur centers are divided by oxidation-reduction
potentials 0.25-0.30 volts apart, implying that the actual
oxidative step is carried out within the structure of the
site I segment and not between the substrate and the first
acceptor site. Moreover, the oxidation-reduction potential of
iron-sulfur center I is significantly modified by the addition
of ATP (18,19), which is equivalent to the changes mediated
by ATP on the mid-oxidation-reduction potentials of cytochrome
b_T and a_3 in sites II and III, respectively (cf. 6). Nevertheless, interesting as these results appear to be, as far
as suggesting potential sites of energy transduction, it is
clear that the functional contribution or significance of the
different iron-sulfur centers from the NADH-Q region is yet
to be established.

The Approach of Comparative Biology in Yeast Systems for Elucidating the Properties of the NADH-Q Region of Mitochondria

The use of different phenotypic variants of yeast cells
grown under deprived conditions has been of great help in the
efforts to elucidate the mechanism of electron transport and
energy coupling at the NADH-coenzyme Q segment of mitochondria.
The main contributions to this approach have been elegantly
developed by Ohnishi et al. (6) and Garland et al. (20),
indicating the important role of specific low-potential iron-
sulfur centers of the NADH dehydrogenase in both electron
transfer and energy conservation at the so-called "site I"
region of the energy coupling machinery of mitochondria. A
careful comparison of the mitochondria from Saccharomyces
cerevisiae, S. Carlbergensis and Candida utilis showed a close
correlation existent between the occurrence of site I phosphorylation and the appearance of both specific binding sites
for the inhibitor rotenone and the presence of the iron-sulfur
protein responsible for the "g=1.94" EPR signal in the NADH
dehydrognease region (cf. 6). The observation (5,20,21) that
cells of C. utilis grown in iron or sulfur-limited conditions
are made simultaneously deficient in site I phosphorylation,
rotenone or piericidin-sensitivity and the "g=1.94" EPR signal
at the NADH dehydrogenase (this conversion being likewise
reversed by in vivo induction procedures) was of central
importance to support the above conclusion. Moreover, by
growing cells of C. utilis in the presence of cycloheximide,

an inhibitor of protein synthesis, the simultaneous disappearance was observed of site I phosphorylation, the EPR signals arising from iron-sulfur centers 1 and 2 and the sensitivity to piericidin (22). In summary, the approach using phenotypic variants of different yeast strains strongly suggests that at least one of the "EPR-active" iron-sulfur proteins at the NADH-Q region plays a role in the conservation of energy at site I of oxidative phosphorylation.

The Reconstitution of the NADH-ubiquinone Segment of Mitochondria in Model Membranes

Racker et al. have been able to functionally reconstitute in vesicular model lipid systems the activity of oligomycin-sensitive ATPase (23), site III of energy conservation (24) and the energy-coupled rhodopsin system of bacteria (25) (Fig. 2). Recently, the reconstitution of the first site of energy conservation in liposomes was also accomplished by the same laboratory (9) (cf. Fig. 2). By mixing the NADH-ubiquinone reductase (complex I) with phospholipids and an oligomycin-sensitive ATPase derived from bovine heart mitochondria, vesicular elements capable of phosphorylating ADP during oxidation of NADH by ubiquinone I were obtained (9). Such vesicles displayed an absolute requirement for phosphatidyl ethanolamine and a partial one for phosphatidylcholine being sensitive to rotenone, uncoupling agents and oligomycin. An NADH dehydrogenase which retains full activity with ferricyanide as acceptor, being devoid of rotenone-sensitive NADH-ubiquinone reductase activity, has also been isolated from complex I (26). The enzyme can be reconstituted in lipid vesicles with either cardiolipin, phosphatidylcholine, phosphatidyl ethanolamine or a mixture of the three phospholipids to give rates of rotenone-sensitive NADH-ubiquinone reductase approaching those in complex I. Moreover, the preparation catalyzes oxidative phosphorylation sensitive to rotenone when it is embedded in the above lipid mixture in the presence of an oligomycin-sensitive ATPase.

The Use of Specific Inhibitors of Electron Transport at the NADH-ubiquinone Region of the Mitochondrial Oxidation Chain.

There is a large body of literature indicating phenomenological aspects of the inhibtion of electron transport at the NADH-cytochrome b segment of the oxidation chain by a sizable number of agents seemingly unrelated in their chemical or physical properties. Thus, it has been reported that oxygen uptake and coupled ATP synthesis is prevented when mitochondria

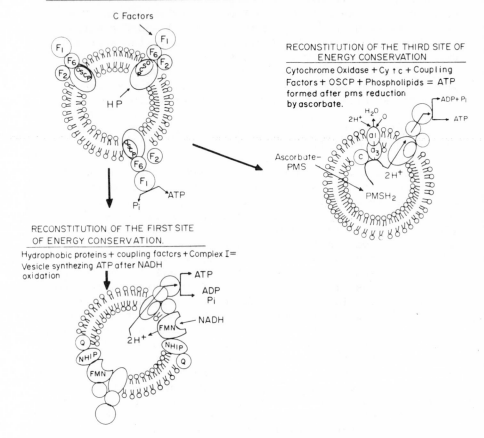

Fig. 2. Reconstitution experiments of energy-conserving reactions in model vesicular membranes.

oxidizes NADH-dependent substrates by: alkylguanidine derivatives (27), amytal (28), acetaldehyde (29), detergent compounds
such as sublytic concentrations of Triton X-100 (30), lecithinases (31), metal chelators such as O-phenantroline (32),
rhein (33), cychloheximide (34), progesterone and diethylethylbestrol (35), various classes of hydrophobic organic compounds
of the aliphatic and aromatic series (36), rotenone (37),
piericidin (38) and some synthetic macrocyclic polyethers (39).
From all the described inhibitors, both rotenone and piericidin have been evaluated at more depth than any other drug.
From these studies, it has been suggested that piericidin A or
rotenone sensitivity appear to be closely related to the presence of site I energy conservation and to some iron-sulfur

centers (6-8). However, three yeast systems have been obtained
where site I energy conservation is present and the accompanying electron transport is piericidin A-insensitive (20,21,40).
Thus, it is apparent that electron transfer through the piericidin, a sensitive site, is not obligatory for the occurrence
of site I energy conservation, while there is no example yet
known where rotenone-sensitive respiration does not yield
phosphorylation at site I. From available evidence it appears
that the rotenone inhibitory site is in very close proximity
to iron-sulfur center 2 on its oxygen side (6-8), a site where
both lipids and proteins are involved in non-covalent forces,
which holds this inhibitor at its specific binding site. Aside
from diethylethylbestrol and piericidin, which seem to share
the same inhibitory site as rotenone (8,35), there is no
equivalent information pertaining to the large group of
inhibitors indicated above. On the other hand, our work (39)
has shown that macrocyclic polyethers such as dibenzo-18-
crown-6 (compound XXVIII), or dicyclohexyl-18-crown-6 (compound
XXXI) already characterized as a mobile carrier molecule for
K^+ in model membranes (41,42) inhibits electron transport at
the NADH-ubiquinone segment of intact mitochondria, submitochondrial sonic vesicles and Hatefi complex I without inhibit-
ing electron flow at the cytochrome $\underline{b}$ or $\underline{a}$ segments. The
inhibition of electron transport by the polyethers was found
to be unrelated to its ability for mediating net K^+ translocation across lipid membranes (39). Moreover, when several
macrocyclic polyether analogs (43) were subsequently studied
at our laboratory, it was found (Table I) (44) that the synthetic ionophores inhibit electron flow at the NADH-Q segment as
a function of having: a) an internal ring diameter of 2.6-
3.2 Å and a high capacity to chlatrate K^+, although the transmembrane K^+ flux mediating the model carriers was unrelated
to their inhibitory action per se: b) two hydrophobic sub-
stituents (dibenzo>dicyclohexyl groups) to facilitate their
lipid solubility at the hydrocarbon membrane phase; c) specific
stereoisomeric requirements, disclosed by quantitative variations found in the inhibitory power between cis and trans
isomers; d) a homogeneously dispersed electronegative cloud in
the molecular cavity which surrounds the chlatrated K^+, since
substitution of oxygens by nitrogens or sulfurs in the cation-
criptating crown made the ionophores inactive to inhibit electron flow; e) ability to partitionate from the aqueous to the
lipid membrane phase, determined by the concentration of K^+
present in water; and f) ability to form monomer-dimer transition complexes with K^+ in the membrane interphase. Most
interesting was the fact that although added K^+ was not required for the active polyethers to inhibit oxygen uptake in

TABLE I

Effect of Macrocyclic Polyethers on the Coupled Oxidation of Glutamate in Intact Mitochondria

Polyether Tested	Percent Inhibition of Oxygen Uptake Stimulated by ADP	
	K^+	Li^+
Cyclohexyl-15-crown-5 (VII)	0	0
Cyclohexyl-18-crown-6 (XIII)	0	0
18-crown-6 (XXVII)	0	0
Dibenzo-18-crown-6 (XXVIII)	85	26
Dicyclohexyl-18-crown-6 (XXXI-AB)*	67	30
Dicyclohexyl-18-crown-6 (XXXI-B)	42	30
Dibenzo-21-crown-7 (XXXIII)	51	33
Dicyclohexyl-21-crown-7 (XXXIV)	9	4
Dibenzo-24-crown-8 (XXXV)	12	0
Dibenzo-30-crown-10 (XXXVIII)	8	0
Dibenzo-18-crown-6 (Nitrogen subst. in 2 positions of the crown) (cpd. E)	0	0
Dibenzo-18-crown-6 (sulphur subst. in 2 positions of the crown) (cpd. Z)	0	0

The reaction mixture for these experiments contained: 10 mM glutamate (TEA) pH 7.4; 800 µM ADP-tris, pH 7.4; 3 mM inorganic ortho-phosphate (TEA) pH 7.4; 10 mM triethanolamine (Cl⁻) pH 7.4; 25 mM of the indicated alkali metal cation; 200 mM sucrose and 1.3 mg mito-chondrial nitrogen in 5.0 ml medium at 25°C. The indicated polyethers were added in 10 µl volume (1:4) of dimethylformamide-ethanol at a final concentration of 5.5×10^{-5} M.

mitochondrial membranes, the extent of their inhibitory action was affected by the alkali ion present in the medium. In fact, the extent of inhibition was similar in the absence or presence of K^+, while a considerable decrease of this inhibition was observed when K^+ was substituted by Li^+ or Na^+ (Table I). These results emphasize again that the ability to complex alkali ions is associated with the respiratory inhibition med-iated by the polyethers. Attempts were subsequently made to localize the site of interaction of the inhibitory macrocyclic molecules at the sequence of electron carriers from the NADH-cytochrome b segment. For this purpose we followed the ap-proach devised by Bois and Estabrook (45) and Gutman et al., (7) to measure the oxidation-reduction of iron-sulfur centers at the NADH dehydrogenase from beef heart ETP particles. As noticed by these authors, when measuring the absorbance changes

at 470 minus 500 nm, a cycle of initial bleaching due to NADH
oxidation followed by recolorization of the chromophore after
NADH is exhausted is obtained (Fig. 3A). The residual bleaching, which approaches 30% in our study, is due to the superimposed reduction of cytochrome b, while the total corrected
bleaching can be ascribed to the reduction of iron-sulfur
centers from the NADH dehydrogenase (45). As illustrated in
Fig. 3B, both the extent of recolorization and the cycle time
of the absorbance changes measured are significantly affected
by rotenone. It is apparent that the component that remains
bleached accounts for 66% of the initial total bleaching,
while the cycle time shown two phases: one slow before NADH
is exhausted, and one fast when the substrate has been completely oxidized. The slow phase is the result of two opposing processes: reduction of the chromophore by NADH and its
slow re-oxidation by the respiratory chain. These results
show that inhibition of electron flow by rotenone is incomplete
and that some residual oxidase activity always remains active.

The identity of the "permanently bleached chromophore" that
remains in such a state in the presence of rotenone was clari-
fied by recent EPR studies carried out at $13^{\circ}K$ by Gutman et al.
(46). By correlating the effects of ATP and rotenone both on
the oxidation-reduction state of iron-sulfur center 2 and the
absorbance changes at 470 minus 500 nm, it was concluded that
rotenone blocks electron flow not directly on center 2, but
in a very close proximity of center 2, on its oxygen site (46).
Following this approach, an experiment was devised to locate
the site of interaction of macrocyclic polyether XXVIII in the
NADH-ubiquinone region. Figure 3C shows the control absorbance
changes at 470-500 nm obtained in the presence of the solvent
where the polyehter is dissolved and those with two different
concentrations of the macrocyclic molecule. It is clear that
the cycle time and extent of recolorization of the chromophore
disclosed at 470-500 nm are affected by the polyether in the
same manner as that shown by rotenone, piericidin or diethylethylbestrol (45,8,35). Thus, it may be concluded that the
polyethers inhibit electron flow immediately after iron-sulfur
center 2, on its oxygen side. Experiments of competition
between different inhibitory drugs are in progress at our
laboratory to elucidate the existence of a very sensitive site
utilized for the interaction of a vast number of substances
whose only apparent common link resides in their high lipophilicity (36). Further experiments are also required to
decide whether the inhibitory form of the polyethers in the
membrane is the charged carrier species with a chlatrated K^+;
the K^+ uncomplexed form, freely available to chlatrate K^+, or

else, to complex an organic cationic species bound to the membrane.

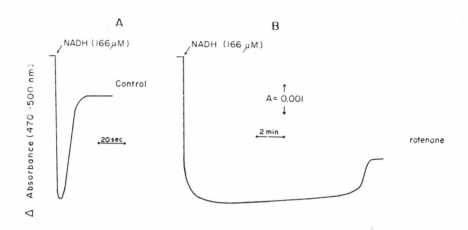

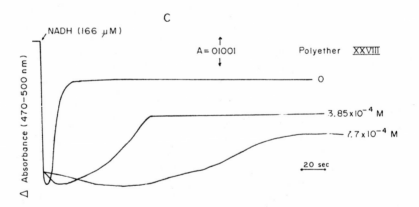

Fig. 3. *Kinetics of absorbance changes at 470 minus 500 nm induced by NADH in electron transport particle (ETP) from mitochondria. The reaction was followed in the DW2 Aminco spectrophotometer in a medium containing 0.1 M KPi, pH 7.4 and 4 mg protein/ml at 25°C. The reaction was started with the indicated addition of NADH. A) untreated electron transport particle, B) electron transport particle treated with 12 nmoles rotenone/mg protein for 30 min at 0°C before assay, C) electron transport particle treated with the indicated concentrations of macrocyclic polyether XXVIII in same conditions as in B.*

Comparative "K⁺-independent Effects" of Ionophores on the Electron Transport Mechanisms of Mitochondria and Photosynthetic Membranes

In 1962, Baltscheffsky and Arwidsson (47) reported that valinomycin inhibited ATP formation by 50 per cent in the presence of succinate in chromatophores of R. rubrum. Further work by Gromet-Elhanan (48) showed that valinomycin as well as nonactin inhibit NAD$^+$ photoreduction to the same extent as photophosphorylation in chromatophores. Similar to the polyethers effects observed in mitochondrial electron transport, he also found that valinomycin and nonactin do not require added K$^+$ or K$^+$-mediated translocation to exert their inhibitory action.

Similarly, Keister and Minton (49) and Telfer and Barber (50) observed that valinomycin inhibits photophosphorylation in R. rubrum chromatophores, presumably by primarily inhibiting electron transport, and that the ionophore blocks coupled electron transport in chloroplasts. Both effects are independent of added K$^+$ or of K$^+$ transport mediated by valinomycin. McCarty also reported (51) that valinomycin inhibits both ferricyanide reduction and the coupled phosphorlyation in chloroplasts and subchloroplast particles. In these systems, however, this type of inhibition rather rules out an uncoupling effect. Moreover, the results obtained by Gromet-Elhanan (48) and Keister and Minton (49) preclude the suggestion that valinomycin acts as an energy-transfer inhibitor. Thus, it is plausible to suggest that K$^+$-ionophores such as valinomycin and nonactin, parallel to the macrocyclic polyethers in mitochondria, are exerting their "K$^+$-independent effects" in photosynthetic membranes as primary inhibitors of electron transport.

With the above facts as a reference framework, a working hypothesis can be advanced in the sense that membranes of mitochondria, chloroplasts and photosynthetic bacteria posess common properties at specific control sites of electron transfer in oxidation chains, which can be disclosed through the inhibitory action of K$^+$-ionophores in the absence of K$^+$ translocation mediated by these compounds. Fast changes in the electronegativity of such sites may offer critically vulnerable loci to their chelation/or modification of their surface charge either by the free uncomplexed ionophores in the former case or by the charged chlatrated carriers in the latter alternative.

FUTURE PERSPECTIVES

Much has to be done in research before a definite answer is given to the central question of what is the mechanism by which electron transport is coupled to ATP synthesis at the NADH-ubiquinone segment of mitochondrial membranes. A detailed study of both the biogenesis of the polypeptide chains from the NADH-dehydrogenase and the in vivo assembly of the iron-sulfur centers which reside at such enzyme is critically needed. The unequivocal identification of the primary proton carrier of the segment, as well as the complementary development of experimental criteria to make the separate generation of electrical and chemical components of the H^+ gradient, kinetically compatable with the individual pocket reactions occurring at the electron transport system of the NADH-Q region, are also necessary.

Also of great importance is the isolation in crystal form of membrane electron carriers from the NADH-Q segment with or without lipids, along with the development and application of techniques such as high resolution NMR and X-ray diffraction, in order to characterize their native molecular structure and conformation. Following this line, the parallel application of new techniques of bilayer formation (52) and chemical manipulation of charged translocators (53) should be expected in order to understand their individual electrochemical properties. Finally, a word should be mentioned with respect to the potential use of biochemical genetics in the further elucidation of the segment under discussion. It is accepted as a truism that the main contributions of biochemical genetics to oxidative phosphorylation, as a whole, appear to be represented so far not by what has been resolved but by what has been suggested (54). Yet, significant progress has been made in the field with respect to improving our knowledge on the genetic nature of the cytochrome chain and ATPase molecular components (Tzagoloff, this volume). Contradictory reports with respect to the obtention of yeast mutants to respiratory inhibitors such as rotenone (cf 6) indicate that at present the NADH-ubiquinone segment of the mitochondrial oxidation chain still remains elusive to the attack of biochemical genetics. However, it is possible to envisage that the future use of different species of yeasts, bacteria and fungus (54) in these studies should bring new unexpected scientific problems and new vistas of research.

ACKNOWLEDGMENTS

The authors acknowledge the expert collaboration and discussions of Miss Maria Eugenia Vargas in the present work.

REFERENCES

1. Orme-Johnson, N.R., Orme-Johnson, W.H., Hansen, R.E., Beinert, H. and Hatefi, Y. (1971) Biochem. Biophys. Res. Commun. 44, 446.
2. Ohnishi, T., Asakura, T., Yonetani, T. and Chance, B. (1971) J. Biol. Chem. 246, 5960.
3. Albracht, S.P.J. and Slater, E.C. (1971) Biochim. Biophys. Acta 245, 503.
4. Orme-Johnson, N.R., Hansen, R.E. and Beinert, H. (1974) J. Biol. Chem. 249, 1922.
5. Light, P.A., Ragan, C.I., Clegg, R.A. and Garland, P.B. (1968) FEBS Letters 1, 4.
6. Ohnishi, T. (1973) Biochim. Biophys. Acta 301, 105.
7. Gutman, M., Singer, T.P., Beinert, H. and Casida, J.E. (1970) Proc. Nat. Acad. Sci. US 65, 763.
8. Singer, T.P. and Gutman, M. (1971) Adv. Enzymol. 34, 79.
9. Ragan, C.I. and Racker, E. (1973) J. Biol. Chem. 248, 2563.
10. Mitchell, P. (1966) Chemiosmotic Coupling in Oxidative and Photosyntgetic Phosphorylation, Glynn Research Ltd., Bodmin, Cornwall, England.
11. Mitchell, P. (1968) Chemiosmotic Coupling and Energy Transduction, Glynn Research, Ltd., Bodmin, Cornwall, England.
12. Witt, H.T. (1971) Quart. Rev. Biophys. 4, 365.
13. Crofts, A.R. (1974) in Perspective in Membrane Biology (Estrada-O, S. and Gitler, C., eds.) pp. 373-398, Academic Press, New York.
14. Slater, E.C. (1971) Quart. Rev. Biophys. 4, 35.
15. Boyer, P.D. (1968) in Biological Oxidations (Singer, T.P. ed.) pp. 193-235, Interscience Publ., New York.
16. Williams, R.J.P. (1969) in Current Topics in Bioenergetics (Sanadi, D.R., ed) Vol. 3, pp. 79-156, Academic Press, New York.
17. Skulachev, V.P. (1973) Abst. 9th Intl. Congr. Biochem., Stockholm, No. 45bl.
18. Ohnishi, T., Wilson, D.F., Asakura, T. and Chance, B. (1972) Biochem. Biophys. Res. Commun. 46, 1631.
19. Ohnishi, T., Wilson, D.F. and Chance, B. (1972) Biochem. Biophys. Res. Commun. 49, 1087.

20. Garland, P.B. (1970) Biochem. J. 118, 329.
21. Haddock, B.H. and Garland, P.B. (1971) Biochem. J. 124 , 155.
22. Ohnishi, T., Panebianco, T. and Chance, B. (1972) Biochem. Biophys. Res. Commun. 49, 99.
23. Kagawa, Y. and Racker, E. (1971) J. Biol. Chem. 246, 5477.
24. Racker, E. and Kandrach, A. (1971) J. Biol. Chem. 246, 7069.
25. Racker, E. and Stoeckenius, W. (1974) J. Biol. Chem. 249, 662.
26. Ragan, C.I. and Racker, E. (1973) J. Biol. Chem. 248, 6876.
27. Hollunger, G. (1955) Acta Pharmacol. Toxicol., Kbh, 11 Suppl. 1.
28. Ernster, L., Jalling, O., Low, H. and Lindberg, O. (1955) Exp. Cell Res. Suppl. 3, 124.
29. Cederbaum, A.I., Lieber, C.S. and Rubin, E. (1974) Arch. Biochem. Biophys. 161, 26.
30. Chappell, J.B. (1963) J. Biol. Chem. 238, 410.
31. Badano, B.N., Boveris, A., Stoppani, A.O.M. and Vidal, J.C. (1973) Mol. Cell Biochem. 2, 157.
32. Butow, R.A. and Racker, E. (1965) J. Gen. Phys. 49, 149.
33. Kean, E.A. (1968) Arch. Biochem. 127, 528.
34. Garber, A.J., Jomain-Bam, M. , Salganicoff, L., Farber, E. and Hanson, W. (1973) J. Biol. Chem. 248, 1530.
35. Otamendi, M. and Stoppani, A. (1974) Arch. Biochem. Biophys. 165, 21.
36. Yuguzhinsky, L.S., Smirnova, E.G., Ratnikova, L.A, Kolesova, G.M. and Krasinskaya, I.P. (1973) Bioenergetics 5, 163.
37. Ernster, L., Dallner, G. and Azzone, G.F. (1963) J. Biol. Chem. 238, 1124.
38. Hall, C., Wu, M., Crane, F.L., Takahashi, N., Takamura, S. and Folkers, K. (1966) Fed. Proc. 25, Abs. #192.
39. Estrada-O, S. and Cárabez, A. (1972) Bioenergetics 3, 429.
40. Clegg, R.A., Ragan, C.I., Haddock, B.A., Light, P.A., Garland, P.B., Swann, J.C. and Bray, R.C. (1969) FEBS Letters 5, 207.
41. Eisenman, G., Ciani, S.M. and Szabo, G. (1968) Fed. Proc. 27, 1289.
42. Tosteson, D.C. (1968) Fed. Proc. 27 1269.
43. Vargas, L.M.E. (1974) Thesis, Instituto Politécnico Nacional, Mexico.
44. Estrada-O, S., Vargas, M.E. and Gómez-Lojero, C. (1975) Proc. 5th Int. Congr. Biophys., Copenhagen, in press.
45. Bois, R. and Estabrook, R.W. (1969) Arch. Biochem. Biophys. 129, 362.

46. Gutman, M., Singer, T.P. and Beinert, H. (1972) Biochem.
     11, 556.
47. Baltscheffsky, H. and Arwidsson, B. (1962) Biochim.
     Biophys. Acta 65, 425.
48. Gromet-Elhanan, Z. (1970) Biochim. Biophys. Acta 223, 174.
49. Keister, D.L. and Minton, N.J. (1970) Bioenergetics 1, 367.
50. Telfer, A. and Barber, J. (1974) Biochim. Biophys. Acta
     33, 343.
51. McCarty, R.E. (1969) J. Biol. Chem. 244, 4292.
52. Montal, M. and Mueller, P. (1972) Proc. Nat. Acad. Sci.
     US 69, 3561.
53. Montal, M. and Korenbrot, J.I. (1973) Nature 246, 219.
54. Novak, L. (1974) Biochim. Biophys. Acta 346, 101.
55. Wikström, M. (1973) Biochim. Biophys. Acta 301, 155.

ENERGY COUPLING IN RECONSTITUTED COMPLEXES FROM MITOCHONDRIA

Peter C. Hinkle

INTRODUCTION

The study of energy coupling in purified segments of the respiratory chain and the ATPase complex of mitochondria has finally allowed the problem of the mechanism of oxidative phosphorylation to be pursued at the enzyme level. This was made possible by the development by Kagawa and Racker (1,2) of a method for reconstitution of these enzyme complexes into liposome membranes. This method consists of mixing the enzyme, liposomes and cholate and removing the cholate slowly by dialysis. In addition, in some cases simply sonicating a mixture of the liposomes and enzyme complex also produces coupled vesicles (3). The reconstituted complexes are active in translocating protons across the liposome membrane in an electrogenic process coupled to electron transfer or ATP hydrolysis. The rates of electron transfer or ATP hydrolysis are also stimulated several fold by uncouplers, a phenomenon analogous to respiratory control in mitochondria. When a respiratory complex and the ATPase were reconstituted together oxidative phosphorylation was observed (4-6).

I have been particularly interested in the role of proton translocation in the mechanism of oxidative phosphorylation. Mitchell proposed in his chemiosmotic hypothesis (7,8) that proton translocation is the reaction that the three coupling regions of the respiratory chain and the ATPase complex have in common which allows coupling between them. In other hypotheses the proton translocation is considered to be on a side pathway of energy transduction. Thus it was relevant to study proton translocation in the isolated segments of the respiratory chain to see whether this function could be separated from energy coupling phenomena in at least one of the coupling regions.

We have now succeeded in applying the cholate dialysis technique to all three coupling regions of the chain; NADH-CoQ reductase (6,9), reduced CoQ-cytochrome c reductase (10,11), and cytochrome c oxidase (4,5,12-14) as well as the partially purified ATPase complex (1,2) and have found coupled proton translocation associated with each complex.

Cytochrome c oxidase

This terminal enzyme complex of the respiratory chain consists of about six polypeptide chains and the electron transfer components cytochrome a, cytochrome a$_3$ and enzyme-bound copper. The important questions of which polypeptides correspond to the different functional groups and whether all of the polypeptides are needed for activity are still not clear. Incorporation of the isolated complex into liposomes by the cholate-dialysis procedure formed vesicles which showed electrogenic proton translocation and stimulation of the rate of electron transfer by uncouplers. An experiment showing the latter phenomenon, which we refer to as respiratory control, is shown in Table I.

Table I

RESPIRATORY CONTROL IN CYTOCHROME OXIDASE VESICLES

Respiration was measured with a Clark oxygen electrode, with cytochrome oxidase vesicles (20 µg protein) in 50 mM KP$_i$ pH 7.0, 20 mM Na Ascorbate and 1.2 mg cytochrome c in a final volume of 1.2 ml.

Additions	oxygen uptake ng atoms/min	RCR
None	53	--
100 µM 1799	205	3.9
4 µM CCCP	124	2.3
0.5 µg Valinomycin	59	1.1
0.4 µg Nigericin	59	1.1
Val. + Nig.	196	3.8
0.06% Triton X-100	205	3.9

Cytochrome c was added to the vesicles during the assay so that it was only external. As shown in table I the uncoupler 1799 stimulated respiration about four fold. We have obtained up to ten fold stimulation in certain preparations. Another uncoupler, CCCP, did not stimulate fully alone but did in the presence of valinomycin (not shown). This small synergism between CCCP and valinomycin is presumably due to an effect of lipid soluble cations (val-K$^+$) on the permeability of the membrane to lipid soluble anions (CCCP$^-$) which has been observed in other systems.

Valinomycin, which carries potassium ions across the membrane, did not stimulate electron transfer even though the K$^+$ concentration was about 100 mM. This is different from observations with intact mitochondria where valinomycin alone causes uncoupling at high K^+ concentrations (but not below 5 mM K^+) presumably because of swelling and damage to the membrane. Cytochrome oxidase vesicles are about 400-1000 Å in diameter, however, and may be less sensitive to osmotic damage caused by K^+ uptake. Addition of both valinomycin and nigericin, however, completely uncoupled cytochrome oxidase vesicles because nigericin can exchange K^+ for H^+ making the combined activities equivalent to electrogenic proton permeability. This effect has also been observed in mitochondria at low K^+ concentrations and in submitochondrial particles (15).

The detergent Triton X-100 also caused complete uncoupling. The fact that a similar rate is obtained with uncouplers and with a detergent indicates that the enzyme complex is predominantly oriented in the membrane with the cytochrome c binding site exposed to the external medium which corresponds to the sidedness of intact mitochondria. Recent studies by Mr. R. Carroll in E. Racker's laboratory indicate that the sidedness of the reconstituted system can be changed by altering the conditions of reconstitution. This is one of the most interesting aspects of the reconstitution process and will be studied extensively. Right now it is not possible to predict what sidedness a particular system will have, but our approach has been to use one impermeant substrate, eg. cytochrome c, to select for complexes with a certain sidedness in studies of coupling phenomena.

Studies of proton translocation in cytochrome oxidase vesicles were carried out with a recording pH meter with a sensitivity of 0.1 pH units full scale. A glass anaerobic cell was used so that the suspension could be equilibrated anaerobically before transport was initiated with a small pulse of oxygen. Reduced Naphthoquinone Sulfonate was used as the reductant for cytochrome c because it forms two hydrogen ions when oxidized, unlike ascorbate which forms only one. Figure 1 shows a typical oxygen-pulse experiment to measure proton efflux driven by respiration. Addition of oxygen caused a rapid acidification of the medium followed by a slow return to the original pH when respiration stopped. The coupling ratio for proton translocation can be determined from the number of protons transported, calculated from the buffering power of the suspending medium which is measured in each experiment with small additions of HCl, and the amount of oxygen added. This ratio is very close to two (1.9 ± 0.2) over a wide range of oxygen pulse size, ionic composition of the medium, pH values and temperature.

Valinomycin and K^+ are usually used as the permeant counter ion, although other permeant ions may also be used.

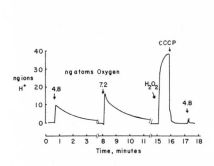

Fig. 1 <u>Proton translocation in cytochrome oxidase vesicles</u>

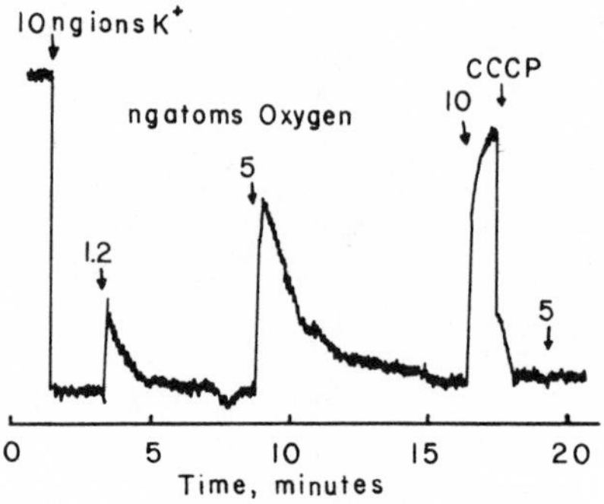

Fig. 2 <u>Potassium ion uptake by cytochrome oxidase vesicles</u>

Figure 2 shows a recording from a K^+-sensitive glass electrode in a similar oxygen pulse experiment to measure potassium ion uptake by the vesicles. Although K^+ uptake was technically more difficult to measure, the coupling ratio for uptake is also close to two.

According to the chemiosmotic hypothesis the energy coupling in cytochrome oxidase is a result of its proposed ability to carry electrons across the membrane from cytochrome c on the outside to the inside where oxygen is reduced to water picking up two protons per electron pair from the internal aqueous phase. The pH change inside has been measured directly by trapping phenol red inside the vesicles during reconstitution (13). Studies of the proposed transmembrane nature of cytochrome oxidase, however, have indicated that the protein is in contact with the medium on both sides of the membrane (16), but have not shown that electrons are actually carried across the membrane or that the protons taken up inside mitochondria are the same protons that form water in the reduction of oxygen.

Reduced Coenzyme Q - cytochrome c reductase (complex III)

Studies have recently been carried out with complex III (10,11) which are analogous to those described with cytochrome oxidase. Complex III is the cytochrome b-cytochrome c_1 region of the respiratory chain which carries electrons from coenzyme Q to cytochrome c. It was isolated thirteen years ago by Hatefi, Rieske and their co-workers, and consists of about eight polypeptide chains some of which have been identified as cytochrome c_1, cytochrome b, iron-sulfur protein and antimycin binding protein (cf. 17). When this complex is mixed with phospholipids and cholate and the cholate removed by dialysis, liposomes are formed which show uncoupler stimulation of electron transfer, coupled proton efflux, and uptake of K^+ in the presence of valinomycin.

Rates of electron transfer from reduced CoQ_2 to cytochrome c by the reconstituted complex in the presence of various uncouplers or ionophores are shown in Table II.

Table II

EFFECT OF IONOPHORES ON THE RATE OF REDUCTION OF CYTOCHROME C BY COMPLEX III VESICLES

The reduction of cytochrome c was measured with an Aminco DW-2 spectrophotometer at 550-540 nm with 1 µg protein complex III vesicles, 8 nmoles cytochrome c, 13 nmoles reduced CoQ_2 in 1.0 ml of 25 min KP_i pH 7.5 at 22°.

Additions	cytochrome c reduction µmoles/min/mg	"RCR"
None	3.1	--
1799, 40 µM	22.5	7.3
CCCP, 10 µM	34.1	11.0
Nigericin, 50 ng	4.2	1.4
Valinomycin, 50 ng	6-4	2-1.3
Val. + nig.	31.7	10.2

The uncouplers 1799 or CCCP or a mixture of valinomycin and nigericin gave a ten-fold increase in the rate of electron transfer. In this case valinomycin alone gave a transient stimulation of the reaction which lasted about five seconds and then returned to the controlled rate. This effect, which may not have been observed with cytochrome oxidase vesicles because of the slow response time of the oxygen electrode, is presumably caused by the transient collapse of the $\Delta\bar{\mu}H^+$ when valinomycin lowers the membrane potential. As the pH gradient is established, however, the $\Delta\bar{\mu}H^+$ rises until the rate of electron transfer is inhibited. In the presence of both valinomycin and nigericin complete uncoupling is obtained, as was shown in cytochrome oxidase vesicles.

Studies of proton translocation in complex III vesicles

were carried out similarly to those described with cytochrome oxidase vesicles except that reduced CoQ_2 was the substrate and pulses of ferricyanide were given as the oxidant
to oxidize cytochrome c present in the medium. The $H^+/2e$
ratio was two within experimental error for proton translocation with an additional two protons formed by the overall
oxidation of hydroquinone by ferricyanide. The uptake of two
K^+ per electron pair was also observed in the presence of
valinomycin.

There are two ways to arrange the $b-c_1$ region across the
membrane to fit the "loop" structure of the chemiosmotic
hypothesis. The first, favored by Mitchell, is to place CoQ
between cytochrome b and cytochrome $\underline{c}_1$ so that the cytochrome
$\underline{b}$('S) and CoQ form a proton-translocating loop. In this view
added quinone analogues do not react at the same site as the
natural CoQ_{10}, but reduce cytochrome $\underline{b}$ instead. The other
possibility, proposed by Skulachev, is that an undiscovered
hydrogen carrier exists between cytochrome $\underline{b}$ and cytochrome
$\underline{c}_1$ to form the loop. Whatever the mechanism of proton translocation in this region of the respiratory chain is eventually shown to be, we can at least conclude that the
components involved are in preparations of complex III.

NADH – Coenzyme Q reductase (complex I)

The conditions for reconstitution of complex I were
first studied by the cholate-dialysis technique by adding
both complex I and the F_0-F_1 ATPase and measuring oxidative
phosphorylation (6). Unlike the other reconstituted systems,
complex I requires mixtures of pure phosphatidylcholine and
phosphatidylethanolamine. Recently, we have studied vesicles
made with complex I alone, and measured proton translocation,
uptake of the permeant anion tetraphenylboron, and the stimulation of electron transfer by uncouplers and ionophores (9).
Since NADH was added only in the assay, the sidedness of the
system was like submitochondrial particles, and proton uptake
following addition of NADH and CoQ_1 was observed which
returned to a higher pH than at the beginning because of the
one proton taken up in the overall reduction of quinone by
NADH. The $H^+/2e$ ratios for translocation were about 1.4.
These measurements were technically difficult, however,
because the low buffering power of the purified lipids limited the maximum amount of proton translocation to about
10 ng ions/mg protein which is only 2% of that found with
the other reconstituted systems.

The stimulation of NADH oxidation by uncouplers or ionophores is shown in Table III.

Table III

EFFECT OF IONOPHORES ON THE RATE OF NADH OXIDATION BY COMPLEX I VESICLES

NADH oxidation was measured spectrophotometrically with 20 µg protein complex I vesicles, 125 µM NADH, 150 µM coenzyme Q_1 in 1.0 ml of 20 mM KP_i, pH 8.0 at 30°.

Additions	NADH oxidation µmoles/min/mg
None	0.35
1799, 40 µM	0.87
Valinomycin, 0.2 µg	0.38
Nigericin, 0.4 µg	0.38
Val. + Nig.	0.86
rotenone, 1 µg	0.15
rotenone + 1799	0.15

The stimulation by uncouplers and synergism between valinomycin and nigericin discussed previously are also observed with complex I vesicles. The "respiratory control ratio" caused by uncouplers was only 2.5, but this is probably low because of some rotenone insensitive damaged enzyme which is not coupled. When the rotenone insensitive activity is subtracted from the total the control ratio is 3.6.

Mitochondrial ATPase complex (F_0-F_1)

This complex was reconstituted using the uncoupler-sensitive ATP-$^{32}P_i$ exchange reaction to detect coupling activity (1). In addition to catalyzing this exchange, however, the system was found to show a two-fold stimulation of ATPase activity by uncouplers or by the combination of valinomycin and nigericin in a medium containing potassium ions. Studies of proton translocation with a recording pH meter also showed proton uptake driven by ATP hydrolysis (2), although the coupling ratio H^+/ATP was not determined. Skulachev has also studied this system and found that ATP hydrolysis can drive the uptake of lipid soluble anions indicating that the proton uptake is electrogenic (18).

Some interesting studies of the lipid specificity for reconstitution of the ATPase complex have been reported (19,20). It was found that the ATP-$^{32}P_i$ exchange required both phosphatidylethanolamine (PE) and phosphatidylcholine (PC) in a ratio of about 4:1. This mixture was not required

for the reconstitution process itself, however, since PE
vesicles could be reconstituted by cholate dialysis and the
PC added later by incubation with PC liposomes and the PC
transfer protein. Recent studies (20) have also shown that
the requirement for PE can be replaced by a mixture of acetyl
PE and stearylamine. It appears that both the free amino
group and the charge of the lipids are important variables.
Further studies on the role of lipids in the reconstituted
systems will be interesting.

Oxidative Phosphorylation

The combination of respiratory complexes with the ATPase
complex has been investigated by Dr. Racker. It was found
that cytochrome oxidase would drive phosphorylation if it
and the ATPase were reconstituted into the same liposome. The
sidedness of the cytochrome oxidase turned out to be the
critical factor for observing oxidative phosphorylation (4,5).
Internal cytochrome c added during reconstitution was active
when reduced by ascorbate and the permeant mediator
N-methylphenazinium methylsulfate (PMS). External cytochrome
c was not only inactive in driving phosphorylation, but
actually reversed the coupling of the oxidation of internal
cytochrome c. The requirement for sidedness is readily explained by the chemiosmotic hypothesis since the ATPase
couples the efflux of protons to ATP synthesis and the "loop"
of the PMS hydrogen carrier and cytochrome oxidase electron
carrier would create an electrochemical gradient to drive
protons outward. If the oxidase complexes with opposite
polarity were also activated by adding external cytochrome c
the resulting proton translocation outwards (electrons moving
inwards and protons absorbed inside) would neutralize the
gradient and inhibit phosphorylation.
Two other systems have also been coupled to the ATPase
complex by reconstitution, NADH-CoQ reductase (6) and bacter-
iorhodopsin (21). Each of these systems catalyzes proton
translocation inward.

Conclusions

The development of methods for incorporating transport
enzymes into liposomes represents a large step forward in
this rather broad field. The reconstitution of transport
systems related to oxidative phosphorylation that I have described is also relevant to our understanding of the mechanism
of energy coupling. The results are most readily explained
by Mitchell's chemiosmotic hypothesis. They could not be
called proof for it, however, because they are also

consistent with the chemical or conformational hypotheses if
it is proposed that each coupling site has a proton pump
which is on a sidepathway of energy transfer. To approach
this problem more directly, Dr. Thayer and I have studied
the kinetics of ATP synthesis driven by an artificially
imposed electrochemical proton gradient in submitochondrial
particles (22). This reaction, which is similar to Jagendorf's pH jump phosphorylation in chloroplasts, can be driven
faster than steady state oxidative phosphorylation, indicating that if proton translocation is on a sidepath, it is a
very fast sidepath.

The prospects for further work with the reconstituted
systems are considerable. A thorough study of the structure
of each system and the kinetics of electron transfer and
proton transport would help considerably in arriving at a
detailed mechanism for energy conservation in the respiratory chain. Many other key questions such as the mechanism
of the proton-translocating ATPase may also be aided by
studies of reconstitution with altered F_1 or F_0. Once the F_0
part of the ATPase is purified it should also be possible to
study it in liposomes where it acts as a proton "carrier" or
"channel".

REFERENCES

1. Kagawa, Y. and Racker, E., J. Biol.Chem. <u>246</u>, 5477
   (1971).
2. Kagawa, Y., Biochim.Biophys. Acta <u>265</u>, 297 (1972).
3. Racker, E., Biochem.Biophys.Res.Commun. <u>55</u>, 244 (1973).
4. Racker, E. and Kandrach, A., J.Biol.Chem. <u>246</u>, 7069
   (1971).
5. Racker, E. and Kandrach, A., J.Biol.Chem. <u>248</u>, 5841
   (1973).
6. Ragan, C.I. and Racker, E., J.Biol.Chem. <u>248</u>, 2563
   (1973).
7. Mitchell, P., Nature <u>191</u>, 144 (1961).
8. Mitchell, P., Biol.Rev. <u>41</u>, 445 (1966).
9. Ragan, C.I. and Hinkle, P.C., J.Biol.Chem., in press.
10. Hinkle, P.C. and Leung, K.H., in <u>Membrane Proteins in
    Transport and Phosphorylation</u>, Azzone, G.F., Klingenberg,
    M.E., Quagliariello, E. and Siliprandi, N., eds.
    Elsevier, New York, 1974, p. 73.
11. Leung, K.H. and Hinkle, P.C., J.Biol.Chem., in press.
12. Hinkle, P.C., Kim, J.J. and Racker, E., J.Biol.Chem. <u>247</u>,
    1338 (1972).
13. Hinkle, P.C., Fed.Proc. <u>32</u>, 1988 (1973).
14. Racker, E., J. Membrane Biol. <u>10</u>, 221 (1972).

15. Chance, B. and Montal, M., in Current Topics in Membranes and Transport, Bronner, F. and Kleinzeller, A., eds. Academic Press, New York, vol. 2, p. 99 (1972).
16. Schneider, D.L., Kagawa, Y. and Racker, E., J.Biol.Chem. 247, 4074 (1972).
17. Gellerfors, P. and Nelson, B.D., Eur. J. Biochem. 52, 433 (1975).
18. Skulachev, V.P., Ann. New York Acad. Sci. 227, 188 (1974)
19. Kagawa, Y., Johnson, L.W. and Racker, E., Biochem. Biophys. Res. Commun. 50, 245 (1973).
20. Knowles, A.F., Kandrach, A., Racker, E. and Khorana, H.G., J. Biol. Chem. 250, 1809 (1975).
21. Racker, E. and Stoeckenius, W., J. Biol. Chem. 249, 662 (1974).
22. Thayer, W.S. and Hinkle, P.C., J. Biol. Chem., in press.

PART II

BIOGENESIS

The existence of non-Mendelian, and therefore believed to be cytoplasmic (extranuclear), mutations affecting mitochondrial functions has been known for over twenty-five years. Attempts at understanding of the molecular basis of this phenomenon are, however, of much more recent vintage, and much remains to be done. What is known is that mitochondria contain their own characteristic and distinct DNA; that this DNA is used as a subsidiary repository for cellular genetic information; that it can be replicated, damaged and repaired; that it can be transcribed into RNA of all three classes (transfer, ribosomal, and messenger RNA), and that these RNAs form part of a characteristic and specific translational system. Although mitochondria have the capacity for autonomy in their replication, the extent of this autonomy is strictly limited. Only a few polypeptides localized in the inner mitochondrial membrane and essential for its structure and function are formed. This synthesis occurs in a temporal sequence that is tightly linked and integrated with other events in the cell.

The presentations in this section address themselves to many of the most current and actively explored contemporary problems in the field: the nature, structure and function of the polypeptide gene products specified by mitochondrial DNA; their interaction with one another and with other essential mitochondrial functions; their role and regulation during mitochondrial assembly, and the nature of mutations relevant to mitochondria and their function in general.

APPLICATION OF CONTINUOUS CULTURE
IN THE STUDY OF MITOCHONDRIAL MEMBRANE
BIOGENESIS IN YEAST

by A. W. Linnane, S. Marzuki, G. S. Cobon, and
P. D. Crowfoot

INTRODUCTION

The study of the biogenesis of mitochondria
in yeast is facilitated by the ease with which the
function of yeast cells can be manipulated.   Both
the level of development and the composition of
yeast mitochondria may be varied by alterations in
such parameters as the culture conditions, the use
of specific inhibitors of mitochondrial functions,
and by mutation.   For example, growth of cells
under anaerobic conditions (1), or in the presence
of high levels of inhibitors of the mitochondrial
protein synthesising system, such as chloramphenicol (2), reduces mitochondrial function and
development.   Similarly, loss or alteration of
mitochondrial DNA by mutation (ρ^0 or ρ^- mutants)
result in the loss from the mitochondrial membranes
of those components, such as cytochromes aa$_3$, and b
which are totally or partially synthesised by the
mitochondrial protein synthesising system (3).
Such mutants must be grown on fermentable substrates, as these cells are incapable of oxidative
metabolism.   However, cells grown in the presence
of high concentrations of fermentable substrates
are catabolite repressed, which often complicates
the interpretation of the results.   In those
instances where glucose repression may obscure the
primary results of interest, it can be overcome by
culturing the cells in a glucose chemostat.   We
propose herein to describe this procedure and its
usefulness in the study of mitochondriogenesis.

The use of chemostat cultures is of course
routine in many areas of microbiology where
metabolic control phenomena are under investigation.   However, a novel second application of
chemostat culture is illustrated by the use of a
fatty acid auxotroph.   Such an organism has a
requirement for a molecule which is essential for

195

the structural integrity of all cell membranes. If the organism is grown in batch culture in the presence of an excess of unsaturated fatty acid then it is similar to the wild type (prototroph). However, the unsaturated fatty acid content of the cellular membranes may be manipulated by growing the cells in a batch culture containing growth limiting amounts of the lipid component (4,5). Under batch culture conditions with limited lipid the cellular membranes are in a state of flux, as the unsaturated fatty acid content is continually decreasing as the medium is depleted of unsaturated fatty acid, which makes the effect of any specific level of lipid content difficult to study.

In this chapter we describe how the chemostat culture technique can be used to advantage to obtain any desired steady state level of membrane unsaturated fatty acid content. Cells may be grown on fermentable substrates free from catabolite repression, and may be maintained in an exponential phase of growth at a given membrane unsaturated fatty acid level for several cell generations.

METHODS USED

Yeast Strains:

The strains of Saccharomyces cerevisiae used were:
L410 (α ura his ρ^+); E5 (α ura his ρ^o), a mitochondrial DNA less respiratory deficient mutant derived from L410; KD115 (α ole ρ^+), an unsaturated fatty acid auxotroph; and KD115-2 (α ρ^+), a revertant of KD115.

Theoretical Considerations:

No attempt is made here to give an exhaustive account of the theoretical background of continuous culture, which can be found in several excellent articles (6,7,8) and books (9,10).

In a continuous culture, increase in the cell density dx/dt is determined by the growth rate of the cells, and by the rate at which the culture is diluted by the inflowing medium.

Thus,
$$\frac{dx}{dt} = \mu x - Dx \qquad (1)$$

where x is the cell density (mg cell dry wt/ml), μ is the specific growth rate (h^{-1}) and D is the dilution rate (h^{-1}).

In the steady state,

$$\frac{dx}{dt} = \mu x - Dx = 0$$

and thus
$$\mu = D \qquad (2)$$

Applying Michaelis-Menten kinetics, according to Monod (6),

$$\mu = \frac{\mu_{Max} \, S}{K_s + S} \qquad (3)$$

Where μ_{Max} is the maximum specific growth rate of an organism on a given medium, S is the concentration of the growth limiting substrate, and K_s is the substrate concentration which gives half of the μ_{Max}.

It should be appreciated that equation (3) although satisfactory for most situations, as an empirical description of μ, does not fully express the complex kinetics of a chemostat culture (11).

In the steady state, substituting (2) in (3)

$$D = \frac{\mu_{Max} \, S}{K_s + S} \qquad (4)$$

or

$$S = \frac{K_s \, D}{\mu_{Max} - D} \qquad (5)$$

Since both K_s and μ_{Max} are intrinsic properties of a microorganism, it is obvious from equation (5) that in the steady state, the concentration of the growth limiting substrate in the culture is determined only by the dilution rate D. Thus the actual concentration of the growth limiting substrate can be varied in the culture by altering the dilution rate D.

Establishment of Chemostat Growth Conditions:

Fermenting vessels may be fabricated in the laboratory from glass or stainless steel, or commercial apparatus may be used. A description of typical apparatus, together with precautions which need to be taken in setting up a culture, may be found in references (6-10). In our laboratory glass fermentors which incorporate a simple side arm overflow outlet are used.

Cells are grown according to the chemostatic principle, so it is necessary to monitor various culture parameters as well as the cell density. It is essential when first using a particular strain or medium that the desired growth factor is indeed limiting, i.e. the steady state cell density depends on the concentration of that factor in the input medium. For example, if it was desired to grow cells in a glucose limited culture then the concentration of glucose in the input medium would be varied, and the steady state cell density (at the same dilution rate in each case) determined. If the steady state cell density increases linearly with the glucose concentration in the input medium, then the culture is glucose limited. If this is not the case, then some other factor is limiting, and must be increased in concentration in the medium.

When auxotrophic mutants are grown under conditions where the required factor is limiting or reduced, the culture must be checked for revertants, especially when cultures are maintained in steady state for long periods. In addition for yeast, the cultures must be checked for the proportion of petite mutants which under certain conditions accumulate in the culture to an abnormal extent (12).

EXPERIMENTAL & DISCUSSION

Effect of alterations in culture dilution rate on cell metabolism:

An example of a glucose limited culture is shown in Fig. 1, where the cell density and the ethanol and glucose concentrations in a culture are plotted as a function of the dilution rate of the culture. The steady state cell density remained constant at about 5.5 mg cell dry weight/ml when the dilution rate was varied between 0.025 h^{-1} and 0.2 h^{-1}. Over this range of dilution rate, the metabolism of the cells was both fermentative and oxidative, as shown by the low level of ethanol accumulated in the culture medium and the near complete utilization of the glucose in the medium. Slight decreases in the cell density have been observed as the dilution rate is decreased. Thus, as shown in Fig. 1, the cell density decreased from about 5.6 mg/ml to about 5.3 mg/ml as the dilution rate was decreased from 0.2 to 0.025 h^{-1}. This has been attributed by Fiechter and Von Meyenburgh (13) to an increase in the proportion of the total energy requirement needed for maintenance of cellular functions, and therefore not available for growth.

At a high dilution rate (>0.2 h^{-1}) (Fig. 1), a change toward a purely fermentative metabolism was observed. Thus, an accumulation of ethanol in the culture medium together with a concomitant drop in the cell density to about 1.5 mg cell dry weight/ml are observed. The actual concentration of glucose in the culture remained below 0.5 mM until the dilution rate was about 0.3 h^{-1}, and then increased sharply. This increase was accompanied by a decrease in the steady state cell density. The critical dilution rate of 0.35 h^{-1} is equal to the μ_{Max} of the yeast strain when grown on glucose; at a dilution rate above 0.35 h^{-1}, a steady state could not be established, as the dilution rate exceeded the μ_{Max}. This example illustrates that it is possible, simply by varying the dilution rate of the culture, to grow the cells under conditions whereby fermentative processes alone occur or both fermentative and oxidative processes

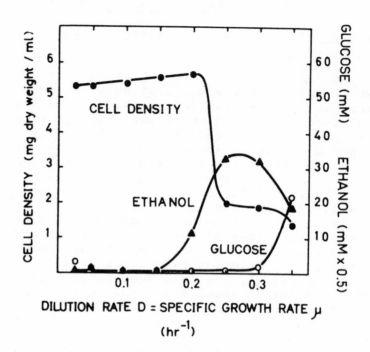

<u>Fig. 1:</u>   <u>Effect of dilution rate on glucose
chemostat culture parameters.</u>

<u>Saccharomyces cerevisiae</u> KD115-2 (α OLE ρ^+)
was grown aerobically at 28°C in a glucose limited
chemostat culture on a medium described previously
(18). Initially the dilution rate was set at
0.025 h^{-1}. Samples were taken at about 12 h
intervals and the culture glucose, and ethanol,
concentrations and the cell density were determined.
When a steady state was established, the dilution
rate was increased to 0.05 h^{-1}. Samples were again
taken until a new steady state was established.
The same procedure was repeated at other dilution
rates.   The results presented are in each case
the average of three steady state samples.

occur. Examination of the cell cytochrome spectra confirmed that the mitochondrial functions were derepressed below a dilution rate of 0.2 h^{-1} and repressed at a dilution rate of about 0.3 h^{-1}.

In this communication, two examples of the application of chemostat culture in the study of mitochondrial membrane biogenesis will be presented, viz. the effect of glucose derepression on the lipid synthetic capacity of petite cell mitochondria and unsaturated fatty acid depletion of cells on mitochondrial protein synthesis.

1) Respiratory Deficient (Petite) Mutants Grown Under Catabolite Derepressed Conditions.

Respiratory deficient mutants are unable to obtain energy by the process of oxidative phosphorylation, and consequently must be grown on fermentable substrates such as glucose. However, as mentioned earlier, high concentrations of glucose in the medium induce catabolite repression, and it then becomes uncertain whether any effects observed are directly due to the cytoplasmic petite mutation or to catabolite repression.

As shown in Fig. 1, respiratory competent cells may be grown in a glucose limited chemostat culture such that the glucose level is maintained below that level which induced catabolite repression (14). For respiratory deficient cells the situation is slightly more complicated as, since these cells are incapable of oxidative metabolism, the dilution rate at which the cells are no longer subject to catabolite repression cannot be predicted from the cell density. Thus when the petite mutant E5 is grown at various dilution rates, as described in the legend to Fig. 1, and a diagram of culture parameters similar to that of Fig. 1 is constructed, a monophasic cell density curve is obtained. However, the dilution rate at which the alteration in metabolism occurs can be anticipated to be close to that at which the change-over occurs in respiratory competent cells grown under the same conditions. This can be confirmed by examining the relative concentrations

201

in the cells or mitochondria of components which
are subject to catabolite repression.   These
components include cytochrome c, ubiquinone and
several citric acid cycle enzymes (14, 15).   It
is convenient in practical terms to measure the
cytochrome levels.   Fig. 2 shows the cytochrome
content of a petite mutant and its respiratory
competent parent strain grown under catabolite
derepressed conditions in the chemostat, and under
repressed conditions in batch culture.   The
results show that at a dilution rate of less than
about 0.2 h^{-1}, the cells are derepressed for
cytochrome c synthesis, the level of cytochrome c
approximating the level in the parent strain grown
under the same conditions.

As an example of the application of the
chemostat technique we are examining the nature of
the biochemical lesions resulting from the total
loss of mitochondrial DNA.   In particular we
consider herein the in vitro lipid synthetic
capacity of mitochondria.   It is known that petite
cells contain all of the phospholipids found in
wild-type cells, and that glucose repression alters
the cellular phospholipid composition of both wild
type and petite cells (16).   However, as some
phospholipids are synthesised by both the mitochondria and the microsomes (17), it is possible
that deletion of mitochondrial DNA could result in
loss of the ability to synthesise a given phospholipid in the mitochondria, while the microsomal
synthesis remained unaffected.   However, as shown
in Table 1 we have established by examination of
the in vitro lipid synthetic capacity of derepressed petite mitochondria that these mitochondria are
able to synthesise all of the phospholipids normally
synthesised by wild type mitochondria.   It may be
concluded therefore, that none of the enzymes
involved in the synthesis of phospholipid from
sn-glycerol-3-phosphate in mitochondria is coded
for by mitochondria DNA, or synthesised by the
mitochondrial protein synthesising system.

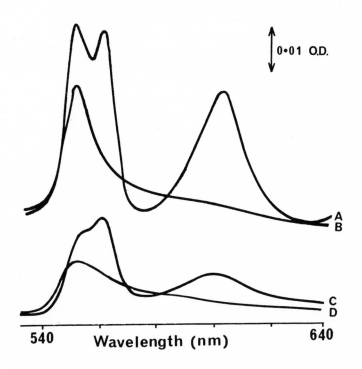

<u>Fig. 2:</u> <u>Cytochrome spectra of mitochondria from</u>
<u>repressed and derepressed ρ^+ and ρ^o</u>
<u>cells.</u>

Strains L410 (ρ^+) and E5 (ρ^o) were grown
in glucose chemostat cultures under derepressed
conditions and under repressed conditions in batch
culture.   Mitochondria were prepared, and
cytochrome spectra determined as described
previously (17).   All preparations were measured
at a protein concentration of 5 mg/ml.

A, L410 derepressed;   B, E5 derepressed;

C, L410 repressed;   D, E5 repressed.

TABLE 1: In vitro lipid synthesis by mitochondria from derepressed ρ^0 and ρ^+ cells.

Lipid Synthesised	ρ^0 Mitochondria	ρ^+ Mitochondria
Phosphatidic acid	14	24
Phosphatidylglycerol	11	26
Diphosphatidylglycerol	2	5
Phosphatidylinositol	40	22
Phosphatidylserine	6	5
Phosphatidylethanolamine	17	5
Phosphatidylcholine	-	-
Di and Tryglycerides	10	12

Saccharomyces cerevisiae strains L410 (ρ^+) and E5 (ρ^0) were grown under derepressed conditions in the glucose chemostat. Mitochondria were prepared and incubated with sn-(2-^{3}H)glycerol-3-phosphate for 30 mins (17). The results are presented as mol of product/100 mol of glycerolipid (^{3}H) glycerol in each species.

2)   Use of Chemostat Culture Technique to Manipulate Steady State Mitochondrial Membrane Unsaturated Fatty Acid Level.

(a)   The Effects of Unsaturated Fatty Acid Depletion on Oxidative Phosphorylation

The unsaturated fatty acid auxotroph KD115 has been extensively utilised in our laboratory to investigate the role of unsaturated fatty acids in membrane structure and function. In earlier studies, the cells were grown in batch culture to early stationary phase, the level of unsaturated fatty acid in the medium at inoculation determining the final membrane unsaturated fatty acid content (4,5). It was found that the efficiency of oxidative phosphorylation dropped in parallel with the membrane unsaturated fatty acid level, so for studies at very low membrane unsaturated fatty acid levels the use of glucose as carbon source was obligatory. Thus there were three factors potentially contributing to phenomena observed at low membrane unsaturated fatty acid levels:

(1)   catabolite repression,

(2)   extensively depleted cells were harvested in stationary phase,

(3)   changes in the membrane unsaturated fatty acid content.

When cells are grown under derepressed conditions in the glucose chemostat, the first two complicating factors can be avoided. Moreover, the cellular steady state unsaturated fatty acid content could be maintained at any desired point between the wild type level of 75% to as low as 10% of the total fatty acids, by altering the concentration of the fatty acid supplement in the input medium (18, 19).

It was found that even a small decrease in the steady state cellular unsaturated fatty acid content to 60%, for example, resulted in a decrease in the molar growth yield compared with the fully supplemented culture. This decrease in molar growth yield was shown to be a consequence of a reduced efficiency of oxidative phosphorylation in such cells. There was no accumulation of glucose or its products (ethanol) in the medium, viz. the ethanol was being oxidised but the energy yield was reduced. It was found that the efficiency of oxidative phosphorylation continued to decrease as the steady state unsaturated fatty acid content was reduced further, until at about 34% unsaturated fatty acids, oxidative phosphorylation was fully uncoupled.

Thus by using the technique of chemostat

culture, it has been possible to examine changes
in molar growth yield which can be correlated
with the efficiency of oxidative phosphorylation
at each level of unsaturated fatty acids. In this
way we have obtained data from the parameters of
the continuous culture which confirm previous
in vitro measurements and establish that, in vivo,
the efficiency of oxidative phosphorylation is
directly related to the membrane lipid composition.

> (b) The Effect of Unsaturated Fatty Acid
> Depletion on Mitochondrial Protein
> Synthesis

In previous studies we were unable to
demonstrate in vitro protein synthetic activity
in mitochondria isolated from cells extensively
depleted of unsaturated fatty acid in batch culture.
Further investigation revealed that these mitochondria apparently lacked ribosomes. However,
the mitochondria contained almost normal levels of
cytochromes aa_3, and b, which are synthesised in
part by the mitochondrial protein synthesising
system (20,21). This apparently anomolous
situation in which neither protein synthetic
activity nor intact ribosomes could be detected
in isolated mitochondria, while some of the products of mitochondrial protein synthesis appeared to
be present in appreciable amounts, raised several
questions. We considered two possibilities to
explain the results.

The presence of the products of the
mitochondrial protein synthesising system raises
the first possibility that mitochondrial protein
synthesis may operate normally in vivo at low
levels of unsaturated fatty acids. The observed
lack of mitochondrial protein synthesis in vitro
could then be due to disruption of the fragile
lipid depleted mitochondria during their isolation,
with a consequential loss from the organelle of the
mitochondrial ribosome. Indeed electron micro-
scopic examination of the isolated organelles
containing 15% unsaturated fatty acids showed that
they were extensively damaged.

Alternatively, a functional mito-

chondrial protein synthesising system may be
absent from cells containing low levels of
unsaturated fatty acids.   When cells are depleted
of unsaturated fatty acids in batch cultures, the
membrane unsaturated fatty acid content steadily
decreases as the cells grow.   The mitochondrial
protein synthesising system may operate normally
only until a critical membrane unsaturated fatty
acid level is reached.   Relatively few cell
generations occur between reaching this critical
unsaturated fatty acid level and cessation of
growth, so if the mitochondrial ribosomes under-
went a more rapid turnover than the products of
the mitochondrial protein synthesising system
(such as cytochrome aa$_3$), the depleted mitochondria
would contain no detectable ribosomes while still
retaining some products of the system.

In order to resolve this question, cells
were grown in chemostat cultures at low steady
state membrane unsaturated fatty acid levels for
a number of generations, so that components formed
at high unsaturated fatty acids but no longer
synthesised at low unsaturated fatty acid levels
could be diluted out.   The steady state cellular
membrane unsaturated fatty acid content could be
lowered from the wild type level of 75% to 34%
without affecting the synthesis of proteins by the
mitochondrion (18).   However, as the cellular
steady state unsaturated fatty acid content was
lowered further, to about 28% or below, the total
mitochondrial cytochrome level fell to less than
10% of normal (Table 2).   Cytochrome aa$_3$ was not
detectable even when the samples were examined at
the temperature of liquid nitrogen, indicating
that at low levels of unsaturated fatty acids the
products of the mitochondrial protein synthesising
system are not formed.   The presence of these
compounds in cells depleted of unsaturated fatty
acid in batch culture is therefore attributable to
insufficient dilution of the cytochromes due to
the relatively few cell divisions which occurred
between depletion of the unsaturated fatty acid
below about 30% and cessation of cell growth at
about 15% unsaturated fatty acid content.

It is of interest that the total amount of

TABLE 2: Mitochondrial Cytochrome Levels of
Strain KD115 Grown at Various
Unsaturated Fatty Acid Levels

Steady State Unsaturated Fatty Acid (% Total Fatty Acid)	Mitochondrial Cytochrome (nmoles/mg protein)		
	cc_1	b	aa_3
75	0.44	0.68	0.25
60	0.31	0.62	0.23
50	0.40	0.52	0.21
44	0.47	0.60	0.23
34	0.37	0.52	0.25
28	0.05	0.07	Undetectable
22	0.04	0.05	Undetectable

The unsaturated fatty acid auxotroph strain
KD115 was grown to different steady state
levels of membrane unsaturated fatty acid.
Mitochondria were prepared and cytochrome
spectra measured, as described previously
(18).

cytochrome c synthesised at low levels of unsaturated fatty acids (about 28% or less) is also
greatly reduced. Cytochrome c is synthesised
by the cytoplasmic protein synthesising system so
that the reduction in cytochrome c content cannot
be accounted for simply by the absence of mitochondrial protein synthesis. Indeed, as the cells
are capable of growth by fermentation, the cytoplasmic protein synthesising system must be fully
functional. Moreover, the effect is not a direct

result of the absence of cytochromes $\underline{b}$ and $\underline{aa}_3$, as derepressed petite mutants contain almost normal levels of cytochrome $\underline{c}$ (see Fig. 2). A number of explanations are possible, among which it may be suggested that due to changes in the mitochondrial membrane lipid composition (and hence its molecular architecture), cytochrome $\underline{c}$ cannot be readily integrated into the mitochondrial membrane. It is then conceivable that the inability of the membrane to incorporate cytochrome $\underline{c}$ leads to a feed back control of cytochrome $\underline{c}$ synthesis in the cytoplasm. However, at this stage of our invest-igations this interpretation is uncertain, and this aspect of the problem warrants further study.

(c)   The Effect of Unsaturated Fatty Acid Depletion on Mitochondrial Ribosome Synthesis

The absence of mitochondrially synthesised cytochromes from cells grown in the chemostat culture for several generations at low levels of unsaturated fatty acids (about 28% or below) indicates that the mitochondrial protein synthesising system is not active under such conditions. Subsequently, we demonstrated that morphologically intact mitochondria could be isolated from the cells and that these mitochondria contain no ribosomal subunits (18) but it was shown by RNA analysis that mitochondrial transcription occurs. The unsaturated fatty acid depleted organelles contain about 50 μg total RNA/mg protein, a level similar to that found in mitochondria isolated from normal cells (19). Furthermore, the RNA has the characteristic guanine plus cytosine content of mitochondrial RNA (19).

Sucrose density gradient analysis showed that the isolated RNA sediments as a broad low molecular weight band (8-12$\underline{S}$); the intact ribosom-al 21$\underline{S}$ and 16$\underline{S}$ species were not observed (19). At this time it is not certain whether the low molec-ular weight RNA occurs in vivo or is formed during the isolation of the mitochondria.

SUMMARY   AND FUTURE PROSPECTIVES

In this paper, two specific examples of the
use of chemostat culture techniques in the study
of mitochondriogenesis have been discussed.   The
procedure is particularly useful for the study of
catabolite repression.   In the present study, the
in vitro phospholipid synthetic capacity of mitochondria isolated from respiratory deficient cells
was studied under conditions free from the complic-
ations of catabolite repression.   It was found
that these mitochondria had a similar phospholipid
synthetic capacity in vitro to mitochondria from
wild type cells, indicating that none of the enzymes
involved in mitochondrial lipid synthesis are products of the mitochondrial protein synthesising
system.

The chemostat culture technique can be
adapted to the study of several independent, though
interacting, systems.   As an example, conditions
under which simultaneous control of both the degree
of catabolite repression and the cellular membrane
fatty acid composition may be studied were described.   The effect of lowered mitochondrial membrane
unsaturated fatty content on the mitochondrial
protein synthesising system was reported.

It has been shown that at low cellular levels
of unsaturated fatty acids, mitochondrial ribosomes
are not formed although mitochondrial RNA synthesis
occurs.   Moreover, the cytoplasmic protein synthesising system, which is responsible for the formation of the mitochondrial ribosomal proteins, is
active.   The question then arises as to why
ribosome assembly does not occur under these
conditions.

As a first approach to this problem, we are
using antibodies against mitochondrial ribosomal
proteins to determine the localization of these
cytoplasmically synthesised proteins within the
unsaturated fatty acid depleted cell.

REFERENCES

1. Watson,K., Haslam,J.M. & Linnane,A.W. 1970, J.Cell. Biol. 46, 88-96.
2. Lamb,A.J., Clarke-Walker,G.D. & Linnane, A.W. 1968, Biochim. Biophys. Acta. 161, 415-427.
3. Linnane,A.W. & Haslam,J.M. 1970 in Current Topics in Cellular Regulation (B.L.Horecker & E.R.Stadtman, Eds) pp 101-172, Academic Press, New York & London.
4. Proudlock,J.W., Haslam,J.M. & Linnane, A.W. 1971. Bioenergetics 2, 327-349.
5. Haslam,J.M., Proudlock,J.W., & Linnane,A.W., 1971. Bioenergetics 2, 351-370.
6. Monod,J. 1950, Annls.Inst.Pasteur Paris, 79, 390-410.
7. Tempest,D.W. 1970, Adv.Microbial Physiol.4, 223-250.
8. Tempest,D.W., 1970, in Methods in Microbiology, Vol.2 (J.R.Norris & D.W.Ribbons, eds) pp 259-276. Academic Press, London-New York.
9. Malek,I. & Fencl,Z. 1966, Theoretical and Methodological basis of continuous culture of microorganisms. Publishing House of Czecho-slovak, Acadamy of Science, Prague & Academic Press, New York & London.
10. Kubitschek,H.E. 1970, Introduction to Research with continuous cultures. Prentice-Hall, N.J.
11. Uden, N.van., 1971 in The Yeasts, (A.H.Rose & J.S.Harrison, eds) Academic Press, London & NY.
12. Marzuki,S., Hall,R.M., & Linnane,A.W. 1974. Biochem.Biophys.Res.Commun., 57, 372-378.
13. Feichter,A. & Von Myenberg,M.K. 1968, in Proc. 2nd Int.Symp.Yeast,Bratislava. pp 387-399. Prague and Academic Press, New York & London.
14. Perlman,P.S. & Mahler,H.R., 1974. Arch. Biochem. Biophys., 162, 248-271.
15. Lamb,A.J., 1969, Ph.D. Thesis, Monash University Clayton, Australia.
16. Jakovcic,S., Getz,G.S., Rabinowitz,M.,Jakob,H., & Swift,M. 1971, J. Cell Biol. 48, 490-502.
17. Cobon,G.S., Crowfoot,P.D. & Linnane,A.W., 1974. Biochem. J. 144, 265-275.
18. Marzuki,S., Cobon,G.S., Haslam,J.M. & Linnane, A.W., 1975. Arch. Biochem. Biophys. In Press.

19. Marzuki, S., Cobon, G.S., Crowfoot, P.D. & Linnane, A.W. 1975. <u>Arch. Biochem. Biophys</u>. In Press.

20. Schatz, G. & Mason, T.L. 1974. <u>Ann. Rev. Biochem</u>. <u>43</u>, 51-87.

21. Tzagoloff, A., Rubin, M.S. & Sierra, M.F. 1973. <u>Biochim. Biophys. Acta</u>, <u>301</u>, 71-104.

MITOCHONDRIAL ASSEMBLY: ATTEMPTS AT RESOLUTION OF COMPLEX FUNCTIONS

Henry R. Mahler*

Studies devoted to the problem of mitochondriogenesis occupy an ever increasing fraction of the effort of bio-chemists and cell biologists.[1-10] This effort has resulted in the proof of the supposition long held by certain geneti-cists that these organelles possess a separate genophore, distinct from that of the nucleus, and the demonstration of their capability for expressing the information encoded in its DNA. At the same time it has become apparent that these contributions in the *formation* of the organelle are strictly limited in scope and extent, and in the limit of ρ^0 mutants of *Saccharomyces cerevisiae*, entirely dispensable. Cells of such strains contain mitochondria that are topologically, and to a certain extent functionally, homologuous to the ones found in the wild type organism.[11] The intramitochondrial system was thus shown to be essential only to the formation of a fully competent organelle, capable of respiration and linked energy transduction. It participates in the expres-sion of mitochondrial competence and function rather than the *assembly* of the essential framework undergirding this pro-cess[12]. Conversely, studies on mitochondriogenesis and the contribution of mitochondrial gene expression to this pro-cess can reveal much that has been hidden concerning mito-chondrial structure and function.

1. Nature of translational products and statement of the general problem

In the course of the formation of a functional organelle, the mitochondrial contribution is probably limited to no more than ten polypeptides.[5-10,13-17] These entities are required for the synthesis of three enzyme complexes located on or in the cristae of the inner membrane and retain this structural integration into these complexes during their extraction and purification[5-10,18-21] (Table 1). The complexes are cyto-chrome oxidase, $CoQH_2$:cytochrome c reductase and the terminal

* Recipient of Research Career Award K06 05060 from the National Institute of General Medical Sciences, National Institutes of Health; research supported by Research Grant GM 12228 from this Institute.

TABLE 1

FRACTION OF INNER MEMBRANE TRANSLATED INTERNALLY

COMPONENT			SUBUNITS		TOTAL (%
Complex Portion		Fraction of Membrane (%)	Translated internally (Number) M_R x 10^{-3}	Fraction of total (%)	
$c : O_2$	aa_3	11	~40, ~30, ~20 (3)	50	5.5
$CoQH_2: c$	b	2.8	~30 (1)	50	1.4
ATP Synth	CF_o	10.5	~29,22,19,~ 7.5 (4)	100	10.5
Composition based on data of Harmon, Hall and Crane[34]					17.4
Independent estimate based on particle weight of complexes					19.0

CF_o ≡ Membrane attachment site of oligomycin-sensitive ATPase

ATP synthetase system [oligomycin-sensitive (OS) ATPase]. A number of inferences appear justified as a result:

a) These identified products account both qualitatively and quantitatively for observations with site specific inhibitors: therefore very few, if any, additional products remain unassigned in nature and localization.

b) At least in animal mitochondria these major products are probably the only ones: taken together with the other known mitochondrial gene products (RNA), they pretty nearly saturate the coding capacity of a single strand of their mtDNA.

c) In the steady state all these major products appear to be synthesized in roughly the same amount and, perhaps, at the same rate: therefore there must exist a mode for the co-ordinate regulation of these processes and, if a second class of minority products should exist, it must be subject to a different regulatory device.

d) All of the functional entities to which mitochondria make a contribution are equally dependent on one from the nucleocytosolic (NC) system of gene expression: their biogenesis is a co-operative venture, requiring the simultaneous and coordinate participation of both the NC and the mitochondrial (mt) system. Therefore, devices must exist to accomplish this coordination and its integration with other cellular events.

These inferences in turn raise a number of questions, among them the following:

a) Can one dissect the two classes of contributions with regard to their function, and if so discern a structural distinction for this difference?

b) Is there evidence for interactions between these mitochondrial contributions of stoichiometric or regulatory significance in the biogenesis of the complexes and of the organelle?

c) Can one adduce evidence for mutual interdependence between entities containing mt gene products and the mt genome itself? What is their relationship to entities produced by the NC system?

It is to these questions that we in the Mitochondrial Biogenesis Group of Indiana University have addressed ourselves to recently, using *S. cerevisiae* as our experimental organism. Our modest contributions towards providing some tentative answers concerning the first three will constitute the remainder of this presentation.

2. What is the function of the polypeptides synthesized internally?

We attempted to solve this problem explicitly first by controlled resolution of cytochrome oxidase. More recently as a result of some unexpected findings implicating the OS-ATPase in some reactions of mtDNA we have indirectly gained some novel insights into the functions of this enzyme complex as well.

a. Resolution of cytochrome oxidase - Apparently homogeneous preparations, some retaining very high levels of enzymatic activity have recently been obtained from heart muscle[23,24] and the fungi *S. cerevisiae*[25-28] and *Neurospora crassa*.[19,29] Upon complete dissociation and analysis by electrophoresis (sometimes supplemented by electrofocusing) they have been shown to be multimeric proteins containing either seven (fungi) or six (animals) non-identical polypeptide subunits. We too have been studying such preparations from both beef heart and yeast, and have asked whether it is possible to subject them to successive resolution, removing one (or more) polypeptides at a time without denaturating the enzyme. In this Dr. S.H. Phan has been successful by means of the procedure outlined in Table 2. Some of the characteristics of the three classes of enzyme, the unresolved (Class A: Y_7 and B_6), partially resolved (Class B: Y_5 and B_5) and completely resolved ones (Class C: Y_4 and B_4) are described in Tables 2-5.

TABLE 2

PROCEDURES FOR OBTAINING VARIOUS FORMS OF CYTOCHROME OXIDASE

PROCEDURE	PREPARATION[*]		CLASS
1. *Extract* Keilin-Hartree particles with *cholate*, replace cholate by *Tween 20*, fractionate with *Am$_2$SO$_4$*	Y$_7$	B$_6$	A
2. *Chromatography* on *L-Leucine-sepharose:* elute with gradient (0.5 M - 0.75 M) Tris (pH 8.0) - 0.5% Tween 20	Y$_5$	B$_5$	B
3. Recycling *chromatography* on *Sephadex G-200* in presence of *0.1% SDS*	Y$_4$	B$_4$	C

Y, B ≡ enzymes from yeast and beef heart, respectively

The subunit composition of the various preparations is summarized in Table 3. The data are based on densitometry of stained gels after a) electrophoresis in SDS, and b) electrofocusing in urea; as well as on electrophoretic analysis of proteins dissociated in SDS and labeled with [^{125}I] by chemical iodination. Applications of this technique to Y$_7$ and Y$_5$ are shown in Fig. 1A and B. Although bands IV and V cannot be resolved under these conditions, they can be separated in other gel systems (≥16% acrylamide) and by isoelectric focusing. As prediced from studies by Poyton and Schatz[28], enzymes of Class C, which are much less hydrophobic in their subunits than the standard (Class A) enzyme, turn out not to require detergents for their solubilization and to be soluble in aqueous buffers. Interestingly enough so is enzyme B$_5$. This is in accord with the much higher content of polar residues of subunit III of the beef heart compared to the fungal enzymes (J.R. Rowland, unpublished). The polypeptides removed by the various chromatographic processes can be recovered in separate fractions; their removal leads to proteins (Classes B and C) with altered physical properties, specifically with regard to their size and mass (Table 4). However, as shown in Table 5, this removal of proteins in going from Y$_7$ to Y$_5$ and Y$_4$ is accompanied by *increases* in the specific content of the two prosthetic groups heme a and Cu, and the ratio A$_{420}$/A$_{280}$, but is without any significant effect on any of the other characteristic spectroscopic

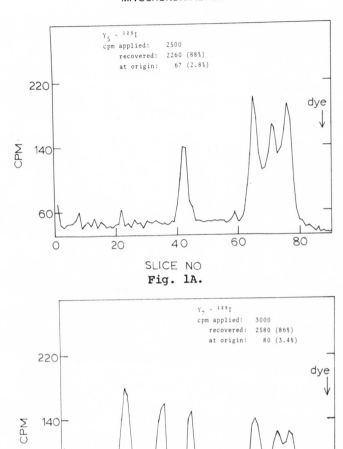

Fig. 1A.

Fig. 1B.

properties. Thus dissociation and removal of polypeptides I-III does not lead to either the loss of, or any extensive change in, the environment around the heme prosthetic group. Qualitatively similar results for the environment around the Cu-binding site - which is also conserved - have been obtained in EPR studies by Prof. H. Beinert of the University of Wisconsin.

TABLE 3

SUBUNIT COMPOSITION OF CYTOCHROME OXIDASES

PREPARATION	COMPOSITION (particle masses in Kdaltons)[a]	POLARI[
Y_7	I(42.4), II(34.1), III(24.7); IV,V(14.6), VI(12.3), VII(10.6)	A
Y_5	III ; IV,V , VI , VII	A
Y_4	IV,V , VI , VII	P
B_6	I(47.5), III(20.4); IV,V(14.5), VI(13.0), VII(11.0)	A
B_5	III ; IV,V , VI , VII	P
B_4	IV,V , VI , VII	P

[a] Values for yeast enzymes are extrapolations derived from Ferguson plots

[b] A = apolar, requires amphipathic environment
P = polar, soluble in purely aqueous media

TABLE 4

PHYSICAL PROPERTIES OF CYTOCHROME OXIDASES

PARAMETER	Y_7	Y_5	Y_4	B_6	B_5	B_4
Stokes radius (A)	75	70	60	70	65	
$s_{20,w}$ (S)	7.8	7.1	3.4	16.5	5.5	
M_R (corr) x 10^{-3}	260	170	107	205	124	
Min M_R (from heme a)	125	70.9	55	109	77.5	59
(from Cu)	112	68.0		92.5	75.2	
(from subunits)	153	76.8	52.1	121	73.4	53

CONCLUSION: Under conditions of physical measurements all enzymes are dimeric

TABLE 5

PROPERTIES OF YEAST CYTOCHROME OXIDASES

PARAMETER	Y_7	Y_5	Y_4
Heme (nmoles/mg)	8.02	14.1	18
Cu (ngatoms/mg)	8.91	14.7	n.d.
P-lipid (%)	5.46	2.00	n.d.
γ/α (Fe^{2+})	9.1	9.0	9.0
A_{420}/A_{444} (Fe^{2+})	0.40	0.39	0.39
A_{605}/A_{550} (Fe^{2+})	3.20	3.20	3.22
A_{280}/A_{420} (Fe^{2+})	3.16	2.84	2.76

All values are the means for at least three preparations, with std. deviations <10%

Is this true also of the enzymatic properties? The data of Table 6 indicate that this is in fact the case. The binding constant for cytochrome *c* is completely unaffected, and the turnover number suffers either virtually no ($Y_7 \rightarrow Y_5$), or, less than a 50% reduction ($Y_5 \rightarrow Y_4$) coincident with the quantitative separation from the rest of the molecule of polypeptides I + II, and III respectively. Similarly the total activity, heme and Cu present in Y_7 is recovered to the extent of >75% in Y_5 and >50% in Y_4. We conclude that prosthetic groups are attached to components of, and catalytic activity is retained by, a particle comprising only subunits IV-VII present in dimeric form.

b. Nature and function of two subunit classes - It thus appears that as in the case of ATPase[18,30] the subunits of cytochrome oxidase fall into two classes[19,28]: The first comprises a group of less hydrophobic polypeptides, specified and synthesized wholly by the NC system and concerned with catalyzing the oxidation of ferrocytochrome c by oxygen. The second consists of a group of polypeptides of greater hydrophobicity, synthesized in the mitochondria, and responsible for the integration of the polypeptide of the first class - singly or in combination - into their site of attachment in the inner membrane. Such integration appears to be essential for providing the *proper* spatial and catalytic environment during assembly of the complex (e.g. attachment of heme prosthetic groups and proper subunit configuration), membrane

TABLE 6

ENZYMATIC PROPERTIES OF CYTOCHROME OXIDASES

PARAMETER	Y_7	Y_5	Y_4	B_6	B_5	B_4
K_M (μM)	49.1	51.4	48.0	42.5	51.5	43.9
V (μmole x min^{-1} x mg^{-1})	164	216	180	63.8	108.5	79
V (mmole x min^{-1} x μmol a)[a]	20.4	15.3	13	8.1	8.7	4.7
[b]	44.9	33.6	28			

[a] pH 7.2, 50 mM phosphate (standard assay)

[b] pH 6.6, 35 mM phosphate (optimal conditions)

All values are means for three preparations, standard deviations <10%

integration (i.e. siting relative to other complexes) and function (i.e. link between oxidation and energy transduc tion) . However since cytoplasmic *petites* are capable of forming inner membranes which contain subunits IV-VI[31] some aspects of this dependence cannot be absolute. Furthermore the results described raise the question whether the hydro-phobic polypeptides of the second class - which are believed to span the inner membrane[31,32] - are actually required for the functionally significant, vectorial separation of cyto-chromes a and a_3.[33-37] One possibility is that the resi-dual phospholipids and detergent still present in Y_4 and B_4 may be able to substitute for this function and can gener-ate the required anisotropy. Another possibility is that the protein has already undergone a profound structural reorientation in the course of its detachment from the mem-brane and subsequent purification. If so this alteration must have occurred and the anisotropy lost in the isolation of Class A enzymes and thus maintanence of proper orient-ation does not depend on the presence or absence of the hydrophobic polypeptides. Finally, the change may be too subtle to be appreciated by the methods employed and become manifest only upon conversion of the protein into function-al vesicles.[35,37] If this is the case a comparison of Y_7, Y_5 and Y_4, the latter supplemented or not by the separated subunits should prove instructive.

c. Interaction of OS-ATPase with some novel and characteristic ligands - We have long been interested in the mode of action of the phenanthridinium dye ethidium bromide (EtdBr), a highly specific probe for mitochondrial structures and functions both *in vivo* and *in vitro*. In the course of these studies we found[38,39] that isolated mitochondria of *S. cerevisiae* are capable of incorporating EtdBr into mtDNA in novel, probably covalent linkage with a stoichiometry of one molecule of dye per 100 bases. This reaction appeared to require a particular, probably energized, state of the particle (see Fig. 4,) and so we decided to probe the polypeptide environment of the reaction site by means of a specific photoaffinity label.[40,41] We therefore synthesized the diazido derivative (dEtdBr) of EtdBr labeled with [3]H (Fig. 2), showed it to be the equal to the parent compound in mutagenic effectiveness *in vivo* and to compete for the same site in the binding reaction. We then added the labeled derivative to isolated mitochondria, converted it to the dinitrene *in situ* by uv-irradiation and asked which, if any, of the mitochondrial polypeptides become labeled under these conditions. To our surprise such covalent insertion appears restricted to a single polypeptide of the inner membrane with a $M_R \simeq 7800$. In this property as well as its solubility in CHCl$_3$–MeOH (Fig. 3 A

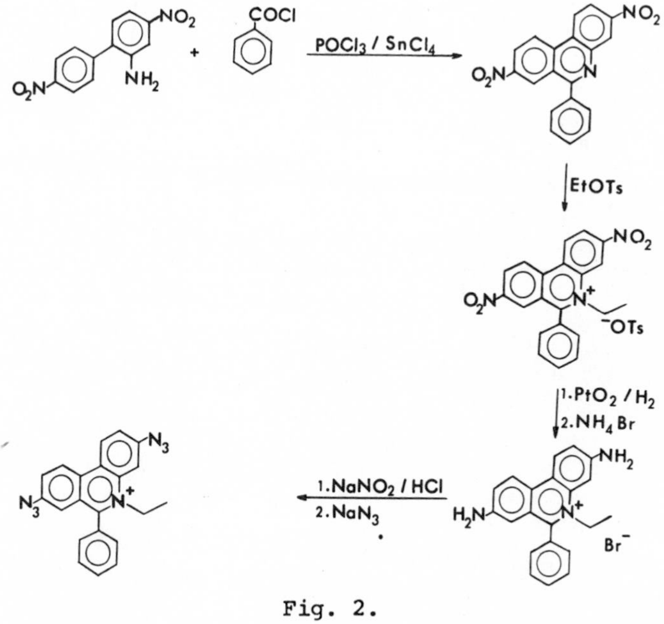

Fig. 2.

221

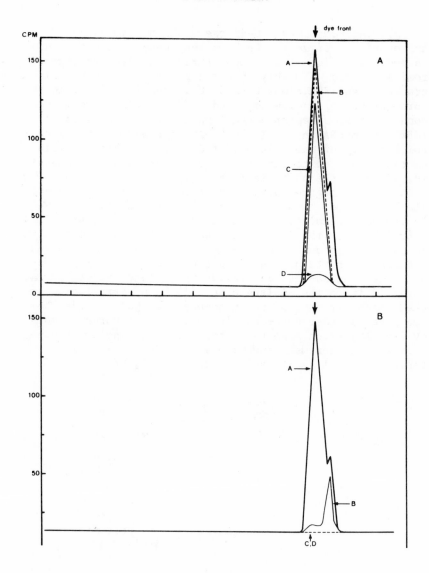

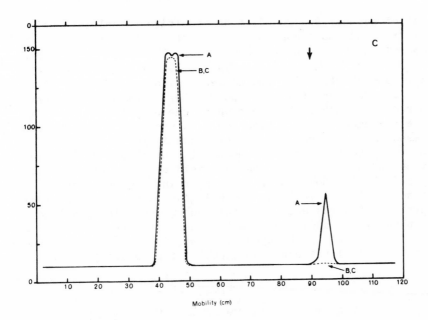

Fig. 3. *Analysis of dEtdBr-binding protein by gel electrophoresis.*
(A) Mitochondria (5 mg protein/ml), from IL16 grown on RHT +
3% glycerol, in MSM buffer (1 ml) were incubated with 5 μCi of
^{3}H-dEtdBr. After washing and dissociation, the sample (100-
200 μg) was loaded on the gel and electrophoresis performed
(trace A). Another sample (trace B) was run after the UV-treated
mitochondria had been washed with acetone. This acetone-washed
sample was then extracted with 2:1 chloroform-methanol and both
extract (trace C) and remaining material (trace D) analyzed.
(B) Previous to the addition of ^{3}H-dEtdBr, mitochondria under
the conditions described under (A), were treated with either
20 μg/ml oligomycin (trace A), 5 μM azidonitrophenol (trace C)
or 5 μM diazidoeuflavine (trace D). Mitochondria were also
isolated from cells grown in the presence of 4 mg/ml chloram-
phenicol. They were then treated under the conditions described
above (A) and results of binding are shown as trace B.
(C) Mitochondria from strain 73/1 were dissociated and loaded
on the gel. Results are shown at trace A. Identical samples
were washed with acetone (trace B) or extracted with 2:1 chloro-
form-methanol (trace C) previous to loading on the gel. In the
latter case, only the organic extract was analyzed. In all
cases, gels were sliced into 1 mm slices, solubilized by treat-
ment with H_2O_2 and counted.

and B) this component resembled the smallest, most hydrophobic subunit of the yeast OS-ATPase[18,30,42,43] and we
therefore set out to look for additional evidence in favor
of this supposition. The synthesis of the component, like

223

that of subunit 9, proved to be sensitive to chloramphenicol[38,44] and, more convincingly it is quantitatively precipitated with an antiserum to OS-ATPase[43,44] kindly provided by Dr. Alex Tzagoloff (Table 7). It thus appears that EtdBr and dEtdBr are bound to yeast mitochondria at a site that is identical to, overlaps with or is closely adjacent to the hydrophobic (F_0) portion of the OS ATPase. More direct evidence for this hypothesis is provided by experiments[45] assessing the effect of complex formation with isolated, purified protein either by means of equilibrium dialysis or by measuring changes of quantum yield of these fluorescent dyes.[47,48] As shown in Table 8 their interaction with the intact OS ATPase leads to a significant enhancement of their intrinsic fluorescence, and this enhancement is magnified further by the addition of ATP. These effects are specific with respect to i) the type of dyes; ii) the nucleoside phosphate and iii) the proteins capable of generating them.

TABLE 7

EXTRACTION AND IDENTIFICATION OF PROTEIN-BOUND dEtdBr

FRACTION	CPM
A (Mitochondria)	1940
B (Submitochondrial particles)	1710
C (Extract)	1680
D (Gel peak)	1700
E (Immune precipitate)	1700

Mitochondria (5 mg/ml protein) were incubated with radioactive dEtdBr. An aliquot was removed (A) and total incorporation measured. They were then converted to submitochondrial particles (B) and extracted with Triton X-100 under conditions for ATPase solubilization. The extracted material (C) was split into two fractions: one (D) was loaded directly on a poly-acrylamide gel and the other (E) was precipitated with antibody to oligomycin-sensitive ATPase. The gel was sliced and counted and total counts under the single peak determined. All counts presented are normalized to the same volume of starting material.

TABLE 8

FLUORESCENCE ENHANCEMENT BY VARIOUS LIGANDS

Component / Ligand	dEtdBr 355→420	EtdBr 322→585	Euflavine 305→510	desethyl EtdBr 318→560	ANS 360→432
None	1.00	1.00	1.00	1.00	1.00
Buffer	0.79	0.78	0.83		
OS ATPase (0.5 mg/ml)	2.32	2.15	2.18		
F₁ (0.5 mg/ml)	0.91	0.95	0.98		
Subunit 9 (0.1 mg/ml)	1.03	0.91	0.92		
BSA (0.5 mg/ml)	0.84	0.83	0.88		
OS ATPase (0.5 mg/ml)	2.05	2.15		1.20	1.60
OS ATPase + ATP (5 mM)	2.55	2.50		1.20	1.04

All ligands at a concentration of 0.5 µM; buffer = 0.01 M Tris, pH 7.5;
1% Brij 58; 3% glycerol.

Wavelength are those used for excitation and emission, in nm.

As concerns i) most of the relevant data are summarized in the table: EtdBr, dEtdBr, euflavine and its diazido-derviative all are molecules carrying a delocalized positive charge, which are known from other experiments[39,45] to compete for the same site, and they produce qualitatively similar effects. On the other hand removal of the positive charge in the phenanthridium dye,[46] or the use of a negatively charge fluorophore, such as aminonaphthalene sulfonate[49] abolishes the effect. We can then ask whether we can use this phenomenon to probe the interaction of the OS ATPase with some other characteristic ligands known or suspected to interfere with this molecule in its functional capacity. The results of a large number of experiments are summarized in Table 9: Dio-9, a characteristic inhibitor[50,51] of F_1, as well as oligomycin and venturicidin which inhibit the enzyme by binding (reversibly) to its F_0 component[52,55] do not interfere with the binding of the fluorophore to the enzyme, but prevent the structural rearrangement in the latter induced by ATP, monitored by the incremental fluorescence enhancement resulting from its addition. Two ligands are known to become

TABLE 9

COMPETITION BETWEEN (d)EtdBr AND OTHER LIGANDS FOR OS ATPASE (1 μM)

Ligand	Reaction Concentration	Fluorescence Enhancement by Protein alone	Protein plus ATP	Photoaffinity Labeling
Dio 9	20 μg/ml	0	+	
Oligomycin	10 μg/ml	0	+	
Venturicidin	2 μg/ml	0	+	
DCCD	1 μM[a]	+	+[c]	+
ANP[b]	1 μM[a]	+	+[c]	+
Colicin K	1 μM	+	+[c]	
FCCP	1 μM	±	±	
1799	1 μM	+	+[c]	

[a] stoichiometric with the amount bound covalently

[b] 2-azido-4-nitrophenol (after covalent insertion)

[c] no further effect on addition of ATP

A zero indicates no effect, a + a significant (>80%) block

covalently attached to the OS ATPase: the well-known inhibitor DCCD, long known to form a covalent link to a hydrophobic component of the enzyme,[54-57] and the nitrene produced
photochemically from 2-azido-4-nitrophenol (ANP), an analogue
of the uncoupler 2,4-dinitrophenol.[41] The formation of a
covalent adduct with either compound, a process that occurs
with stoichiometry of one mole of ligand per mol of ATPase
leads to a) a inhibition of enzymatic activity; b) a decrease
in fluorescence enhancement and c) a reduction of the amount
of labeled dEtdBr that can become photochemically inserted
into subunit 9. The last two effects are strictly proportional to the amount of ligand bound covalently. Uncouplers
such as FCCP and 1799 at low concentrations produce effects
that appear reminiscent of those observed with the reversible
inhibitors. However, colicin K which we had earlier shown to

226

be an uncoupler of mitochondrial oxidative phosphorylation[45]–
as it is of the bacterial process[58] – apparently is capable of
interfering with the usual interaction between enzyme and
fluorophore.

As concerns ii) we have been able to show that among the
nucleoside triphosphates tested the effectiveness is restricted to ATP and ITP. ADP and the β,γ-methylene analogue of
ATP[59], while incapable of producing enhancement themselves,
can compete with ATP for its binding site. These are the
specificities expected and previously established for OS-
ATPase[54,55] and ATP-synthetase.[60] Finally, regarding iii),
the data in Table 8 indicate the specificity of the reaction
for the intact OS ATPase molecule. The inability of F_1 to
substitute for it is not unexpected, but since subunit 9 constitutes the eventual target of the photochemical attack by
dEtdBr its complete inability to function as an acceptor requires some comment. What we believe this to be due to is
the following: The binding interactions require a hydrophobic core or domain in or on the enzyme[62,63] of which subunit 9 forms only a part, albeit the one most accessible to
the photochemically generated nitrenes. There is some additional evidence for this hypothesis: complex formation is
abolished when the enzyme is treated with phospholipase A;[61]
and its occurrence in turn interferes with the encounter
quenching of perylene, a process known to require a hydrophobic environment.[64]

Apparently DCCD and ANP share this binding domain with
the phenanthridinium and acridinium dyes. If we now assume
that the modulating effect by ATP as measured by additional
fluorescence enhancement probes an alteration in the same
binding domain then two inferences emerge: a) The modu-
lation, probably involving the unmasking of additional
negative charges around the binding site, may be referrable
to additional conformational constraints on the protein-bound
fluorophore. This supposition is confirmed by attendant increases in fluorescence polarization[65-68] which, under appropriate conditions, can approach values close to 0.5, the
theoretical maximum, which is also attained in the absence of
ATP upon complete immobilization of the fluorophore by covalent attachment. b) Binding of the substrate, presumably to
its primary site on F_1, affects the structure of a relatively
distant hydrophobic domain, of which subunit 9 forms a part
and which subsumes the DCCD, oligomycin and venturicidin

binding sites. Thus the two principal reactive sites of the OS ATPase are capable of some form of functional communication, and the reaction described here can serve as a useful model for the behavior of the energy conservation sites[47,63, 67] in intact membranes.

3. *Interactions between mitochondrial gene products*

Previous studies, especially in Schatz's laboratory by Ebner *et al.*[31,68] and by Goffeau *et al.*,[69] had shown that nuclear mutations could prevent the proper integration of mitochondrially synthesized polypeptides either into just one, or more than one, of the enzyme complexes of the inner membrane. Is there evidence for similar cross-talk or regulatory interplay of products with a mitochondrial specification? Some evidence in favor of such a possibility has just been reported by Colson *et al.*[69a] Our own studies in this direction have made use of a new class of mitochondrial mutants.

a. *A new class of* res⁻ *mutants with a mitochondrial mode of inheritance* - These findings were the result of investigations on a novel class of respiratory deficient (*res⁻*) mutants isolated by a selective procedure designed to enhance the probability of their being of mitochondrial origin.[70] Tables 10 and 11 summarize the characteristics[8,70,71] of their genotype and phenotype, respectively, and compare them with those of two well documented classes of *res⁻* mutants[3-6,68]: suppressive *petite* (ρ^-) strains with a mitochondrial, and certain *pet* strains with a chromosomal (nuclear), mode of inheritance. It can be seen from Table 10 that while the new mutants satisfy all the criteria for being of mitochondrial origin, and in this regard resemble suppressive ρ^- strains, they differ from them in their capability of transmitting mitochondrial genetic markers. Their phenotype is rather reminiscent of certain pleiotropic *pet* mutants such as one studied by Goffeau *et al.*[69] in *Schizosaccharomyces pombe*. Of particular relevance are two additional sets of observations, also summarized in Table 11: the presence of an altered ATPase and of mitochondrial protein synthesis in the mitochondria of at least one representative of this class (UF 73/1).

b. *Nature of the altered product* - It thus appeared of interest to identify the consequences of the mutational alteration in order to determine whether, as we had hoped, we were dealing indeed with a mitochondrial point mutation. To do this we applied the photoaffinity labeling technique described earlier to mitochondria of strain 73/1 and compared the results with those obtained with its wild type[39] (Fig. 3C). We

TABLE 10

GENOTYPE OF VARIOUS *res⁻* MUTANT CLASSES

OPERATION	*TYPE*	*res⁻UF*	ρ^- (*suppressive*)	*pet*
1) Complementation with ρ^0		−	−	+
2) Meiotic segregation of *res⁻* (tetrad analysis)		0:4	0:4	2:2
3) Mitotic segregation of *res⁻* character in diploids		+	+	−
Extent		>90%	<1-100%	
4) Elimination of *res⁻* coincident with ρ (mtDNA)		+	+	−
5) Ability to transmit mt markers		−	+	+

found that most of the radioactivity in the mutant had been incorporated into a large polypeptide ($M_R \simeq 40,000$) rather than the small one characteristic of the wild type. The two polypeptides are probably related since they share two characteristic properties: solubility in organic solvents and biosynthesis by the mitochondrial system (i.e. its inhibition by chloramphenicol). It thus appears that a mutational change in mitochondrial DNA has produced a lesion in a product of mitochondrial translation. This lesion may have resulted in a product which is more prone to aggregation, one that is less easily cleaved during the obligatory conversion of a precursor into product, or has produced some reorientation leading to photoaffinity labeling of a different component. All of these possibilities are consistent with both, the presence of a small proportion of the total in the wild type position, and certain suggestive observations in the literature.[30,60,72-78] What we are currently doing is to exclude some of these possibilities and establish the relevance, if any, of this change to the pleiotropic phenotype, particularly with respect to the lesion in cytochrome oxidase.

TABLE 11

PHENOTYPE OF VARIOUS *res*⁻ MUTANT CLASSES

TYPE CHARACTER	*res*⁻UF^a	ρ⁻ (incl. ρ⁰)	pet
Enzymes			
Functional respiratory chain	–	–	–
Cytochrome oxidase	–	–	–
aa₃	–	–	–
b	+	–	+
NADH:cytochrome c reductase	low	–	low
ATPase detachment of F₁	normal	easy	easy
oligomycin sensitivity	–	–	– or +
Mitochondrial translation			
Formation of fMet–puro	+	–	+
Incorporation insens to CHX	+	–	+
sens to CAP	+	–	+
Petite phenocopy by CAP	+	–	
Altered product(s)	+		+

a Results on ATPase and protein synthesis are for one strain only (UF 73/1)

4. *Interactions between mitochondrial gene products and its genetic system*

a. Reaction of ethidium bromide with mtDNA - In *S. cerevisiae* EtdBr can bring about the quantitative conversion of ρ⁺ cells to ρ⁻ clones upon transient exposure of the former in buffer.[79,80,81] As already mentioned the use of radioactive EtdBr (Compound IV in Fig. 2) permitted the demonstration that as a result of such exposure the molecule becomes incorporated into mtDNA in a novel, probably covalent, fashion[82] to the extent of one EtdBr per 100 bases. Isolated mitochondria are able to carry out the same reaction,[38] forming a product that is stable in buffer but becomes rapidly degraded upon addition of an energy source, provided either by oxidative phosphorylation or by ATP added to the medium (Fig. 4). This degradative reaction (DNAse) is accompanied by activation of an ATPase,[39] and isolated mitochondria are therefore capable of bringing about reactions 1) through 3).

1) $mtDNA + mEtdBr \longrightarrow [mtDNA' \cdot EtdBr_m]$

2) $[mtDNA' \cdot EtdBr_m] + nATP \xrightarrow[\text{ATPase}]{\text{nuclease(s)}}$ fragments +

       $nATP + nPi$

3)=1)+2) $mtDNA + mEtdBr' + nATP \longrightarrow$ fragments $+ nADP +$
       nPi ($m \simeq 100$ nucleotides/EtdBr; $n = 6$ ATP/
       phosphodiester bond in DNA)

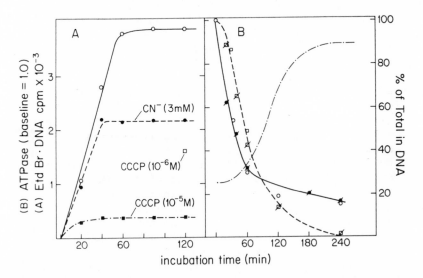

Fig. 4.   (A) Incorporation of $[^3H]EtdBr$ into mtDNA. Mitochondria
were incubated in buffer in the presence of EtdBr, samples were
removed at the times shown and assayed for alkali stable,
trichloroacetic acid-precipitable activity. Inhibitors at the
concentrations shown, were added simultaneously with EtdBr.
(B) Degradation of modified mtDNA and activation of ATPase in
response to addition of succinate. Three separate experiments
are shown: $O, \square$ represents a double label set (^3H for EtdBr,
^{14}C for DNA), while $\emptyset$ used labeled EtdBr but unlabeled cells,
and ⌀ used labeled cells but unlabeled EtdBr; pre-incubation
was as in A for 90 min, reaction was started by addition of
disodium succinate (10 mg/ml). ATPase is shown as dash-dotted
line, with the baseline activity equal to 0.4 μmoles P_i x min^{-1}
x mg^{-1}.

Of particular interest is reaction 2) which can be regarded
as both an ATP-dependent degradation of (modified) mtDNA, as
well as an ATPase dependent on the presence of this (modified)
mtDNA. A variety of experiments involving specific inhibitors of the mitochondrial ATP synthetase and ATPase, as well

231

as with mutants with mutationally altered ATPases, have con-
vinced us that the ATPase implicated in reaction 2) is the
mitochondrial enzyme itself[8],[39],[71] This fact suggests a
close and obligatory link between a key function of the mitochondrial genome, namely its mutagenic alteration (at least
by one particular agent) and a part of the energy transducing
device, an entity that requires the expression of that genome
for its synthesis.[26],[52],[53],[83]

   b. Reconstruction with soluble enzymes - If the above
supposition is correct, it should be possible to reconstruct
a soluble system capable of catalyzing reaction 2) and containing the oligomycin sensitive ATPase as an essential component. Recent data obtained by Dr. Bastos and shown in
Table 12 indicate that we have been successful in this en-
deavor. The system contains as catalysts the OS-ATPase
complex (*not* F_1) and a mitochondrial protein isolated on
DNA-cellulose columns, acting on a labeled covalent modification product (mtDNA'·EtdBr) (previously synthesized and
extracted from mitochondria), plus ATP. It is highly sensitive to inhibitors of both the F_1 (Dio 9) and F_0 portion
(oligomycin) of the ATPase, but relatively insensitive to
uncouplers (CCCP and Colicin K). Coincident with the
degradation of the substrate there is a 100% stimulation of
the base line hydrolysis of ATP by the enzyme system: both
the DNA binding protein and the modification product are required; no stimulation is observed when either is omitted.
As shown also, the system catalyzes only the excision of the
EtdBr (plus perhaps some bases in the immediate vicinity) but
not the extensive degradation of the remaining polynucleotide
tracts. We have thus obtained evidence that reaction 2)
really consists of two components

excision 2a) $[mtDNA'·EtdBr_m]+ x$ ATP $\longrightarrow$ [mtDNA']+m[EtdBr-
$$X(?)] + x \text{ ADP} + x \text{ Pi}$$

hydrolysis 2b)   [mtDNA']+ y ATP (?) $\longrightarrow$ fragments + ...

Confirmatory evidence for a similar sequence in intact mitochondria is provided by the detailed kinetics of reaction 2)
(Fig. 4B) in which considerable EtdBr is liberated, while
most of the bases are still present in an acid-insoluble form.
Additional studies using sedimentation in alkaline sucrose
gradients show that the EtdBr being excised was actually concentrated in one part of the molecule, perhaps on only one
strand. Furthermore, reactions 2a) and 2b) can be distinguished - even in cells or isolated mitochondria - by their
differential sensitivity to -SH reagents;[84] 2a) is insensitive
while 2b) is completely inhibited.

TABLE 12

PROPERTIES OF SOLUBLE SYSTEM FOR DEGRADATION OF Mt[^{14}C]DNA·[^{3}H]EtdBr

SYSTEM		% RELEASED		ATP HYDROLYZED
		EtdBr	Bases	(nmoles)
Complete*		70-75[a]	2	1.25
Omit	ATPase	2		0.10
	DNA-Binding protein	2		0.60
	All protein	2		
	ATP	7		-
	DNA	-		0.70
Add	CCCP (10^{-5} M)	52		
	Colicin K (10^{-6} M)	61		
	Oligomycin (20 μg/ml)	17		
	Dio-9 (20 μg/ml)	17		

* Modification product (1 μC[^{3}H]; 0.1 μC[^{14}C-Ade]; 0.3 nmoles EtdBr, 30 nmoles nucleotide P), 0.5 mg of oligomycin-sensitive ATPase, 3 mg of DNA binding protein in 2.0 ml of MTKE buffer, 60 min, 30°.

a Four separate experiments

c. Some novel functions of mitochondrial ATPase - The various studies just described suggest a novel and hitherto unexpected function for this enzyme complex: as an essential and direct participant in a system responsible for the excision of EtdBr from mtDNA. Can we find additional evidence for such a supposition? Strong support is provided by genetic studies: we had previously demonstrated that a certain haploid strain *uvs ρ72* originally isolated and studied by Dr. E. Moustacchi at Orsay[85] appears greatly more susceptible to mutagenesis by EtdBr than does N123, its wild type parent, and that this susceptibility can be correlated with an enhanced rate of degradation of the mtDNA after modification by the mutagen.[39,81] In a collaborative effort with Prof. P. S. Perlman at Ohio State who constructed the required strains by crossing ρ72, N123 and their DNA^0 (ρ^0) derivatives to appropriate testers (T). We asked two questions: What is the molecular basis of this enhanced rate of degradation and

what, if any, is the contribution of the mitochondrial genetic system to it? The results are summarized in Table 13.
We see that first of all the increased capacity (approx
twice that of wild type) for degradation is specified by the
mitochondrial genome. Secondly, this capacity - which is
found both in the haploid and the diploid which contains the
same information - is retained by the soluble system and is
a reflection of an alteration in its *ATPase* rather than its
DNA binding protein ("Enzyme"). Thus the ATPase itself controls the excision reaction.

Is there a structural correlate to this functional link?
In other words does the OS-ATPase become associated with some
component in the "Enzyme" fraction to form a DNA-excision
complex, analoguous to various ATP dependent DNAses of bacterial sources, all of them molecules which also exhibit
associated ATPase activity (for references, see[39])? Two
types of experiments bear on this question. In one we have
placed the "Enzyme" and the ATPase in glycerol gradients
singly or in combination and then tested for activity in the
complete system: the former activity which by itself remains
at the top of the gradients becomes associated and co-sedi-
ments with the ATPase. In the second we have added dEtdBr
to mitochondria whose DNA had been prelabeled with [3]H-

TABLE 13

MOLECULAR BASIS OF HYPERACTIVITY OF *uvs* ρ5

	Diploid	A	B	C	D
	Genotype	$ρ72/T^0$	$(ρ72)^0/T$	$N123/T^0$	$(N123)^0/T$
Experiment I					
Degradation of EtdBr- MtDNA in cells		++	+	+	+

		1	2	3	4	5	6	7	8
Experiment II									
source of	ATPase	72	A	B	C	D	A	B	C
	Enzyme	72	A	B	C	D	B	A	C+A
Degradation by reconstituted soluble system		++	++	+	+	+	++	+	+

234

adenine, isolated the covalent modification product and added it in the dark to a soluble system which contained an ATPase prelabeled with ^{14}C-leucine. We then exposed it immediately to uv and placed it on sucrose gradients where we found labeled protein (presumably subunit 9) to cosediment with the labeled DNA.

What is the significance of all this? EtdBr, although a superior mutagen, is not a molecule that can in any way be considered to constitute an ordinary challenge for mitochondria or their DNA. But enzyme (systems) capable of introducing nicks into DNA at appropriate places have been postulated to form an essential part of damage repair systems for DNA and, even more importantly, of systems responsible for its function in genetic recombination.[86-90] Perhaps, EtdBr when inserted into DNA, with part of the molecule (perhaps the phenyl ring) still capable of insinuating itself into a hydrophobic pocket of the ATPase mimics some configuration essential for this vital mitochondrial function, particularly if this process is in any way dependent on the resident ATPase.

SPECIFIC EFFECTS OF ETHIDIUM ON MtDNA IN CELLS OF *S. cerevisiae*

1. *Inhibits* replication and transcription

2. Transient exposure in buffer, quantitatively *converts* progeny to ρ^-

3. Subsequent exposure to *energy supply* results in *progressive fragmentation*

4. Events are subject to *prevention* and *cure* by e.g. euflavin or Antimycin A

New finding: transient exposure leads to *covalent fixation* of Ethidium

5. *Conclusions*

These studies suggest a complex and multidimensional reciprocal interplay between the genes and gene products of the two cellular systems involved in the assebly of the mitochondrial inner membrane. They have also afforded some novel insights into and functions for some of its most important constitutents. Further exploration should prove a continued and challenging area for future investigations.

6. Acknowledgements

I am indebted to all members of the mitochondrial biogenesis group at Indiana University for their splendid experimental and conceptual contributions and for providing
an intellectually stimulating environment within which to
carry out these investigations. In particular I wish to
thank K. Assimos, R. Bastos, and S. H. Phan for permitting
me to include their unpublished observations in this presentation. I am also grateful to Profs. E. Moustacchi,
P. S. Perlman and P. P. Slonimski for stimulating discussions and for providing us with some of the yeast strains;
and to Drs. P. Heytler, R. J. Guillory and S. Silver for
some of the inhibitors used in this investigation.

REFERENCES

[1] Ashwell, M. and Work, T.S. (1970) *Ann. Rev. Biochem. 39,*
251-290.

[2] Borst, P. (1972) *Ann. Rev. Biochem. 41,* 333-376.

[3] Sager, R. (1972) *Cytoplasmic Genes and Organelles.* Academic Press, New York.

[4] Linnane, A.W., Haslam, J.M., Lukins, H.B. and Nagley, P.
(1972) *Ann. Rev. Microbiol. 26,* 163-198.

[5] Mahler, H.R. (1973) *CRC Crit. Rev. Biochem. 1,* 381-460.

[6] Kroon, A.D. and Saccone, C. (1974) eds., *The Biogenesis
of Mitochondria.* Academic Press, New York.

[7] Schatz, G. and Mason, T. (1974) *Ann. Rev. Biochem. 43,*
51-87.

[8] Mahler, H.R., Bastos, R.N., Flury, U., Lin, C.C. and Phan,
S.H. (1975) in *Genetics and Biogenesis of Mitochondria
and Chloroplasts.* (P.S. Perlman, C.W. Birky, Jr. and
T.J. Byers, eds.) Ohio State University Press, in
press.

[9] Attardi, G., Constantino, P., England, J., Lynch, D.,
Murphy, W., Ojala, D., Posakony, J. and Storrie, B.
(1975) in *Genetics and Biogenesis of Mitochondria and
Chloroplasts.* (P.S. Perlman, C.W. Birky, Jr. and
T.J. Byers, eds.) Ohio State University Press, in
press.

[10] see contributions by Beattie, Griffiths, Linnane and
Tzagoloff in this volume

[11] Perlman, P.S. and Mahler, H.R. (1970) *J. Bioenergetics
1,* 113-138.

[12] Perlman, P.S. and Mahler, H.R. (1974) *Arch. Biochem. Biophys. 162,* 248-271.

[13] Weislogel, P.O. and Butow, R.A. (1971) *J. Biol. Chem. 246,*
5113-5119.

14 Swank, R.T., Sheir, G.I. and Munkres, K.D. (1971) *Bio-chemistry 10*, 3924-3930.

15 Ibrahim, N.G., Stuchell, R.N. and Beattie, D.S. (1973) *Eur. J. Biochem. 36*, 519-527.

16 Rogers, P.J., Yue, S.B. and Stewart, P.R. (1974) *J. Bact. 118*, 523-533.

17 Jones, L.R., Mahler, H.R. and Moore, W.J. (1975) *J. Biol. Chem. 250*, 973-983.

18 Tzagoloff, A., Rubin, M.S. and Sierra, M.F. (1973)*Biochim. Biophys. Acta 301,*71-104.

19 Sebald, W., Machleidt, W. and Otto, J. (1973) *Eur. J. Biochem. 38*, 311-324.

20 Weiss, H. and Ziganke, B. (1974) *Eur. J. Biochem. 41*, 63-71.

21 von Ruechen, A., Werner, S. and Neupert, W. (1974) *FEBS Lett. 47*, 290-294.

22 Poyton, R.O. and Groot, G.S.C. (1975)*Proc. Nat. Acad. Sci. USA 72*, 172-176.

23 Kuboyama, M., Yong, F.C. and King, T.E. (1972) *J. Biol. Chem. 247*, 6375-6383.

24 Volpe, J.A. and Caughey, W.S. (1974) *Biochem. Biophys. Res. Commun. 61*, 452-459.

25 Shakespeare, P. and Mahler, H.R. (1971) *J. Biol. Chem. 246*, 7649-7655.

26 Rubin, M.S. and Tzagoloff, A. (1973) *J. Biol. Chem. 248*, 4269-4274.

27 Mason, T.L., Poyton, R.O. Wharton, D.C. and Schatz, G. (1973) *J. Biol. Chem. 248*, 1346-1354.

28 Poyton, R.O. and Schatz, G. (1975) *J. Biol. Chem. 250*, 752-761.

29 Rowe, M.J., Lansman, R.A. and Woodward, D.O. (1974) *Eur. J. Biochem. 41*, 25-30.

30 Tzagoloff, A., Akai, A. and Rubin, M.S. (1974) in *The Biogenesis of Mitochondria* (A.D. Kroon and C. Saccone, eds.) Academic Press, New York, pp. 405-421.

31 Ebner, E., Mason, T.L. and Schatz, G. (1973) *J. Biol. Chem. 248*, 5369-5378.

32 Eytan, G.D. and Schatz, G. (1975) *J. Biol. Chem. 250*, 767-744.

33 Schneider, D.L., Kagawa, Y. and Racker, E. (1972) *J. Biol. Chem. 247*, 4074-4079.

34 Harmon, H.J., Hall, J.D. and Crane, F.L. (1974) *Biochim. Biophys. Acta 344*, 119-155.

35 Hinkle, P.L. (1974) *Ann. N.Y. Acad. Sci. 227*, 159-165.

36 Wikström, M.K.F. (1974) *Ann. N.Y. Acad. Sci. 227*, 146-158.

37 Skulachev, V.P. (1974) *Ann. N.Y. Acad. Sci. 227*, 188-202.

38 Mahler, H.R. and Bastos, R.N. (1974) *Proc. Natl. Acad. Sci. USA 71*, 2241-2245.

39 Bastos, R.N. and Mahler, H.R. (1974) *J. Biol. Chem. 249*, 6617-6627.

40 Knowles, A.F. (1971) *Accts. Chem. Res. 5*, 155-160.

41 Hanstein, W.G. and Hatefi, Y. (1974) *J. Biol. Chem. 249*, 1356-1362.

42 Tzagoloff, A. and Meagher, P. (1971) *J. Biol. Chem. 246*, 7328-7336.

43 Sierra, M.R. and Tzagoloff, A. (1973) *Proc. Nat. Acad. Sci. USA 70*, 3155-3159.

44 Tzagoloff, A. and Meagher, P. (1972) *J. Biol. Chem. 247*, 594-603.

45 Bastos, R.N. and Mahler, H.R. (1975) *J. Biol. Chem.*, in press.

46 Mahler, H.R. (1973) *J. Supramol. Struct. 1*, 449-460.

47 Azzi, A. and Santato, M. (1971) *Biochem. Biophys. Res. Commun. 44*, 211-217.

48 Gitler, C., Rubalcava, B. and Caswell, A. (1969) *Biochim. Biophys. Acta 193*, 479-481.

49 Rubalcava, B., Martinez de Munoz, D. and Gitler, C. (1969) *Biochemistry 8*, 2742-2747.

50 Guillory, R.J. (1964) *Biochim. Biophys. Acta 89*, 197-207.

51 Schatz, G. (1968) *J. Biol. Chem. 243*, 2192-2199.

52 Griffiths, D.E. and Houghton, R.L. (1974) *Eur. J. Biochem. 46*, 157-161.

53 Griffiths, D.E., Houghton, R.L., Lancashire, W.E. and Meadows, P.A. (1975) *Eur. J. Biochem. 51*, 383-402.

54 Senior, A.E. (1973) *Biochim. Biophys. Acta 301*, 249-277.

55 Pedersen, P.L. (1975) *Bioenergetics 6*, 243-275.

56 Cattell, K.J., Lindop, C.R., Knight, I.G. and Beechey, R.B. (1971) *Biochem. J. 125*, 169-177.

57 Stekhoven, F.S., Waitkus, R.F. and van Moerkerk, Th. B. (1972) *Biochemistry 11*, 1140-1150.

58 Plate, C.A., Suit, J.L., Jettem, A.M. and Luria, S.E. (1974) *J. Biol. Chem. 249*, 6138-6143.

59 Yount,R.G., Babcock, D., Ballantyne, W. and Ojala, D. (1971) *Biochemistry 10*, 2484-2489.

60 Hatefi, Y., Stigall, D.L., Galante, Y. and Hanstein, W.C. (1974) *Biochem. Biophys. Res. Commun. 61*, 313-321.

61 Somlo, M. and Krupa, M. (1974) *Biochem. Biophys. Res. Commun. 59*, 1165-1171.

62 Cunningham, C.C. and George, D.T. (1975) *J. Biol. Chem. 250*, 2036-2044.

63 Knowles, A.F., Kandrach, A., Racker, E. and Khorana, H.G. (1975) *J. Biol. Chem. 250*, 1809-1813.

64 Klip, A. and Gitler, C. (1974) *Biochem. Biophys. Res. Commun. 60*, 1155-1162.

[65] Weber, G. (1953) *Adv. Protein Chem. 8*, 415–459.

[66] Brewer, G.J. (1974) *Biochemistry 13*, 5038–5045.

[67] Azzi, A., Fabbro, A., Santato, M. and Gherardini, P.L. (1971) *Eur. J. Biochem. 21*, 404–410.

[68] Ebner, E., Mennucci, L. and Schatz, G. (1973) *J. Biol. Chem. 248*, 5360–5368.

[69] Goffeau, A., Landry, Y., Foury, F., Briquet, M. and Colson, A-M. (1973) *J. Biol. Chem. 248*, 7097–7105.

[69a] Colson, A.-M., Goffeau, A., Briquet, M., Weigel, P. and Mattoon, J.R. (1975) *Molec. gen. Genet. 135*, 309–326.

[70] Flury, U., Mahler, H.R. and Feldman, F. (1975) *J. Biol. Chem. 249*, 6130–6137.

[71] Mahler, H.R., Bastos, R.N., Feldman, F., Flury, U., Lin, C.C., Perlman, P.A. and Phan, S.H. (1975) in *Membrane Biogenesis: Mitochondria, Chloroplasts and Bacteria* (A.A. Tzagoloff, ed.) Plenum Press, New York, pp. 15–61.

[72] Tzagoloff, A. and Akai, A. (1972) *J. Biol. Chem. 247*, 6516–6523.

[73] Michel, R. and Neupert, W. (1973) *Eur. J. Biochem. 36*, 53–67.

[74] Werner, S. (1974) *Eur. J. Biochem. 43*, 39–48.

[75] Küntzel, H. and Blossey, H-C. (1974) *Eur. J. Biochem. 47*, 165–171.

[76] Wheeldon, L.W., Dianoux, A-C., Bof, M. and Vignais, P.V. (1974) *Eur. J. Biochem. 46*, 189–199.

[77] Rogers, P.J. and Stewart, P.R. (1974) *J. Bact. 119*, 653–660.

[78] Hanstein, W.G. (1975) *Fed. Proc. 34*, 595, abst. #2126.

[79] Slonimski, P.P., Perrodin, G. and Croft, J.H. (1968) *Biochem. Biophys. Res. Commun. 30*, 232–239.

[80] Perlman, P.S. and Mahler, H.R. (1971) *Nature New Biol. 231*, 12–16.

[81] Mahler, H.R. (1973) in *Molecular Cytogenetics* (B. Hamkalo, ed.) Plenum Press, New York, pp. 181–208.

[82] Bastos, R.N. and Mahler, H.R. (1974) *FEBS Lett. 6*, 27–34.

[83] Shannon, C., Enns, R., Whellis, L., Burchiel, K. and Criddle, R.S. (1973) *J. Biol. Chem. 248*, 3004–3011.

[84] Durphy, M., Manley, P.N. and Friedberg, E.C. (1974) *J. Cell Biol. 62*, 695–706.

[85] Moustacchi, E. (1971) *Mol. gen. Genet. 114*, 50–58.

[86] Dujon, B., Slonimski, P.P. and Weill, L. (1974) *Genetics 78*, 415–437.

[87] Perlman, P.S. and Birky, C.W., Jr. (1974) *Proc. Nat. Acad. Sci. USA 71*, 4612–4616.

[88] Radding, C.M. (1973) *Ann. Rev. Genetics 7*, 87-112.

[89] Grell, R.F. (ed.) (1975) *Molecular Mechanisms in Genetic Recombination*, Oak Ridge Symposium 1974. Plenum Press, New York.

[90] Meselson, M.S. and Radding, C.M. (1975) *Proc. Nat. Acad. Sci. USA 72*, 358-361.

CYTOPLASMIC AND NUCLEAR MUTATIONS AFFECTING MITOCHONDRIAL FUNCTIONS*

Alexander Tzagoloff

INTRODUCTION

It is now well established that the biosynthesis of some mitochondrial enzymes depends on a functional system of mitochondrial protein synthesis. Two complexes of the respiratory chain, coenzyme QH_2-cytochrome <u>c</u> reductase and cytochrome <u>c</u> oxidase, and the oligomycin-sensitive ATPase have each been shown to contain one or more mitochondrially translated polypeptides as well as other polypeptides that are synthesized on cytoplasmic ribosomes.

Although most of the protein constituents of mitochondria are made in the cytoplasm and are encoded in nuclear DNA, information about the mitochondrial translation products is still incomplete. For example, it is not known whether they are gene products of mitochondrial DNA.

As a part of a study on mitochondrial biogenesis, we have attempted to isolate mutants of yeast with specific lesions in each of the two respiratory complexes and in the ATPase complex. Such mutants could serve as useful tools for genetic and biochemical investigations of the mechanism of assembly of the enzymes and of the organization of genes on the mitochondrial DNA. The selection method used for finding both nuclear and cytoplasmic mutants as well as some of the genetic and biochemical characteristics of the strains that have been isolated will be reviewed in this communication.

Biosynthetic Origin of the Electron Transfer Chain Complexes and of the ATPase

The electron transfer chain of mitochondria consists of four electron transfer enzymes which jointly carry out the oxidation of NAD-linked substrates and of the tricarboxylic acid cycle intermediate, succinate (1). The energy of oxidation of NADH and succinate is utilized for the synthesis of ATP and this coupling reaction is thought to be catalyzed by the ATPase complex (2-4). These five complexes and their catalytic constituents are shown in the diagram of Fig. 1.

---

*The research was supported by Public Health Research Grants HL 13003 and GM 18868 from the National Institutes of Health, United States PUblic Health Service.

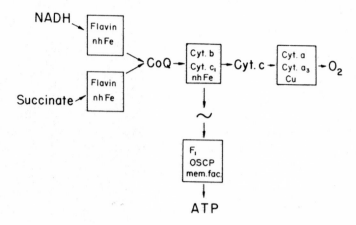

Fig. 1. Enzyme complexes of oxidative phosphorylation. Complex I: NADH-coenzyme Q reductase; complex II: succinate-coenzyme Q reductase; complex III: coenzyme QH$_2$-cytochrome c reductase; complex IV: cytochrome oxidase; complex V: oligomycin-sensitive ATPase. The squiggle symbol is the high energy intermediate or state generated at the complex I, III, and IV spans. The known components of each enzyme are shown in the enclosed squares. The ATPase complex consists of a catalytic ATPase unit, F$_1$, a membrane factor composed of four hydrophobic proteins and OSCP, a protein which functions in the binding of F$_1$ to the membrane factor.

Yeast and <u>Neurospora crassa</u> have been used to study the biosynthesis of the mitochondrial system of enzymes that are involved in oxidative phosphorylation. Up to now such studies have had two major goals. The first has been to obtain the individual enzyme complexes in sufficiently pure state to allow a characterization of their composition and structural organization. The second goal has been to define the genetic and biosynthetic origin of their constituent proteins.

The information obtained from such studies for the three enzyme complexes that are dually derived from mitochondrial and cytoribosomal protein synthesis is summarized in Table I. Based on experiments utilizing specific inhibitors of either mitochondrial or cytoribosomal protein synthesis, four subunits of the ATPase (5) and three subunits of cytochrome oxidase (6-8) have been found to be made in mitochondria. The mitochondrial products of the ATPase correspond to the four proteins of the membrane factor (5) and one of the mitochondrial products of cytochrome oxidase appears to carry the heme prosthetic group of the enzyme (9). The cytochrome <u>b</u> component of coenzyme QH$_2$-cytochrome <u>c</u> reductase has also been determined to be a mitochodrially synthesized protein.

Although the localization of the structural genes for these proteins has not been established, some of the cytoplasmic mutants reported here support the idea that mitochondrial translation products are also coded by mitochondrial DNA.

TABLE I

Biosynthetic Origin of Mitochondrial Proteins

	ENZYME		
	ATPase	Cyt. Oxidase	$CoQH_2-c$
Total subunits	10	7	?
Mit. products	4	3	1
Identity	Mem. fact.	Cyt. $\underline{a}$	Cyt. $\underline{b}$
Cytoplasmic products	6	4	?
Identity	F_1, OSCP	?	Cyt. $\underline{c}_1$

Selection Procedure

Saccharomyces cerevisiae is a facultative anaerobic yeast that grows vegetatively on energy derived from glycolysis even in the absence of functional mitochondria. This circumstance allows an initial enrichment of mutants in which mitochondrial functions are defective by selecting for cells that grow on sugars but not on non-fermentable substrates such as glycerol.

The preselection of strains which cannot grow on glycerol yields a class which consists predominantly of ρ^o and ρ^- mutants, i.e., mutants which either lack mitochondrial DNA or have gross deletions in the DNA that affect the ability of the cells to carry out mitochondrial protein synthesis. Also present in this class are a large number of pet$^-$ strains that have nuclear mutations which cause a loss of mitochondrial protein synthesis.

Since it can be predicted that any mutant which is specifically affected in only one of the enzyme complexes known to contain mitochondrial translation products, must still retain a functional system of mitochondrial protein synthesis, the second step in the screeing procedure is to select by an in vivo assay for those strains which have maintained this activity. This step eliminates the large class of cytoplasmic

and nuclear petites that are defective in mitochondrial translation and consequently have pleiotropic enzymatic deficiencies.

The final step in the screen is to examine the strains
that meet the above two criteria, namely lack of growth on
glycerol and retention of mitochondrial protein synthesis, for
the presence or absence of coenzyme QH_2-cytochrome c reductase,
cytochrome oxidase and ATPase.

Out of approximately 7,000 strains that have been screened
by this method, a substantial number of nuclear and cytoplasmic
mutants have been found which are defective in one of the three
enzymes only (10-12). In Table II, data are presented on the
frequency of the different types of mutants obtained after
mutagenesis with ethylmethane sulfonate, nitrosoguianidine
and manganese.

TABLE II

Frequency of specific nuclear and cytoplasmic mutants

Mutagen	No. strains examined	Cyt. Ox.		$CoQH_2$-c		ATPase	
		Nucl.	Cyto.	Nucl.	Cyto.	Nucl.	Cyto.
EMS	1,500	6	1	6	0	1	0
NTG	1,500	15	0	12	0	2	0
$MnCl_2$	4,000	10	58	0	5	0	1
Total	7,000	31	59	18	5	3	1

Criteria for Distinguishing Nuclear and Cytoplasmic Mutants

Several criteria have been used to determine whether the
mutated gene is located on mitochondrial or nuclear DNA.

1. Mutants are mated to a haploid strain of the opposite
mating type in which mitocondrial DNA has been deleted by
extensive treatment with ethidium bromide (13). This tester
is designated as ρ^o in Fig. 2. If the diploid formed from
the cross is capable of growth on glycerol, the mutant is
preliminarily classified as nuclear. This is based on the
reasoning that the ρ^o tester having no mitochondrial DNA but
a normal nuclear background should complement a haploid strain
with a defect in a nuclear gene. A mutation in mitochondrial
DNA on the other hand, will not be restored by the tester
since the gene is also missing in the tester.

2. Mutants are mated to a respiratory competent ρ^+ tester
and the diploids are forced to sporulate. The meiotic spore
progeny are checked for growth on glycerol and preservation

Mating Test

I. Nuclear mutation

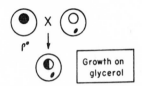

2. Cytoplasmic mutation

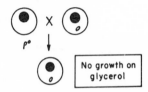

Fig. 2. Complementation test with the ρ^o tester. A mutation in nuclear DNA is represented by the large black circle (nucleus) and in mitochondrial DNA by the black oval-shaped mitochondrion. Wild-type nuclear and mitochondrial DNAs are represented by the clear organelles.

of the enzymatic deficiency. If the defect segregates in a 2:2 Medelian fashion, the mutation is assumed to be nuclear, whereas 0:4 and 0:4 segregations indicate a cytoplasmic mutation (Fig. 3).

3. Nuclear and cytoplasmic mutations can also be discriminated by testing for mitotic segregation of the enzymatic lesion. In this test, the mutants are mated with a ρ^+ tester and the diploids are grown vegetatively. If the mutation is cytoplasmic, the diploids should produce some cells that are genetically pure for the enzymatic phenotype of the original mutant due to segregation of the mitochondrial DNA. This cannot occur if the mutation is in nuclear DNA.

Nuclear Mutants

Table III lists some properties of nuclear mutants in coenzyme QH_2-cytochrome c reductase, cytochrome oxidase and ATPase. Strains with specific lesions in each of the three enzymes fall into different subclasses based on both genetic

245

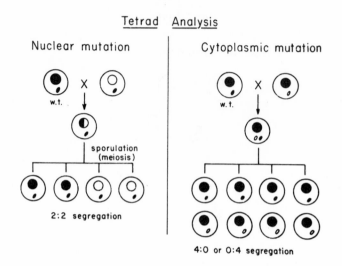

Fig. 3. *Meiotic segregation test. For an explanation of the symbols used, see the legend to Fig. 2.*

TABLE III

Enzymatic properties of nuclear mutants

Strain	Enzymatic Deficiency	Specific Activities				Cytochr.		
		NADH-$\underline{c}$	CoQH$_2$-$\underline{c}$	Cyt. Ox.	ATPase	$\underline{a}$	$\underline{b}$	$\underline{c}$
D273-10B	Wild type	0.35	0.38	0.55	5.4	+	+	+
N5-7	CoQH$_2$-$\underline{c}$	N.D.	<0.01	0.40	2.8	+	−	+
N5-96	CoQH$_2$-$\underline{c}$	N.D.	<0.01	0.30	3.1	+	+	+
N6-78	Cyt. Ox.	0.24	N.D.	<0.01	3.8	−	+	+
N7-13	Cyt. Ox.	0.33	N.D.	<0.01	2.9	+	+	+
E2-126	ATPase	0.09	N.D.	0.09	0	+	+	+

The specific activities refer to μmoles of substrate oxidized or hydrolyzed per min per mg of mitochondrial protein. The ATPase activity refers to units that are sensitive to the inhibitor oligomycin.

complementation and biochemical phenotypes. Among the cytochrome oxidase and coenzyme QH_2-cytochrome c reductase mutants,
some are completely lacking in cytochromes a, a$_3$ and cytochrome
b, respectively. These mutants appear to be of the same general phenotype as some of the pet⁻ mutants reported by Sherman
and Slonimski (14), Subik et al. (15) and Ebner and Schatz (16).
Other strains, however, contain variable levels of the cytochromes even though the corresponding enzymatic activities are
absent. Some ten genetic complementation groups have been
found in both the cytochrome oxidase and coenzyme QH_2-
cytochrome c reductase deficient strains.

The ATPase deficient mutants are also genetically and biochemically different. Two strains have no measurable ATPase
activity in mitochondria or in crude cell homogenates. Preliminary studies indicate, however, that they contain proteins
that cross-react with antiserum to the F_1 ATPase. Another
mutant that was initially characterized by a deficiency in
ATPase only, acquired a pleiotropic phenotype upon subculturing.
This strain synthesizes an enzymatically active F_1 that accum-
ulates in the post-ribosomal supernatant fraction. Analysis
of the mitochondrial products made by this strain indicates
that it lacks a proteolipid that has previously been identified as a subunit of the oligomycin-sensitive ATPase (17). The
nuclear mutation in this strain, therefore, appears to exert
a secondary effect that is expressed in the inhibition of
synthesis by mitochondria of a subunit of ATPase. This situation is analogous to some of the nuclear mutants of cytochrome
oxidase reported by Ebner and Schatz (16) in which a mitochondrial translation product of the enzyme is also under the
control of a nuclear gene.

Cytoplasmic Mutants

A large number of cytoplasmic mutants have been found with
specific deficiencies in either cytochrome oxidase, coenzyme
QH_2-cytochrome c reductase or ATPase (12). In addition,
doubly deficient mutants have also been isolated which lack
combinations of coenzyme QH_2-cytochrome c reductase and cytochrome oxidase or cytochrome oxidase and ATPase. These will
be referred to as mit⁻ mutants to distinguish them from the
pet⁻ mutants. The spectral properties and enzymatic phenotypes of the different classes of mit⁻ mutants are summarized
in Fig. 4.

The enzymatic properties of representative mit⁻ mutants
are presented in Table IV. With the exception of one strain,
all of the mutants deficient in cytochrome oxidase activity
lacked spectrally detectable cytochromes a and a$_3$. The
coenzyme QH_2-cytochrome c reductase mutants lacked cytochrome

Enzymatic Phenotypes of Mutants

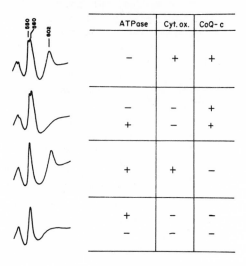

	ATPase	Cyt. ox.	CoQ- c
	−	+	+
	−	−	+
	+	−	+
	+	+	−
	+	−	−
	−	−	−

Fig. 4. Spectral properties and enzymatic deficiencies of different mit⁻ strains. Pluses and minuses refer to the presence or absence of the enzymatic activity.

TABLE IV

Enzymatic properties of cytoplasmic mutants

Strain	Enzymatic Deficiency	Specific Activities				Cytochr.		
		NADH-$\underline{c}$	CoQH$_2$-$\underline{c}$	Cyt. Ox.	ATPase	$\underline{a}$	$\underline{b}$	$\underline{c}$
D273-10B	Wild type	0.35	0.38	0.55	4.45	+	+	+
M5-16	Cyt. Ox.	0.57	N.D.	<0.01	3.93	−	+	+
M15-208	Cyt. Ox.	0.75	N.D.	<0.01	1.20	+	+	+
M6-200	CoQH$_2$-$\underline{c}$	0.02	<0.01	0.53	3.43	+	−	+
M6-239	ATPase	0.10	N.D.	0.10	0	+	+	+
M9-228	Cyt. Ox. CoQH$_2$-$\underline{c}$	0.01	N.D.	<0.01	1.56	−	−	+
M15-153	Cyt. Ox. CoQH$_2$-$\underline{c}$ ATPase	0.01	N.D.	<0.01	0	−	−	+

For explanation, see legend to Table III.

b. Strains that were doubly deficient in both enzymes lacked
a and b type cytochromes. A number of strains were also isolated that were capable of mitochondrial protein synthesis, but
had the same pleiotropic phenotype as ρ^- mutants (M15-153 in
Table IV). The single ATPase mutant that was found was genetically unstable due to a high spontaneous rate of conversion
to ρ^-. This mutant has low levels of respiratory activities
but no oligomycin-sensitive ATPase. In analogy to ρ^- mutants
(18), the cytoplasmic ATPase mutant synthesizes an F_1 type of
ATPase.

Genetic Properties of Mit⁻ Mutants

The cytoplasmic class of specific, doubly deficient and
pleiotropic mutants were tested by the three criteria described in the last section. In all cases, the mutants failed
to be complemented by the ρ^o tester and exhibited non-Mendelian
segregation of the meiotic spore progeny. Furthermore, a
number of strains were tested and found to produce mitotic
segregants which retained the original enzymatic deficiency
(12).

The cytoplasmic mit⁻ mutations showed variable degrees of
stability. Some of the strains did not produce any revertants
while others reverted at frequencies ranging from one revertant per 10^5-10^9 cells. In most strains, the percentage of
ρ^- accumulated in stationary phase cultures ranged from 5-10%.
There were some strains, however, which degenerated to ρ^- at
a high rate.

CONCLUSIONS

The screening procedure based on the selection of mutant
cells that show lack of growth on non-fermentable substrates
but retain the capacity for mitochondrial protein synthesis
has facilitated the isolation of a large number of nuclear
and cytoplasmic strains with specific lesions in cytochrome
oxidase, coenzyme QH_2-cytochrome c reductase and ATPase.

These strains are currently being studied from two stand-
points. The first is to define more precisely, i.e., at the
level of the protein constituents, the target of the mutation.
This information can then be used to study the process by
which each of the three enzymes is assembled much in the same
way that the mutant approach has been used to study the
assembly of complex viral structures. The second aim is to
use the mit⁻ mutants to localize the mutants on the mitochondrial DNA and thereby to enlarge our knowledge of this genome.
Such studies are underway and have already permitted the

249

identification of three different loci in mitochondrial DNA that affect the biosynthesis of cytochrome oxidase and a fourth locus that is concerned with the synthesis of cytochrome b (Slonimski and Tzagoloff, unpublished studies).

REFERENCES

1. Hatefi, Y. (1963) *in* The Enzymes (Boyer, P.D., ed.) Vol. 7, p. 495, Academic Press, New York.
2. Racker, E. (1965) Mechanisms in Bioenergetics, Academic Press, N.Y.
3. Pedersen, P.L. (1975) J. Bioenergetics 6, 243.
4. Mitchell, P. (1974) FEBS Letters 43, 189.
5. Tzagoloff, A. and Meagher, P. (1972) J. Biol. Chem. 247, 594.
6. Sebald, W., Weiss, H. and Jackl, G. (1972) Eur. J. Biochem. 30, 413.
7. Mason, T.L. and Schatz, G. (1973) J. Biol. Chem. 248, 1355.
8. Rubin, M.S. and Tzagoloff, A. (1973) J. Biol. Chem. 248, 4275.
9. Tzagoloff, A., Akai, A. and Rubin, M.S. (1974) *in* The Biogenesis of Mitochondria (Saccone, C. and Kroon, A.D. eds.) p. 405, Academic Press, New York.
10. Tzagoloff, A., Akai, A. and Needleman, R.B. (1975) J. Bacteriol. 122, 826.
11. Tzagoloff, A., Akai, A. and Needleman, R.B. (1975) J. Biol. Chem., in press.
12. Tzagoloff, A., Akai, A., Needleman, R.B. and Zulch, G. (1975) J. Biol. Chem., in press.
13. Goldring, E.S., Grossman, L.J., Krupnick, D., Cryer, D.R. and Marmur, J. (1970) J. Mol. Biol. 52, 323.
14. Sherman, F. and Slonimski, P.P. (1964) Biochim. Biophys. Acta 90, 1.
15. Subík, J., Kuzela, S., Kolarov, J., Kovác, L. and Lachowicz, T.M. (1970) Biochim. Biophys. Acta 205, 513.
16. Ebner, E., Mennucci, L. and Schatz, G. (1973) J. Biol. Chem. 248, 5360.
17. Sierra, M.F. and Tzagoloff, A. (1973) Proc. Nat. Acad. Sci. US. 70, 3155.
18. Schatz, G. (1968) J. Biol. Chem. 243, 2192.

THE CONTROL OF MITOCHONDRIAL MEMBRANE FORMATION

Diana S. Beattie

INTRODUCTION

The process of mitochondrial biogenesis and assembly requires that proteins synthesized on cytoplasmic ribosomes be
integrated into the mitochondrial membrane in conjunction with
the 8-10 hydrophobic proteins synthesized on the unique mitochondrial ribosomes. Furthermore, synthesis of these proteins
is presumably under the control of two distinct and different
genetic systems, one mitochondrial and the other cytoplasmic,
which are both involved in the formation of the mitochondrial
membrane. Hence, a mechanism must exist in the cell to control both the expression of these two genetic systems as well
as the synthesis of proteins at two intracellular sites such
that the different enzyme complexes of the inner mitochondrial
membrane are assembled in a coordinated manner. Several recent
studies have indicated that the formation of three enzyme complexes of the inner membrane cytochrome c oxidase (1) oligomycin-sensitive ATPase (2) and cytochrome b of complex III, (3,4)
require the coordinated synthesis of proteins both in the
mitochondria and in the cytoplasm.

Several controls have been proposed for the synthesis of
mitochondrial proteins on both cytoplasmic and mitochondrial
ribosomes. For example, after addition of inhibitors of mitochondrial transcription and translation to cultures of Neurospora, the activity of different mitochondrial enzymes which
are synthesized in the cytoplasm were observed to increase
significantly. To explain these increases, Barath and Küntzel
(5) have suggested that mitochondrial DNA codes for a repressor
protein(s) which is (are) synthesized in the mitochondria.
These repressor proteins would then leave the mitochondria
and be transported to the cell nucleus where they would act
as repressors on nuclear genes. Blocking either mitochondrial
transcription with ethidium bromide or translation with chloramphenicol would inhibit formation of these repressor proteins
so that synthesis of mitochondrial proteins in the cytoplasm
would no longer be regulated resulting in a continued increase
in the amount of certain proteins.

Likewise, evidence has accumulated to suggest that the
synthesis of mitochondrial membrane proteins on mitochondrial
ribosomes measured both in vitro and in vivo may be stimulated
by proteins synthesized in the cytoplasm (6,7), while the
absence of or a decreased amount of these proteins results in
a decreased rate of mitochondrial protein synthesis (7). Although the mechanism by which cytoplasmic proteins may affect

mitochondrial protein synthesis remains unclear, certain evidence suggests that the stimulation of protein synthesis is specific, i.e., labeling of certain specific mitochondrial proteins is stimulated when proteins synthesized in the cytoplasm have accumulated (4,6). In this context, the recent report (8) that an unidentified protein synthesized in the cytoplasm appears to control the synthesis of a mitochondrially-made subunit of cytochrome oxidase is of considerable interest.

The studies reported in this chapter are an attempt to elucidate more fully the controls of mitochondrial membrane formation. Possible mechanisms for regulating mitochondrial protein synthesis by cytoplasmically-made proteins have been explored by studying isolated polysomes obtained from yeast mitochondria. In addition, the control of synthesis of mitochondrial proteins in the cytoplasm has been studied by measuring the activity of several enzymes in purified mitochondria obtained from cultures of the cellular slime mold, Dictyostelium discoideum, grown in sufficient ethidium bromide to block mitochondrial transcription.

METHODS

The yeast Saccharomyces cerevisae, was grown aerobically on 5% glucose as described previously (9). Mitochondria were prepared either mechanically (9) or by digestion with snail gut enzyme (10). Polysomes were prepared only from mitochondria obtained from spheroplasts, by lysis of the mitochondria with Triton X-100 and centrifugation of the Triton extract in a 20-40% linear sucrose gradient. RNA in the gradient was determined by absorption measurements at 260 nm. Labeled polysomes were obtained by incubating the spheroplasts with cycloheximide for 15 min followed by a 2-5 min pulse with [^{3}H]leucine. Radioactivity in each tube of the graident was determined either by counting an aliquot directly or after precipitation with trichloroacetic acid.

Axenic cultures of Dictyostelium discoideum strain AX-3 were grown as described previously (11). Mitochondria were isolated and purified by sucrose density gradient fractionation (12). The most purified mitochondrial fraction from the gradient contained essentially no contaminating lysosomes and minimal amounts of contaminating peroxisomes as determined by the marker enzymes N-acetylglucosaminidase and catalase. A mitochondrial fraction with the same amount of lysosomal and peroxisomal contamination was also isolated from cells which had been treated with ethidium bromide for 5 days.

RESULTS AND DISCUSSION

Amino acid incorporation in vitro by isolated yeast mitochondria can be supported by either an external ATP-generating
system or by ATP synthesized by the respiratory chain. However, in the assay system developed in our laboratory for
yeast mitochondria (10), the incorporation rates observed in
the presence of succinate and either ADP or ATP were almost
double those obtained with the ATP-generating system and were
inhibited by either antimycin A or oligomycin. One explanation
for this result may be that the transport of ATP across the
yeast mitochondrial membrane is not rapid enough to support
maximum rates of amino acid incorporation. In all experiments,
amino acid incorporation was measured with both the ATP-generating system and a respiratory substrate.

Tzagoloff (6) first reported that mitochondrial protein
synthesis in vivo was increased in yeast cells allowed to
accumulate products of cytoplasmic protein synthesis by growth
of the yeast for various times in chloramphenicol. Perhaps,
this increased labeling of mitochondrial membrane proteins in
vivo may reflect an actual increase in the rate of mitochondrial protein synthesis. To test this possibility, yeast cells
were grown for 13 hr in 5% glucose at which time the culture
was divided into three equal parts. To one was added chloramphenicol (4 mg/ml), to another was added cycloheximide
(20 μg/ml), while the third was used as the control. After
another 3 hr of growth, all three cultures were harvested,
washed and transferred to an equal volume of fresh medium containing 0.8% glucose. A portion of the cells from each culture
were used at this time to prepare mitochondria while the remaining cells were allowed to grow for another one to two hr.
The presence of either inhibitor of protein synthesis in the
growth medium caused significant decreases in the rate of
amino acid incorporation by isolated mitochondria in vitro
despite the extensive washing of both the cells and the mitochondrial pellets prior to the incorporation studies (Fig. 1).
After 1 hr of growth in fresh medium without inhibitor, the
incorporation measured in both the ATP-regenerating and the
ATP-succinate systems was nearly 50% higher in mitochondria
isolated from cells which had been preincubated in chloramphenicol. A higher rate of incorporation was also observed
in mitochondria from cells after 2 hr of growth in fresh
medium; however, these higher rates were not observed when
the cells had been allowed to grow for 5 hr after the preincubation in chloramphenicol. These results suggest that
the products of cytoplasmic protein synthesis which had accumulated during growth in chloramphenicol may actually stimulate

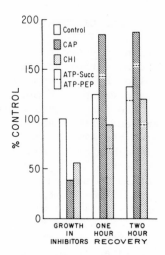

Fig. 1. *Effect of growth in inhibitors on amino acid incorporation. Partially derepressed yeast cells were grown for 3 hr in chloramphenicol or cycloheximide. The cells were washed and transferred to an equal volume of fresh medium containing 0.1% glucose and allowed to grow 1 or 2 hr. Amino acid incorporation was assayed as previously described (7).*

mitochondrial protein synthesis. In contrast, after 1 hr in fresh medium, amino acid incorporation in vitro was derepressed nearly 30% in mitochondria isolated from cells which had been preincubated in cycloheximide. Products of cytoplasmic protein synthesis would be severely depleted by growth of cells in cycloheximide for 3 hr.

The observation that increased rates of mitochondrial protein synthesis in vitro were maintained in isolated mitochondria from cells grown in chloramphenicol suggests several important features about the mechanism of the stimulatory process. First, the cytoplasmically-made proteins which accumulate under these conditions must still be present in the isolated mitochondria. Perhaps, these proteins are part of an enzyme complex and stimulate protein synthesis by augmenting the rate of assembly of the mitochondrially-made proteins into the membrane-bound complex. In support of this suggestion, Poyton and Groot (13) have recently reported that isolated yeast mitochondria can synthesize three subunits of cytochrome oxidase and assemble them into a holoenzyme in the membrane with the four subunits which are made in the cytoplasm. Alternately, the proteins which are present within the isolated mitochondria may act to stimulate either the initiation of protein synthesis, the rate of chain elongation or the release of newly-synthesized proteins from the mitochondrial polysomes.

Another possibility to explain the observed stimulation in vitro in isolated mitochondria is that proteins made in the cytoplasm act as activators for the transcription from mito-chondrial DNA of certain specific messenger RNA's. As a re-sult, more of the ribosomes of the isolated mitochondria would be present in the form of polysomes and greater rates of pro-tein synthesis due to chain elongation would be observed. In support of this suggestion, Ono et al. (8) have recently re-ported that the loss of a nuclear-coded cytoplasmically-made protein causes the loss of a mitochondrial-made subunit of cytochrome oxidase. They have concluded that the nuclear agent (pet 494) which results in a lack of labeling of the mitocondrial-made subunit may act as a "regulatory" mutant.

To gain further insight into which of the above suggestions may explain the control of mitochndrial protein synthesis in yeast, polysomes were isolated from yeast mitochondria as des-cribed under Methods. Absorbancy measurements at 260 nm indi-cated the presence of the mitochondrial monosome at 74S and polysomes consisting of 2-8 monosome subunits. When the Triton X-100 solubilized fraction was treated with EDTA, all the materials absorbing at 260 nm disappeared from the polysome and monosome regions of the gradient and sedimented at approx-imately 30S and 50S, the value for the two subunits of the mitochondrial ribosome. When yeast spheroplasts were incubated for 15 min with cycloheximide, a specific inhibitor of cyto-plasmic protein synthesis, followed by a 2-5 min pulse with [3H]leucine, prior to the preparation of polysomes from iso-lated mitochondria, TCA-precipitable radioactivity was observed mainly in the polysome region of the gradient (Fig. 2). Ribo-nuclease digestion caused degradation of the polysomes to monosomes with release of radioactivity to the top of the gra-dient, while addition of puromycin caused a release of radio-activity from the polysome region to the top of the gradient suggesting that the radioactivity in the polysome region re-presents nascent polypeptide chains (14). Radioactive labeling in the polysome region was completely inhibited by erythromycin indicating that this radioactivity is indeed a measure of mitochondrial protein synthesis.

Yeast cells were grown for 3 hr in chloramphenicol, the specific inhibitor of mitochondrial protein synthesis, grown for 1 hr in fresh medium and then labeled with either [3H]leu-cine (Fig. 2) or [14C]formate (Fig. 3) in the presence of cycloheximide. The polysome to monosome ratio determined by

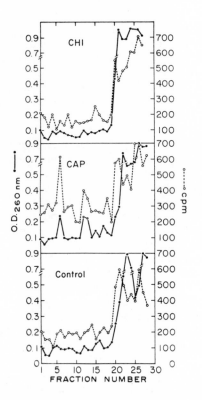

Fig. 2. *Mitochondrial polysome profile from spheroplasts labeled*
in vivo with [³H]leucine. Partially derepressed yeast cells
were grown in inhibitors as described in the legend to Fig. 1.
Spheroplasts were prepared, incubated for 15 min with cyclo-
heximide and then pulse-labeled for 5 min with [³H]leucine.
Mitochondria were isolated by differential centrifugation and
lysed with Triton X-100 as described in Methods. (●——●) O.D.
at 260 nm; (0··0) counts/min due to TCA-precipitable material.

the absorbance at 260 nm increased nearly 2-fold in mitochondria isolated from the cells grown in chloramphenicol compared
to controls. Under these conditions the radioactivity in the
polysome region due to either [³H]leucine or [¹⁴C]formate increased 2.0-fold and 1.5-fold respectively. Conversely, yeast
cells were grown in cycloheximide for 3 hr, in fresh medium
for 1 hr and then labeled with [³H]leucine. Under these conditions the polysome to monosome ratio decreased 60%, while
the radioactivity due to [³H]leucine in the polysome region
showed a corresponding decrease (Fig. 2).

Polysomes were also isolated from mitochondria which had
been incubated with [³H]leucine *in vitro* under the previously
determined optimal conditions. To obtain good polysomes it

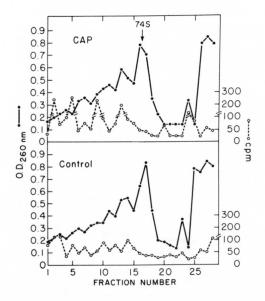

Fig. 3. Mitochondrial polysome profile from spheroplasts labeled in vivo with [14C]formate. Yeast were grown and polysomes labeled as described in the legend to Fig. 2, except that [14C]formate was used. (●—●) O.D. at 260 nm; (O··O) counts/min due to [14C]formate.

was necessary to prepare mitochondria from spheroplasts. After the incubation with the radioactive amino acid, the mitochondria were reisolated, washed twice and then lysed with Triton X-100 to isolate polysomes exactly as described above. Radioactivity of the [3H]leucine was observed predominantly in the polysome region of the gradient (Fig. 4). Mitochondria were also isolated from yeast cells grown previously for 3 hr in either chloramphenicol or cycloheximide as described above, incubated with [3H]leucine and polysomes isolated. The amount of radioactivity in the polysome region was increased nearly 3-fold in mitochondria obtained from cells grown in chloramphenicol and decreased over 75% in mitochondria from cells grown in cycloheximide. It should be noted that these changes are more dramatic than those observed when the cells were labeled in vivo.

The results of both these experiments suggest some possible mechanisms by which proteins made in the cyotplasm may act to

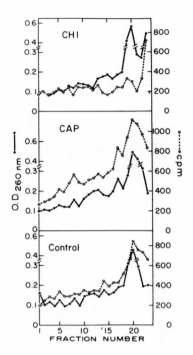

Fig. 4. *Polysome profile from mitochondria incubated in vitro with*
[3H]leucine. Yeast cells were grown in inhibitors as described
in the legend to Fig. 1. Spheroplasts were prepared and mito-
chondria isolated by differential centrifugation. The mito-
chondria were incubated for 15 min with [3H]leucine reisolated,
washed two times and lysed with Triton X-100. Polysomes were
prepared as described in Methods. (●——●) O.D. at 260 nm; (O··O)
counts/min due to TCA-precipitable [3H]leucine.

control mitochondrial protein synthesis. The [3H]leucine
counts in the polysome region of the gradient reflect nascent
polypeptide chains while the [14C]formate counts are a measure
of chain initiation by formyl-methionine. Hence, the increase
in both of these values in mitochondrial polysomes isolated
from cells grown in chloramphenicol suggest that more polypeptide chains are being synthesized. Furthermore, an increased
polysome to monosome ratio is also a measure of greater activity in protein synthesis. One possible explanation for this
increase is that proteins made in the cytoplasm may stimulate
initiation of protein synthesis on mitochondrial ribosomes.
An alternative suggestion is that these proteins cause an
increased synthesis of mitochondrial messenger RNA's resulting
in more polysomes and hence, greater rates of protein synthesis.
The increased labeling of mitochondrial polysomes when isolated

mitochondria are incubated in vitro with [^{3}H]leucine makes the latter hypothesis attractive.

These studies of mitochondrial protein synthesis in yeast have been directed towards an understanding of the factors which regulate mitochondrial protein synthesis. Since the vast majority of mitochondrial proteins are made on cytoplasmic ribosomes, the regulation of their synthesis has also been investigated. For these studies, we have used the cellular slime mold Dictyostelium discoideum grown in the presence of sufficient ethidium bromide to block completely the replication of mitochondrial DNA as well as transcription. In our initial studies (11), we made the unexpected observation that the specific activity of the two mitochondrial membrane-bound flavoproteins, succinate and NADH dehydrogenases in whole cells increased nearly 2-fold in cultures treated with ethidium bromide for 5 days (Fig. 5). Concomitantly, a greater than 50% decrease in activity of cytochrome oxidase and succinate-cytochrome c reductase in whole cells was also observed.

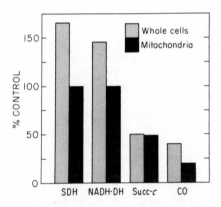

Fig. 5. Effect of growth in ethidium bromide on respiratory enzyme of the slime mold. Enzymes were assayed as described previously (11) in whole cells lysed with Triton X-100 or in mitochondria from the most purified fraction of the sucrose gradient.

It should be noted that the formation of NADH and succinate dehydrogenase has been shown to require proteins synthesized only on cycloheximide-sensitive cytoplasmic ribosomes (9 and not require protein synthesis in the mitochondria. The increased specific activity of these two enzymes in whole cells treated with sufficient ethidium bromide to block further mitochondrial DNA synthesis as well as mitochondrial transcription can be explained by two hypotheses. First, it can be assumed that the formation of new mitochondrial membranes continues despite the cessation of cell growth such that each cell now contains more mitochondria or mitochondrial protein. An increased number of mitochndira per cell (or amount of mitochondrial membrane protein per cell) containing a normal complement of membrane-bound enzymes, which do not require either mitochondrial transcription or translation for their synthesis, would be manifest as a greater specific activity of these mitochondrial membrane-bound enzymes per cell. Alternately, it can be assumed that "abnormal" mitochondrial membranes containing greater amounts of these membrane-bound enzymes are formed when mitochondrial transcription is blocked by ethidium bromide.

To distinguish between these two possibilities, it was necessary to obtain highly purified mitochondria from both control and ethidium bromide-treated cells and to compare the activities of these various enzymes. A mitochondrial fraction containing essentially no contaminating lysosomes and minimal amounts of contaminating peroxisomes was obtained by sucrose density gradient fractionation. Furthermore, the mitochondrial fraction obtained from ethidium bromide-treated cells using the same methods contained exactly the same amount of contamination by lysosomes and peroxisomes. Hence, we were able to compare with confidence the activities of several enzymes and enzyme complexes of the mitochondrial inner membrane in the purified mitochondrial fraction obtained from both types of cells.

After treatment with ethidium bromide, two enzyme complexes of the respiratory chain, cytochrome oxidase and succinate-cytochrome $\underline{c}$ reductase, were reduced by the same amount in the purified mitochondria as in the whole cell (Fig. 5). Spectral analysis confirmed that mitochondria obtained from slime mold cultures treated with ethidium bromide for 5 days contained 82% less cytochrome $\underline{a}$-$\underline{a}_3$ and 35% less cytochrome $\underline{b}$ than mitochondria from control cultures. No significant difference in the activity of either NADH or succinate dehydrogenases were observed in the purified mitochondrial fraction despite the greater activity of both enzymes in whole cells after treatment with ethidium bromide for 5 days. This result suggests that when mitochondrial transcription is completely blocked by ethidium bromide, the synthesis of new mitochondrial membranes continues despite the lack of cell growth. Consequently, the amount of mitochondrial protein per cell has increased by nearly 50%. Currently, electron micrographic studies are in progress to determine whether this increase in mitochondrial protein results from an increased number of mitochondria per cell or an increased size of individual mitochondria.

Similar increases in the activity of mitochondrial enzymes have been reported in Neurospora treated with ethidium bromide. For example, the rifampicin-sensitive mitochondrial RNA polymerase is dramatically increased when mitochondrial transcription or translation has been blocked. In this study, it was not established whether this increase results from more enzyme in the cell or more mitochondrial protein. RNA polymerase, however, is not a membrane-bound enzyme as are the flavoprotein dehydrogenases. Hence, its regulation may occur by a different mechanism.

CONCLUSIONS

The experiments discussed in this chapter provide only a tentative beginning to a greater understanding of the mechanisms which regulate the biogenesis and assembly of the mitochondrial membrane. The importance of the coordinated synthesis of mitochondrial proteins in both the mitochondria and the cytoplasm has been indicated in many recent studies on the biogenesis of the different enzyme complexes of the inner membrane. Hence, the regulation of mitochondrial protein synthesis may be critical for the orderly assembly of a functional mitochondrial inner membrane.

Earlier, we had reported that the accumulation of protein synthesized in the cytoplasm stimulated the synthesis of specific proteins within the mitochondrion measured both in vitro and in vivo. In the present study, we have attempted to examine the mechanism of mitochondrial protein synthesis at the polysomal level to obtain a better insight into the observed stimulation. The data obtained indicate that the polysome to monosome ratio, a measure of protein synthesis, is nearly 2-fold higher in mitochondria obtained from yeast cells in which cytoplasmically-made proteins had accumulated. In addition, more nascent chains are present on these polysomes as indicated by pulse-labeling with either [^{3}H]leucine or [^{14}C]formate. One explanation for these results might be that the protein(s) synthesized in the cytoplasm function as activators for the transcription of specific messenger RNA's from mitochondrial DNA. Alternately, the initiation of protein synthesis on the mitochondrial polysomes may be stimulated by these cytoplasmically-made proteins. Presently, our laboratory is attempting to isolate messenger RNA from mitochondrial polysomes to learn whether different messenger RNA's are present when a stimulation of mitochondrial protein synthesis is observed.

The experiments with the cellular Dictyostelium discoideum confirm previous studies in which increases in activity of different mitochondrial enzymes were reported in Neurospora

treated with either chloramphenicol or ethidium bromide. Two
major theories have been advanced to explain this data. As
mentioned previously, mitochondria may synthesize repressor
proteins which leave the mitochondria and act as repressors
for nuclear genes. An alternative explanation might be that
the drop in total ATP levels which must occur when the activity
of the respiratory chain has been reduced in cells treated with
ethidium bromide may act to trigger a control mechanism which
stimulates the synthesis of certain mitochondrial proteins in
an attempt to remedy the intracellular lack of ATP. Such
possible metabolic controls for mitochondrial biogenesis are
currently under investigation in our laboratory.

While the approach outlined in this chapter towards a
greater understanding of mitochondrial biogenesis is a biochemical one, genetic studies have led to essentially the
same conclusions. Schatz's laboratory has isolated a number
of respiratory-deficient mutants of Saccharomyces cerevisiae
(15). Each mutation obtained was in a single nuclear gene;
however, the proteins synthesized by the mutant mitochondria
in vivo lacked some proteins made by the wild type mitochondria. The results obtained suggested that nuclear mutations
could affect either the synthesis of specific proteins on
mitochondrial ribosomes or the integraton of these proteins
into the mitochondrial membrane. Subsequently, this group
reported that a protein made in the cytoplasm and coded for
in the nucleus was necessary for the synthesis of one mitochondrially-made subunit of cytochrome oxidase. Perhaps this
protein acts to regulate the transcription of a specific
messenger RNA in the mitochondrion or to cause the initiation
of translation of a given protein. Whatever the mechanism of
control, all of these studies emphasize that cytoplasmically-
made proteins play a key role in the regulation of protein
synthesis in mitochondrial ribosomes.

ACKNOWLEDGMENTS

The author wishes to thank N.G. Ibrahim and R.N. Stuchell
who performed the experiments reported in this chapter.

REFERENCES

1. Mason, T.L. and Schatz, G. (1973) J. Biol. Chem. 248, 1355.
2. Tzagoloff, A. and Meagher, P. (1972) J. Biol. Chem. 247, 594.
3. Weiss, H. (1972) Eur. J. Biochem. 30, 469.
4. Lin, L-F. H., Kim, I-C. and Beattie, D.S. (1974) Arch. Biochem. Biophys. 160, 458.

5. Barath, Z. and Küntzel, H. (1972) Proc. Nat. Acad. Sci. (US) 69, 1371.
6. Tzagoloff, A. (1971) J. Biol. Chem. 246, 3050.
7. Ibrahim, N.G., Stuchell, R.N. and Beattie, D.S. (1973) Eur. J. Biochem. 36, 519.
8. Ono, B-I, Fink, G. and Schatz, G. (1975) J. Biol. Chem. 250, 775.
9. Kim, I-C. and Beattie, D.S. (1973) Eur. J. Biochem. 36, 509.
10. Ibrahim, N.G., Burke, J.P. and Beattie, D.S. (1974) J. Biol. Chem. 249, 6806.
11. Stuchell, R.N., Weinstein, B.I. and Beattie, D.S. (1973) Fed. Eur. Biochem. Soc. Letts. 37, 23.
12. Stuchell, R.N., Weinstein, B.I. and Beattie, D.S. (1975) J. Biol. Chem. 250, 570.
13. Poyton, R.O. and Groot, G.S.P. (1975) Proc. Nat. Acad. Sci. (US) 72, 172.
14. Ibrahim, N.G. and Beattie, D.S. Submitted for publication.
15. Ebner, E., Mennucci, L. and Schatz, G. (1973) J. Biol. Chem. 248, 5360.

BIOCHEMICAL GENETIC STUDIES OF OXIDATIVE PHOSPHORYLATION

David E. Griffiths

INTRODUCTION

The biochemical genetics of mitochondrial oxidative phosphorylation have been studied intensively in the past seven years and the subject has been extensively discussed in a recent review by Kovac (1). Parallel studies of the biochemical genetics of oxidative phosphorylation in prokaryotes, due mainly to the work of Gibson (2) and Gutnick (3) have provided new insights into the problem, particularly the relationship to active transport mechanisms. Such studies are giving information on the structure, function and organization of the mitochondrial inner membrane, the bacterial plasma membrane and the ATP-synthetase complex. Information on the sites of synthesis and assembly of mitochondrial inner membrane components, including components involved in membrane transport processes, is now available from the biochemical genetic studies and is of particular value in elucidating the nucleo-cytoplasmic interactions involved in mitochondrial biogenesis.

The major components of the oxidative phosphorylation system can be isolated as five functional complexes of the electron transport chain and ATP synthesis:

COMPLEX I	NADH - ubiquinone reductase (4)	
COMPLEX II	Succinate - ubiquinone reductase (5,6)	
COMPLEX III	Ubiquinol - cytochrome $\underline{c}$ reductase (7,8)	
COMPLEX IV	Cytochrome c oxidase (9,10)	
COMPLEX V	OS ATPase complex or ATP synthetase (11,12)	

In addition, carrier systems for the transport of phosphate (phosphate carrier) and adenine nucleotides (ADP translocase) are closely linked to these complexes in the mitochondrial inner membrane and probably have a structural as well as functional relationship to the above complexes.

The biogenesis of some of the components of these complexes has been discussed in recent reviews by Tzagoloff (13) and Schatz (14). The following major conclusions emerge:

1) The biogenesis of mitochondrial protein complexes involved in energy conservation involves the cooperative action of two genetic systems as these complexes contain subunits synthesized on mitoribosomes (assumed to be mt-DNA coded). Thus nuclear and mitochondrial mutations result in modification of components of the oxidative phosphorylation system.

2) Three subunits of complex IV (cytochrome oxidase) and at least one subunit of complex III (ubiquinone-cytochrome $\underline{c}$

reductase) are mitoribosome synthesized (mt-DNA coded).

3) Four mitochondrial genes or genetic loci have been identified with mitoribosome synthesized components of complexes III and IV (Tzagoloff, personal communication).

4) There is no evidence to indicate the presence of mitochondrially-coded components in complex I and complex II.

5) Four subunits of the ATP synthetase complex are mitoribosome synthesized, and hence are assumed to be coded for by mt-DNA. These include the membrane proteins which are the interaction sites for oligomycin and DCCD.

6) There is tentative evidence that the adenine nucleotide carrier (ADP translocase complex) contains a mitoribosome synthesized component.

7) Study of mitochondrial mutants may lead to the isolation and characterization of previously unknown components of the mitochondrial energy conservation system.

8) Study of nuclear and mitochondrial mutants which lead to a modification of components of the inner membrane may give information as to the mechanism of oxidative phosphorylation and/or the structural organization of the component complexes.

Biochemical genetic studies which are underway in our laboratory deal with the mitochondrial mutants of the yeast Saccaromyces cerevisiae and will be discussed in relation to:

    i)   The isolation and utilization of mitochondrial drug resistant mutants.

   ii)   Mitochondrial genes which specify components of the ATP-synthetase complex.

 iii)   Genetic evidence for a relationship between ATP synthetase and the adenine nucleotide carrier (ADP translocase).

Mitochondrial Drug Resistant Mutants for Analysis of Components of Oxidative Phosphorylation

Studies of the biochemical genetics of oxidative phosphorylation in our laboratory have made extensive use of mitochondrial drug resistant mutants.

The investigation is based on the following premises (15):

1) A drug or inhibitor should have a specific mode of action on mitochondrial energy conservation, i.e., it should have no effect on fermetative growth on glucose, but should inhibit the oxidative phase of growth on glucose or growth on an oxidizable substrate.

2) Drugs which affect mitochondrial energy conservation reactions (inhibitors, uncouplers, ionophores) have specific inhibitor binding sites associated with specific protein

266

subunits of the oxidative phosphorylation complex located in the mitochondrial inner membrane.

3) Drug resistant mutants should exhibit modified sensitivity to the drug at the whole cell, mitochondrial, submitochondrial and purified enzyme level.

4) Demonstration that the mutation is cytoplasmically inherited and located on mt-DNA is good a priori evidence that a mitochondrial gene product which is a component of the oxidative phosphorylation complex has been modified.

The drugs which show promise as specific inhibitors useful in genetic studies of mitochondrial oxidative phosphorylation are listed in Table I. Uncoupling agents, ATPase inhibitors, adenine nucleotide translocase inhibitors and ionophores satisfy some or all of the general criteria outlined above

TABLE I

Mitochondrial drugs useful in genetic studies of oxidative phosphorylation

Inhibitor	Biochemical Locus of Action	Resistant Mutants Isolated and Locus
Oligomycin Rutamycin Ossamycin Peliomycin	O.S. ATPase	Yes, nuclear and mitochondrial
Venturicidin	O.S. ATPase	Yes, nuclear, mitochondrial and cytoplasmic
Trialkyltin salts	O.S. ATPase	Yes, nuclear and cytoplasmic (mitochondrial)
Aurovertin	F_1 ATPase	Yes, nuclear
Dio-9	F_1 ATPase	Yes, nuclear
Bongkrekic acid	ADP translocase	Yes, nuclear and cytoplasmic
Hexachlorophene	ADP translocase	Yes, nuclear (cytoplasmic?)
Rhodamine 6G	ADP translocase?	Yes, cytoplasmic
Valinomycin	K^+ transport	Yes, mitochondrial
Octylguanidine	Energy transfer?	Yes, cytoplasmic
CCCP	Uncoupling agent	Yes, cytoplasmic
TTFB	Uncoupling agent	Yes, cytoplasmic
1799	Uncoupling agent?	Yes, cytoplasmic

(15,16). Detailed biochemistry genetic studies of oligomycin-resistant, venturicidin-resistant and triethyltin-resistant mutants are now available, and correlate well with modification of inhibitor binding sites on the mitochondrial OS ATPase complex.

Oligomycin-resistant, venturicidin-resistant and triethyltin-resistant mutants can be divided into two general classes (Class I and Class II) on the basis of cross resistance to other mitochondrial drugs (17). Class I mutants show cross resistance to aurovertin, Dio-9, venturicidin, triethyltin, uncouplers and other mitochondrial drugs; and these mutants show no resistance or modification at the mitochondrial level, but modification of mitochondrial permeability to drugs may be involved. Genetic analysis of Class I mutants revealed a complex involvement of a nuclear gene and a cytoplasmic element (18).

In contrast, Class II oligomycin-resistant mutants are specifically resistant to oligomycin and structurally related antibiotics, and show no cross resistance to venturicidin, triethyltin, DCCD or uncoupling agents (Table II). All the Class II mutants exhibited typical cytoplasmic inheritance and the resistance determinants are located on mit-DNA. Genetic analysis indicates that two loci (OLI and OLII) located on independent cistrons on mt-DNA are involved (18,19).

TABLE II

Cross resistance of oligomycin resistant, venturicidin resistant and triethyltin resistant mutants

Class	OLIGO	VEN	TET	1799	(MD)*
I	>50	>50	10-20	>20	3-5
II O^R	>50	1	1	1	1
II V^R (V,O)	100	100	1	1	1
II V^R (V,T)	1	>50	>20	10	1
II T^R (T,V)	1	>50	>20	1	1
II T^R (T,V)	1	>50	>20	20	1

*MD = mitochondrial drugs, chloramphenicol, antimycin A, CCCP, etc.

Biochemical studies (17,20) indicate a difference in oligomycin sensitivity which is demonstrable at the mitochondrial, submitochondrial and purified ATPase level, and are consistent with the modification of two of the mitochondrially synthezised components of the OS ATPase complex.

A modification of the oligomycin binding site is indicated which involves modification of an ionizing group of pK 7-9. Current studies are concerned with the isolation of mitochondrially synthesized subunits of the OS ATPase, and comparison of mutant and parental strains by peptide mapping and amino acid sequencing in order to establish a correlation between a mitochondrial mutation and a mitochondrial gene product.

Three types of specific cytoplasmic venturicidin-resistant mutants can be isolated (21). One type is specifically resistant to venturicidin and maps at the OLI locus. Another type, Class II (V,O) mutants, show cross resistance to oligomycin and the resistance allele is located on mt-DNA at a locus which is closely linked to OLI, termed OLIII (22). Class II (V,T) mutants are cytoplasmic mutants which show cross resistance to triethyltin and are similar to cytoplasmic triethyltin mutants.

These resistance alleles are essentially unlinked to other mitochondrial loci and are located on a different molecule of DNA (22,23). There is sufficient genetic evidence to indicate that the binding or inhibitory sites of oligomycin and triehtyltin are not identical and that the triethyltin binding site is located on a different mitochondrial gene product to those which are involved in oligomycin binding. This conclusion is supported by equilibrium binding studies with radioactive triethyltin and studies with an affinity label for this site, di-butyl choromethyltin chloride.

Interaction and cooperative effects between different bindingsites on the mitochondrial ATPase have been demonstrated in studies of the effect of the insertion of the TET^R phenotype into mitochondrial OLY^R mutants (23) and provide an experimental basis for complementation studies at the ATP-synthetase level.

Table III shows the effect of insertion of the T^R mutation on oligomycin sensitivity of the mitochondrial ATPase.

These studies provide strong evidence for separate but interacting reaction sites for oligomycin, triethyltin and venturicidin; and that we are dealing with modification of drug receptor sites on the membrane-bound subunits of the OS ATPase complex.

Mitochondrial Genes which Specify Components of the OS ATPase Complex: The biogenesis of the OS ATPase complex has been intensively investigated by Tzagoloff (13), and has been shown to consist of ten subunits, four of which are tightly associated with the mitochondrial inner membrane and are synthesized on mitoribosomes. These four mitoribosome synthesized subunits contain the sensitivity (inhibitor

TABLE III

INTERACTION OF TRIETHYLTIN AND OLIGOMYCIN SITES

Effect of T^R mutation on oligomycin sensitivity of mitochondrial ATPase

	I_{50} values (µg/mg protein		
Diploids	OLIGO	TET	VEN
KL 110 $O^S T^R$	30	0.3	0.03
KL 111 $O^R T^R$	30	0.3	0.03
KL 113 $O^R T^S$	30	0.3	0.03
KL 114 $O^S T^S$	7	0.6	0.05

ATPase assays were as described previously

binding) sites for oligomycin, venturicidin, DCCD and triethy-
tin as we have demonstrated that the purified O.S. ATPase is
sensitive to these inhibitors and contains the binding sites
for these inhibitors, whereas the F_1-ATPase is not inhibited
and does not bind these inhibitors (20,24). Genetic analysis
indicates that three loci, OLI, OLII and OLIII which are
probably located on two separate genes on mt-DNA specify
oligomycin and venturicidin resistance (21-23). The location
of these resistance alleles on mt-DNA is strongly supported by
evidence for linkage to other mitochondrial loci and ready
loss of the resistance allele on petite induction with ethidium bromide (17-19,22). The cytoplasmic TET[R]VENR and VEN[R]TET[R]
mutations do not appear to be linked to other loci on mt-DNA
and the resistance alleles appear to be located on another
DNA species. The identification of this DNA species is of
major interest in studies of mitochondrial genetics. Recent
experiments (25) have shown that the VEN[R]TET[R] resistance
alleles are retained in ρ^o petites in which mt-DNA has been
deleted by ethidium bromide treatment. However, this cytoplasmic DNA species is closely associated with mt-DNA and it
could be a second mt-DNA molecule and may form part of the
mitochondrial genome. A possible candidate is the 2µ circular
DNA species found to be associated with mitochondrial fractions by several workers and termed omicron-DNA (o-DNA) by
Clark-Walker (26).

Biochemical genetic studies have thus implicated two mitochondrial genes which code for two of the four mitoribosome
synthesized subunits of OS ATPase. A third subunit represented by the triethyltin binding site is determined by an
unknown cytoplasmic determinant which could be another species
of mt-DNA, possibly o-DNA.

Genetic Evidence for a Relationship Between ATP-synthetase and ADP translocase: We have suggested, as a result of studies of TET^R mutants, that the ADP translocase complex contains a cytoplasmically-determined subunit involved in the BA binding site, and that this subunit could be a common subunit of the OS ATPase complex involved in the triethyltin binding site (27). A cytoplasmically-determined subunit in ADP translocase has also been suggested by Haslam et al. (28), but has been questioned by Kolarov and Klingenberg (29) on the basis that a cytoplasmic petite (ρ^-) mutant contains a translocase system with normal sensitivity to atractyloside and bongkrekic acid. As we have demonstrated that components of the OS ATPase complex may be specified by cytoplasmic determinants which are not located on mit-DNA, we suggest that the question remain open. We have investigated the translocase activity and atractyloside binding capacity of mitochondria from parental strains and from ρ^o petites which have retained ($\rho^o V^R T^R$) or lost ($\rho^o V^o T^o$) the $V^R T^R$ resistance alleles.

In both petite strains the translocase activity is not detectable but binding studies with tritiated atractyloside indicate that an ADP sensitive atractyloside site is present, but the affinity for atractyloside is 5-10 fold less than in the parental strain and both petite strains contain less than 15% of the number of binding sites present in the parental strain. The results indicate that a full integreated ADP translocase complex requires the synthesis of a mitochondrially-determined subunit or of mitochondrially-determined components which serve a structural role in the ADP translocase complex.

TABLE IV

ADP translocase in petite mutants

Yeast strain	Translocase	BA Sensitivity	ATR Binding Site
($\rho^+, V^S T^S$)	+	+	+
($\rho^+, V^R T^R$)	+	(+)	+
($\rho^-, V^S T^S$)	+	+	+
($\rho^o, V^R T^R$)	-*		(+)
($\rho^o, V^o T^o$)	-*		(+)

not detectable: (+) modified sensitivity. Translocase assays were as described previously (29).

271

FUTURE PROSPECTS

Studies of the biochemical genetics of mitochondrial drug resistant mutants have proven to be of value in assessing the role and organization of the mitochondrially synthesized hydrophobic subunits of the ATP-synthetase complex. Firm evidence has been obtained that two mitochondrial genes specify two subunits of the ATPsynthetase complex and details of the modification of the primary structure of the relevant poly-peptides obtained from oligomycin-resistant mutants should be available soon.

A major development in these studies is the finding that the triethyltin-resistant and venturicidin-resistant (V,T) mutants are specified by a cytoplasmic element which is not mt-DNA. The possible role of o-DNA is thus of immediate interest but of more general significance is the possibility that the cytoplasmic element (o-DNA?) which specifies the VEN^{R}TET^R phenotype may specify components of other cell membrane systems. The plasma membrane is a possible location for such cytoplasmically-determined gene products. The sensitivity of plasma membrane ATPases to oligomycin, DCCD and to uncouplers raises the possibility that the plasma membrane may contai components in common with the mitochondrial mem-which are specified by a common cytoplasmic determinant. The isolation of cytoplasmically-inherited uncoupler-resistant mutants in this laboratory (16) and the relationship of these mutants to the uncoupler binding protein described by Hanstein and Hatefi (30) may provide a suitable system for testing of this hypothesis.

REFERENCES

1. Kovac, L. (1974) Biochim. Biophys. Acta 346, 101.
2. Gibson, F. and Cox, G.B. (1973) in Essays in Biochemistry Vol. 9, (Campbell, P.N. and Dickens, F., eds.) Academic Press, London.
3. Kanner, B.I. and Gutnick, D.L. (1972) FEBS Letters 22, 197.
4. Hatefi, Y., Haavik, A.G. and Griffiths, D.E. (1962) J. Biol. Chem 237, 1676.
5. Ziegler, D.M. and Doeg, K.A. (1959) Biochem. Biophys. Res. Commun. 1, 344.
6. Davis, K.A.and Hatefi, Y. (1972) Arch. Biochem. Biophys. 149 505.
7. Hatefi, Y., Haavik, A.G. and Griffiths, D.E. (1962) J. Biol. Chem 237, 1681.

8. Rieske, J.S., Baum, H., Stoner, C.D. and Lipton, S.H. (1967) J. Biol. Chem. 242, 4854.
9. Griffiths, D.E. and Wharton, D.C. (1961) J. Biol. Chem. 236, 1850.
10. Fowler, L.R., Richardson, S.H. and Hatefi, Y. (1962) Biochim. Biophys. Acta 64, 170.
11. Tzagoloff, A. and Meagher, P. (1971) J. Biol. Chem. 246, 7328.
12. Hatefi, Y., Stiggall, D.L., Galante, Y. and Hanstein, W.G. (1974) Biochem. Biophys. Res. Commun. 61, 313.
13. Tzagoloff, A., Rubin, M.S. and Sierra, M.F. (1973) Biochim. Biophys. Acta 301, 71.
14. Schatz, G. and Mason, T. (1974) Ann. Rev. Biochem. 43, 820.
15. Griffiths, D.E., Avner, P.R., Lancashire, W.E. and Turner, J.K. (1972) in Biochemistry and Biophysics of Mitochondrial Membranes (Azzone, G.F., Carafoli, E., Lehninger, A., Quagliariello, E. and Siliprandi, N., eds.) p. 505, Academic Press, New York.
16. Griffiths, D.E. (1972) in Mitochondria: Biogenesis and Bioenergetics (Van den Bergh, S.G., Borst, P. and Slater, E.C., eds.) p. 95, North Holland, Amsterdam.
17. Avner, P.R. and Griffiths, D.E. (1973) Eur. J. Biochem. 32, 301.
18. Avner, P.R. and Griffiths, D.E. (1973) Eur. J. Biochem. 32, 312.
19. Avner, P.R., Coen, D., Dujon, B. and Slonimski, P.P. (1973) Mole. Gen. Genet. 125, 9.
20. Griffiths, D.E. and Houghton, R.L. (1974) J. Biol. Chem. 46, 157.
21. Griffiths, D.E., Houghton, R.L., Lancashire, W.E. and Meadows, P.A. (1975) Eur. J. Biochem. 51, 393.
22. Lancashire, W.E. and Griffiths, D.E. (1975) Eur. J. Biochem. 51, 403.
23. Lancashire, W.E. and Griffiths, D.E. (1975) Eur. J. Biochem. 51, 377.
24. Griffiths, D.E. and Griffiths, E.J. (1975) unpublished results.
25. Griffiths, D.E., Lancashire, W.E. and Zanders, E.D. (1975) FEBS Letters 53, 126.
26. Clark-Walker, G.D. (1973) Eur. J. Biochem. 32, 263.
27. Cain, K., Lancashire, W.E. and Griffiths, D.E. (1974) Biochem. Soc. Trans. 2, 218.
28. Haslam, J.M., Perkins, M. and Linnane, A.W. (1973) Biochem. J. 134, 935.

29. Kolarov, J. and Klingenberg, M. (1974) FEBS Letters $\underline{45}$, 320.

30. Hanstein, W.G. and Hatefi, Y. (1974) J. Biol. Chem. $\underline{249}$, 1356.

BINDING AND UPTAKE OF CATIONIC DYES TO A NUCLEAR ETHIDIUM BROMIDE RESISTANT MUTANT OF THE PETITE-NEGATIVE YEAST Kluyveromyces lactis*

Aurora Brunner

INTRODUCTION

Baker's yeast, Saccharomyces cerevisiae, is an excellent biological system for studying mitochondrial mutations. It is a unicellular eukaryote with a short generation time, with stable haploid and diploid stages; cells of opposite sex mate to form zygotes, which under specific conditions sporulate and give rise to four-spored asci. Each spore is haploid which, through dissection techniques (1), can be separated from its sisters and the segregation of its characters studied. Genetically, mitochondrial mutations can easily be detected in these organisms by their non-Mendelian segregation during meiosis and by the mitotic segregation in diploids.

The first mitochondrial mutation described in yeast was the so-called petite mutation (or rho⁻) discovered by Ephrussi et al. (2). This pleiotropic mutation is phenotypically characterized by the appearance of colonies that are smaller than normal. Biochemically they are identified by the lack of cytochromes a + a_3 and b, and by their inability to utilize non-fermentable substrates. This mutation could be observed in S. cerevisiae because it is a facultative anaerobe, i.e., it grows in the absence of oxygen provided a fermentable substrate is present in the culture medium. Numerous reports on the properties of these mutants have appeared in the literature (3-5).

Although the petite mutation arises spontaneously in most strains with a frequency of approximately 1%, Ephrussi et al. (2) observed that acriflavine, and more specifically euflavine (6), was capable of inducing 100% petite mutants at very low concentratons. Slonimski et al. (7) found that another dye, ethidium bromide, at concentrations of about 5 µM was capable of transforming nearly completely a rho⁺ population to the petite condition. It is believed that these effects of euflavine and ethidium bromide are due to their ability to interact with DNA, probably by intercalation.

De Deken in 1961 (8) and Bulder in 1964 (9) by studying the response of several yeasts to acriflavine observed that

---

*I would like to dedicate this paper to the memory of my professor Huguette de Robichon-Szulmajster who introduced me to yeasts.

they could be divided into two groups, those which give rise
to the petite mutation when treated with the dye (petite-
positive) and those that were called petite-negative since
they failed to undergo the petite mutation. However, it was
also observed that the drug affected respiration in both kinds
of yeasts. The effect of ethidium bromide and acriflavine
among several drugs (10) and the effect of ethidium bromide
(11) on the respiration of petite-negative yeasts has also
been studied.

A great deal of work has been done to elucidate the mechanism by which ethidium bromide and acriflavine interfere
specifically with mitochondrial function. Meyer and Simpson
(12) reported that these drugs are more effective in blocking
mitochondrial function than nuclear DNA polymerase of mammalian origin. Also it has been observed that ethidium bromide
inhibits mitochondrial transcription in HeLa cells (13) and
in E. coli (14). A similar action of both ethidium bromide
and acriflavine has been detected in yeast (15-18). All these
effects are probably related to the intercalation of these
drugs into the DNA template.

Mounolou et al. (19) and Moustacchi and Williamson (20)
found that the density or the content of mitochondrial DNA
of petite cells as studied by density gradient centrifugation
was greatly altered; a complete absence of this type of DNA
was observed when S. cerevisiae cells were treated for pro-
longed periods of time with ethidium bromide (21-23). Recently
it has been proposed that ethidium bromide is capable of
interacting with mitochondrial DNA by forming a stable derivative which is further degraded by the action of specific
nucleases in the presence of intramitochondrial ATP. This
reaction which takes place in S. cerevisiae, a petite-positive
yeast, seems to be absent in some petite-negative yeasts
studied by the authors (24). This reaction could explain why
mitochondrial DNA disappears when the cells are treated with
ethidium bromide. Nevertheless, it does not explain the
transitory disappearance of mitochondrial DNA reported for
some petite-negative yeasts (10,25), accordingly the mechanism
by which mitochondrial DNA disappears in a petite-negative
yeast must be looked for at another level.

Although ethidium bromide or acriflavine when used on
petite-negative yeasts induces changes similar to those described for S. cerevisiae, a petite-positive yeast, these
changes seem to be only transitory and when the drugs are
removed the cells reassume their normal behavior (10,25,26).
These reversible effects have also been observed in other
organisms (27,28). A temperature reversible effect of the
mutagenic action of ethidium bromide in S. cerevisiae has
also been described (29).

Studies of the effect of acriflavine on different kinds of yeasts led Bulder to the conclusion that petites were not viable in the so-called petite-negative yeasts. Employing ultra violet light (30) or nitrosoguanidine (31) as mutagens, viable segregational (nuclear) petites have been obtained with Kluyveromyces lactis (a petite-negative yeast); these mutants are phenotypically similar to the vegetative (mito-chondrial) petites of S. cerevisiae. These results suggest that petites are viable in this strain, but for some reason the mitochondrial mutation which could lead to the formation of vegetative petites (rho⁻ mutants) is not stable; whether this is due to the presence of a master copy, as has been suggested (25), or to some other mechanism waits for further demonstration.

Kluyveromyces lactis was described as a petite-negative yeast by Bulder (9). Apart from the fact that diploids are not stable, its life cycle is otherwise identical to that of S. cerevisiae.

Working with a petite-negative yeast offered the advantage of avoiding the pleiotropic mutation elicited by ethidium bromide in petite-positive yeasts. During this work, mito-chondrial and nuclear ethidium bromide resistant mutants, some of which showed a correlated sensitivity to acriflavine, were isolated and studied from both a genetical and a bio-chemical point of view. As the experimental results suggested that the transport of these drugs was altered, the effect of pH on the transport of ethidium bromide and acriflavine in wild type and mutant strains was explored.

MATERIALS AND METHODS

The cells were grown in previously described conditions (32). The method of Mortimer and Hawthorne (1) was used for tetrad dissection.

Binding of ethidium bromide and acriflavine to whole yeast cells. 100 ml liquid YPAD (1% Bacto-yeast extract, 2% Bacto-peptone, 2% glucose and 80 mg/l adenine sulfate) were inocu-lated with a loopful of the strain to be studied. Flasks were incubated in a rotary shaker for 24 hr at $29 \pm 1^{\circ}$. Cells were harvested and centrifuged under sterile conditions, washed twice with distilled water and resuspended in approxi-mately 200 ml of sterile 0.05 M potassium phosphate buffer, pH 7, and starved for 48 hr with one change to fresh buffer after the first 24 hr of starvation. Cells were collected by centrifugation, washed twice with distilled water and re-suspended in two volumes of water. This cell suspension was used for binding experiments. For dry weight determinations, washed yeast cells were layered on pre-weighed aluminum planchets and dried under an infrared lamp.

Ethidium bromide was measured fluorometrically, the excitation and emission wavelengths were 330 and 600 nm, respectively. Acriflavine was measured spectrophotometrically at 451 nm.

RESULTS AND DISCUSSION

K. lactis mutants resistant to ethidium bromide were obtained and selected as previously described (32) on YPADG (1% Bacto-yeast extract, 2% Bacto-peptone, 0.1% glucose, 3% (v/v) glycerol and 80 mg/l adenine sulfate; this medium was made solid by the addition of 2% Bacto-agar) media with 25 or 50μM ethidium bromide. Colonies appeared after several days of incubation. Although our first idea was to use ethidium bromide as a specific mitochondrial mutagen, we later concluded that through this method we were isolating spontaneous mutants that were resistant to this drug. The frequency of this kind of mutation was about 1 in 10^7 cells. Several of these mutants were characterized. Although most of the mutations were of nuclear origin, some presented mutations with nuclear and mitochondrial characteristics.

One of these mutants, KA7-3A (resistant to 10 μM ethidium bromide and 60 μM synthalin), was selected to study in detail whether the mutation was nuclear or cytoplasmic. Diploids from this mutant were obtained by crossing it to a sensitive one. When these diploids were grown on media of YPAG (1% Bacto-yeast extract, 2% Bacto-peptone, 3% (v/v) glycerol and 80 mg/l adenine sulfate) plus 60 μM synthalin, mitotic segregation was observed, pointing to the existence of a mitochondrial mutation. One resistant diploid clone was selected and subcloned in media containing either glucose alone or glycerol plus 60 μM synthalin. When the diploid was subcloned on glycerol plus synthalin (KA10SG) and submitted to sporulation and tetrad analysis it yielded all segregants resistant to synthalin and ethidium bromide. On the other hand, when subcloning was carried out on YPAD media (KA10SD) and diploids sporulated and submitted to tetrad analysis, a nuclear segregation to resistance was observed (Table I). Further studies demonstrated that this drug resistance in this mutant is due to the presence of both nuclear and cytoplasmic mutations in this strain (33). The instability of this mitochondrial mutation is in agreement with the reversibility of mitochondrial changes observed in petite-negative yeasts (10,25,26).

Since acriflavine is similar to ethidium bromide in that it induces the petite mutation in S. cerevisiae (2,7) and intercalates between the DNA bases (34,35), we considered it interesting to test this drug in our mutants. We found that

TABLE I

Ethidium bromide and synthalin resistance in tetrads of KA1OD and KA1OSG

| | No. of Tetrads Analyzed | No. of Tetrads of Type R:S | | | |
| | | 10 µM Ethidium Bromide | | 60 µM Synthalin | |
Diploid		2:2	4:0	2:2	4:0
KA1OD	14	14	0	14	0
KA1OSG	8	0	8	0	8

Drugs were included on YPAG at the final concentrations indicated. Results were scored during four consecutive days, starting 24 hr after plating.

the majority of the ethidium bromide resistant nuclear mutants were sensitive to concentrations of acriflavine in which the wild type grew well. Genetic studies were undertaken in order to dissociate resistance to ethidium bromide from sensitivity to acriflavine. We were unsuccessful and, thus, we concluded that a single mutation or two very closely linked mutations were responsible for this behavior (33). A similar phenomenon has been observed with various other drugs and antibiotics. It has been termed collateral sensitivity by Rank and Bech-Hansen (36) and correlated sensitivity by Celis et al. (33).

Further characterization of the mutation which conferred resistance to ethidium bromide and correlated sensitivity to acriflavine was undertaken. Strain W600B:α,ade$_1$,ade$_2$, leu,EB^S,Acri^R (a gift from Dr. James R. Mattoon) was chosen as the wild type. An ethidium bromide resistant mutant derived from this strain (W600BR) was obtained as mentioned above and crossed to sensitive testers. Thereafter, the diploids were submitted to tetrad analysis to discern the genetic nature of this resistance. Using two different testers (12 tetrads with one, 11 with the other), we found that segregation of resistance gave a Mendelian pattern, indicating a nuclear mutation. Furthermore, all segregants which were resistant to ethidium bromide showed a correlated senstivity to acriflavine. These tests were done on plates at pH 6.7.

However, when the strains were grown on solid YPAG media at pH 4.5 in the presence of either drug, both wild type and mutant strains grew well. Growth on plates seemed inadequate for establishing quantitative differences since local changes in pH or drug concentration are unavoidable situations.

In search of a more quantitative approach to study differences between these two strains, growth curves in liquid media were determined with glucose and glycerol as carbon sources, at different pH values and in the presence of either drug. From the results displayed in Table II, it may be seen that in media with glucose as carbon source (YPAD), there is a difference in the rate of growth between the wild type (W600B) and the ethidium bromide resistant mutant (W600BR), a difference which is independent of pH and the presence or absence of drugs; whereas in media with glycerol as the energy source (YPAG), there is a larger difference in the growth rates of the wild type and the mutant in the presence of the drugs. The difference in the growth rate is also altered by pH. At pH 5.2 the mutant is resistant while at pH 6.6 it is sensitive to ethidium bromide and acriflavine.

TABLE II

Influence of pH, ethidium bromide and acriflavine on the generation time of a wild type and and ethidium bromide resistant mutant in the presence of fermentable and non-fermentable substrates

Mean Generation Time

Strain	pH 5.2 Glucose no additions	10 µM EB	10 µM Acr	Glycerol no additions	10 µM EB	10 µM Acri	pH 6.6 Glucose no additions	10 µM EB	10 µM Acri	Glycerol no additions	10 µM EB	10 µM Acri
W600B												
Exp.1	1 h37'	---	---	1 h34'	∝	1 h32'	1 h46'	1 h48'	1 h34'	1 h27'	∝	2 h7'
Exp.2	---	---	---	1 h42'	∝	1 h38'	1 h35'	---	---	1 h34'	---	1 h50
W600BR												
Exp.1	2 h10'	---	---	2 h56'	2 h50'	2 h53'	2 h11'	2 h15'	2 h25'	3 h	∝	∝
Exp.2	---	---	---	2 h43'	3 h2'	2 h8'	2 h10'	---	---	---	---	---

300 ml nephelometric flasks containing 75 ml liquid YPAD or YPAG with or without drugs we inoculated with a 24 hr-old inoculum, and incubated in a rotary shaker, at 30°C. Readings w taken with a Klett colorimeter provided with the green filter at 30 min intervals once growt had begun. Dashes indicate that the experiment was not done.

The wild type strain is sensitive to ethidium bromide and resistant to acriflavine at both pH values. From these results it may be concluded: first, on plates at pH 6.7 a probable local acidification of the medium brought about by the growth of yeast induces impermeability or insensitivity to ethidium bromide; nevertheless, this local acidification does not prevent the inhibitory effect of acriflavine, which is manifested under these conditions. Second, the mutant always grows at a slower rate independent of the energy source employed. The difference is of 30 min for each doubling with glucose, and increases to nearly an hour in the presence of glycerol. This could be due to a change of permeability to substrates or to a decreased efficiency of the mitochondria in the metabolization of these substrates. Third, since the growth with glucose as a carbon source is not affected by the presence of either ethidium bromide or acriflavine, these drugs seem not to interfere with glycolysis as has already been suggested for this and other petite-negative yeasts (10, 11,24).

In experiments performed by Fukuhara and Kujawa (15) for studying respiratory adaptation, they had to use a pH 6.5 buffer since at lower pH both acriflavine and ethidium bromide had almost no effect. This could be due to an impermeability of these drugs at low pH as we have found with our mutants.

Binding of ethidium bromide and acriflavine to whole yeast cells. Sensitivity to ethidium bromide or acriflavine is observed only when glycerol, a non-fermentable substrate, is present as the carbon source. Since the integrity of the organelle is necessary for the utilization of this substrate these findings indicate that these drugs interfere with some mitochondrial function.

Resistance to these drugs could be due to a change in the permeability of the cell membrane, to a change in the permeability of the mitochondrial membrane or to a change in some protein within the mitochondria which specifically interacts with these drugs and thereby interferes with the normal functioning of the organelle. To further characterize this mutation, the binding of the drugs to whole yeast cells was measured at pH 4.5 and 6. Binding of ethidium bromide depended on an energy source and was greatly influenced by the pH of the medium.

After 15 min incubation with the drug at 30° and in the presence of glucose (Table III), the amount of ethidium bromide bound to the wild type at pH 6 was twice that bound at pH 4.5. In the mutant the percent of ethidium bromide bound at pH 4.5 was not significantly different from that bound at pH 6. It can also be seen that the wild type binds 80% of

TABLE III

Effect of pH on the binding of ethidium bromide and acriflavine by whole yeast cell

Strain	pH	nmoles of EB bound/mg dry weight	% of EB bound[a]	nmoles of Acri bound/mg dry weight	% of Acriflavin bound[a]
W600B	6	19.6 - 1.2 (6)	79.9 - 2.1 (6)	15.2 - 1.7 (2)	92.5 - 9.6 (2)
W600B	4.5	9.2 - 1.9 (4)	38.8 - 4.7 (4)	14.1 - 0.4 (2)	86.1 - 0.9 (2)
W600BR	6	4.1 - 0.9 (6)	18.0 - 2.9 (6)	11.4 - 0.3 (2)	83.4 - 3.5 (2)
W600BR	4.5	2.5 - 2.1 (4)	13.4 - 7.3 (4)	9.7 - 0.2 (2)	71.7 - 6.5 (2)

[a]*The drug mixture without cells was considered as 100%.*
The cell suspension was incubated for 15 min at 30°, an aliquot was then removed and quickly centrifuged in a Beckmann microfuge for 40 sec. The supernatant was removed and ethidium bromide or acriflavine assayed as described. The incubation medium contained in 1 ml final volume: 150 nmoles ethidium bromide or acriflavine, 20 μmoles buffer tartrate-triethanolamine (pH 4.5) or maleate-triethanolamine (pH6) and 2 mg glucose. The reaction was started by the addition of 0.1 ml yeast suspens

the ethidium bromide present in the medium at pH 6, whereas the mutant under the same conditions binds only 18%. The small discrepanices observed when results are expressed as nmoles of ethidium bromide bound/mg dry weight and as percent of the amount ethidium bromide present in the absence of cells are due to the fact that there is a consistent difference among the dry weights of the wild type and the mutant. The same wet weight gives a higher dry weight value for the mutant than for the wild type; this indicates a lower water content in the former.

As shown in Fig. 1, the binding of ethidium bromide is favored by the presence of glucose. If glucose acts as an energy source, ethidium bromide may enter the cell through an energy-dependent transport system, and therefore we are not dealing with a simple binding of the drug to the cell surface. On the other hand, a change in color in ethidium bromide, apparent to the naked eye, is observed when the drug binds to the cell suggesting a change in the hydrophobicity of the medium which surrounds ethidium bromide. Measurement of fluorescence changes of the drug upon binding to whole K. lactis cells have been done with other strains, in this laboratory, and shown also to be glucose dependent.

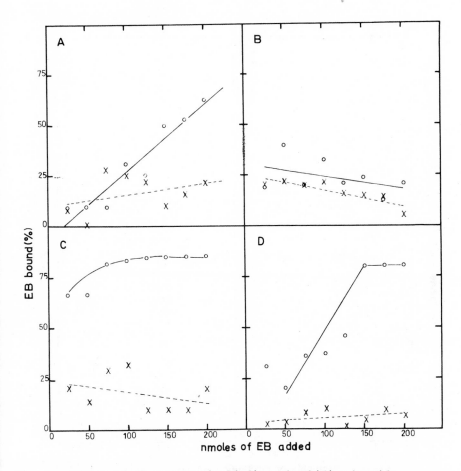

Fig. 1. Effect of glucose on the binding of ethidium bromide.
Whole yeast cells grown and starved as indicated were incubated
at 30° for 15 min, centrifuged and the concentration of ethidium
bromide measured fluorometrically in the supernatant, after the
appropriate dilutions. The incubation mixture contained in
1 ml final volume: 20 μmoles buffer tartrate-triethanolamine
(pH 4.5) or maleate-triethanolamine (pH 6) and the concentrations of ethidium bromide indicated in the figure. The reaction was started by the addition of 0.1 ml yeast suspension.
The concentration of ethidium bromide added was considered as
100%.
    The most probably straight line was calculated for the experi-
ments without glucose in panels C and D; as well as with and
without glucose in panels A and B by the method of least
squares (38). O———O, with 2 mg glucose/ml; x----x without
glucose. A: W600BR, pH 6. B: W600BR, pH 4.5. C: W600B,
pH 6 and D: W600B, pH 4.5.

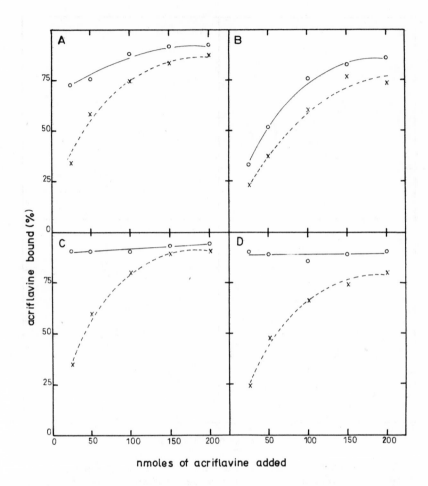

Fig. 2. *Effect of glucose on the binding of Acriflavine. Whole yeast cells grown and starved as indicated, were incubated at 30° for 15 min, centrifuged and the concentration of acriflavine measured spectrophotometrically after appropriate dilutions. The incubation mixture contained in 1 ml final volume: 20 μmoles buffer tartrate-triethanolamine (pH 4.5) or maleate-triethanolamine (pH 6) and the concentrations of acriflavine indicated in the figure. The reaction was started by the addition of 0.1 ml yeast suspension. The concentration of acriflavine added was considered as 100%. O——O with 2 mg/ml glucose; x----x without glucose. A: W600BR, pH 6. B: W600BR, pH 4.5. C: W600B, pH 6 and D: W600B, pH 4.5.*

These results together with those of Table II suggested that the resistance to ethidium bromide could be due to a change in the permeability of the cell membrane. We thought a _priori_ that the reverse situation would be observed with acriflavine since those strains which had become resistant to ethidium bromide were simultaneously made sensitive to acriflavine. However, when binding of acriflavine was measured it was found that: a) the binding of this drug, as shown in Fig. 2, is not absolutely dependent on the presence of glucose especially at high acriflavine concentrations; b) acriflavine, similar to ethidium bromide, is bound to a greater extent by the wild type; c) differences in the binding of this drug to the wild type and mutant cells are not as large as those found with ethidium bromide.

Since the binding to the mutant is lower than the binding to the wild type, the correlated sensitivity found for this drug cannot be explained by an alteration at the level of the cell membrane. The sensitivity to acriflavine must lie at another level. Differences in the mitochondrial permeability or sensitivity towards these drugs must be looked for before any reasonable conclusion can be reached concerning the mechanism of resistance in these mutants.

In this respect, it may be recalled that mitochondrial mutations which affect the permeability of whole yeast cells to imipramine, a cationic drug, have been reported (37). Thus, it is probably that mitochondrial DNA may code for proteins which are common to or regulate the permeability of both the cell and mitochondrial membranes.

ACKNOWLEDGMENTS

I acknowledge the hospitality and helpful advice of Dr. James R. Mattoon of Johns Hopkins University where part of this work was carried out.

REFERENCES

1. Mortimer, R.K. and Hawthorne, D.C. (1969) *in* The Yeasts, Vol. I, (Rose, H.A. and Harrison, J.A., eds.) pp. 397-399, Academic Press, New York.
2. Ephrussi, B., Hottinguer, H. and Chimenes, A. (1949) Ann. Inst. Pasteur 76, 351.
3. Wilkie, D. (1964) *in* The Cytoplasm in Heredity, pp.15-28 Methuen & Co., Ltd., London.
4. Sager R. (1972) *in* Cytoplasmic Genes and Organelles, Academic Press, New York.
5. Jinks, J.L. (1964) *in* Extrachromosomal Inheritance, Prentice-Hall, Inc., New Jersey.
6. Marcovich, H. (1961) Ann. Inst. Pasteur 81, 452.
7. Slonimski, P.P., Perrodin, G. and Croft, J.H. (1968) Biochem. Biophys. Res. Commun. 30, 232.
8. De Deken, R.H. (1961) Exp. Cell Res. 24, 145.
9. Bulder, C.J.E.A. (1964) J. Microbiol. Serol. 30, 1.
10. Luha, A.A., Leslie, E.S. and Whittaker, P.A. (1971) Biochem. Biophys. Res. Commun. 44, 396.
11. Crandall, M. (1973) J. Gen. Microbiol. 75, 363.
12. Meyer, R.R. and Simpson, V.M. (1969) Biochem. Biophys. Res. Commun. 34, 238.
13. Zylber, E., Vesco, C. and Penman, J. (1969) J. Mol. Biol. 44, 195.
14. Richardson, J.P. (1973) J. Mol. Biol. 78, 703.
15. Fukuhara, H. and Kujawa, Ch. (1970) Biochem. Biophys. Res. Commun. 41, 1002.
16. Perlman, Ph.S. and Mahler, H.R. (1971) Nature New Biol. 231, 12.
17. South, D. and Mahler, H.R. (1968) Nature 218, 1226.
18. Mahler, H.R. and Dawidowicz, K. (1973) Proc. Nat. Acad. Sci. (US) 70, 111.
19. Mounolou, J.C., Jakob, H. and Slonimski, P.P. (1966) Biochem. Biophys. Res. Commun. 24, 218.
20. Moustacchi, E. and Williamson, D.H. (1966) Biochem. Biophys. Res. Commun. 23, 56.
21. Goldring, E.S., Grossman, L.I., Krupnick, D., Cryer, R.D. and Marmur J. (1970) J. Mol. Biol. 52, 323.
22. Nagley, P. and Linnane A.W. (1970) Biochem. Biophys. Res. Commun. 39, 989.
23. Nagley, P. and Linnane A.W. (1972) J. Mol. Biol. 66, 181.
24. Nobrega Bastos, R. and Mahler, H.R. (1974) J. Biol. Chem. 249, 6617.
25. Luha, A.A., Whittaker, P.A. and Hammond, R.C. (1974) Mol. Gen. Genet. 129, 311.

26. Kellerman, G.M., Biggs, D.R. and Linnane, A.W. (1969) J. Cell Biol. 42, 378.
27. Soslau, G. and Nass, M.K. (1971) J. Cell Biol. 51, 514.
28. Stuchell, R.N., Weinstein, B.J. and Beattie, D.S. (1973) FEBS Letters 37, 23.
29. Perlman, Ph.S. and Mahler, H.R. (1971) Biochem. Biophys. Res. Commun. 44, 261.
30. Herman, A.I. and Griffin, P.S. (1968) J. Bacteriol. 96, 457.
31. Del Giudice, L. and Puglisi, P.P. (1974) Biochem. Biophys. Res. Commun. 59, 865.
32. Brunner, A., Mas, J., Celis, E. and Mattoon, J.R. (1973) Biochem. Biophys. Res. Commun. 53, 638.
33. Celis, E., Mas, J. and Brunner, A. (1975) Genet. Res. 25, 59.
34. Waring, M.J. (1965) J. Mol. Biol. 13, 269.
35. Lerman, L.S. (1961) J. Mol. Biol. 3, 18.
36. Rank, G.H. and Bech-Hansen, N.T. (1973) Mol. Gen. Genet. 126, 93.
37. Linstead, D., Evans, I. and Wilkie, D. (1974) *in* The Bio-genesis of Mitochondria (Kroon, A.M. and Saccone, C., eds.) pp. 179-192, Academic Press, New York.
38. Croxton, F.E. (1959) *in* Elementary Statistics with Appli-cations in Medicine and the Biological Sciences, p. 114, Dover Publications, Inc., New York.

PART III

MEMBRANE STRUCTURE

One of the most critical problem areas of Bioenergetics
has been the general difficulty of finding good ways to study
membranes. During the last decade, increasing emphasis has
been placed upon the adaptation of the techniques of physics
and chemistry to identify the organization of membranes,
particularly their asymmetric and vectoral properties.

Older approaches, which are continuously being refined,
have involved the use of cell fractionation for purification
of membranes combined with biochemical and morphological
characterization. Electron microscopy still remains as the
only direct way in which membranes can be individually examined,
and some of the chapters in this book consider these approaches.

Most of the newer techniques involve averaging methods,
i.e., give results reflecting the population as a whole but
not the individual membrane profiles or organelles. Among
the newer approaches described in the succeeding chapters
are: the use of chemical probes for asymmetric labeling of
membrane surfaces and for crosslinking of membrane components;
the use of detergents; photoaffinity labeling techniques; and
nuclear magnetic resonance and electron spin resonance techniques, particularly spin labeling. Also, membrane reconstitution is an important approach to elucidating structural
and functional relations in membranes. It was the consensus
that the use of several techniques in combination affords
the most incisive strategy for solving problems of membrane
structural organization.

SPIN LABELING THE HYDROCARBON PHASE OF BIOLOGICAL MEMBRANES

Alec D. Keith

The spin labeling technique is accepted as a general method
of obtaining information about biological membranes. A sizable body of literature now exists using this technique as a
means of analysis of both model systems and the membranes of
intact cells. Several good review articles have been written during the past few years which the reader should consult
for greater details about specific aspects of spin labeling
(1-4).

Although the impact of spin label studies on the general membrane literature has been immense, the general findings can be
stated briefly. Studies on multibilayers and oriented membranes have revealed that spin-labeled fatty acids and sterols
orient in membranous structure in the expected way. Spin
labeled fatty acids are incorporated into phospholipids by a
variety of organisms. Spin label studies have probably cont-
ributed more to the concept of membrane fluidity than any
other approach to the study of membranes. These fluidity considerations relate both to rotational and translational diffusion. Spin labels have been used to monitor the physical
state of membranes, and have been very useful in describing
phase transition and phase separation events. These spin
label detectable events usually correlate with modifications
in membrane-associated biological functions such as transport
activity, growth rate, cold sensitivity, and so on.

Spin labels in common use are organic stable free radicals of
the nitroxide type. Four basic structures and their derivatives constitute most of the spin labels which have so far
been used.

| I | II | III | IV |

The development of these molecules as probes of biological
structure and the syntheses of the probes themselves have involved the efforts of many workers in several nations.

The equation

$$h\nu = g\beta H$$

describes the resonance condition whereby energy absorption

takes place. h is Planck's constant, β is the Bohr magneton,
H is the applied magnetic field, ν is the klystron frequency
and g is a constant defined by the equation. The unpaired
electron is localized in the neighborhood of the nitrogen
nucleus. It is the interaction of the unpaired electron and
the nitrogen nucleus which gives rise to the time-dependent
events which result in useful spectral parameters. More specifically, the interaction of the spin angular momentum of
the unpaired electron with the magnetic moment of the nitrogen
nucleus gives rise to the hyperfine coupling interaction. The
orbital angular momentum of the unpaired electron interacting
with the magnet moment of the nitrogen nucleus results in the
spin-orbit coupling interaction. Careful work has been carried out on single crystals containing spin labels. This work
resulted in measurements of the hyperfine coupling and g-value
tensors. The spin-orbit coupling interaction results in the
g-value tensors.

Spectra taken of spin labels dissolved in liquids show an averaging of both hyperfine and g-value tensors. If the rotational motion is on a sufficiently rapid time scale then the
isotropic hyperfine coupling constant, A_N, is very close to
equaling

$$A_N = 1/3(A_{xx} + A_{yy} + A_{zz}).$$

For the same conditions the isotropic g-value, g_N, comes very
close to equaling

$$g_N = 1/3(g_{xx} + g_{yy} + g_{zz}).$$

Complete motional averaging results in narrow lines which have
very nearly equal integrated intensity. Increasing the solvent viscosity gradually causes loss of motional narrowing.
It is this loss of motional narrowing which results in information about rotational diffusion and, therefore, the restriction of rotational diffusion imposed by the local environment of the spin label.

INFORMATION PARAMETERS
Hyperfine coupling and g-value constants

The isotropic hyperfine coupling constant, A_N, and the isotropic g-value, g_N, both show small but easily measurable changes
resulting from solvent polarity. These dependencies are
compared to   dielectric constant below. Both A_N and g_N
constants depend only on the microenvironment of the spin
label and so are valid for measurements in which the spin
label of interest is solubilized in microstructures. For
example, the small spin label, 5N10, dissolves in concentrated

Table I. Nitrogen hyperfine couplings (gauss) and g values for
2,2,6,6-tetramethylpiperidinal-n-$_{oxyl}$ in different solvents.

Solvent	Dielectric constant	Hyperfine coupling	g value
n-Hexane	1.9	15.2	2.0061
Ethyl ether	4.3	15.3	2.0061
t-Butanol	10.9	16.0	2.0059
n-Butanol	17.8	16.1	2.0059
Ethanol	24.3	16.1	2.0059
Methanol	32.6	16.1	2.0059
Ethylene glycol	37.7	16.4	2.0058
Water	78.5	17.1	2.0056

membrane preparations of E. coli and results in an A_N of 14.3
gauss (the A_N for 5N10 is 16.1 gauss in water and 14.2 gauss
in octadecane). In less concentrated preparations of the same

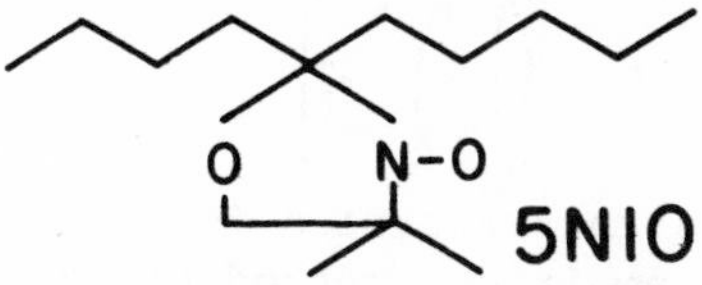

5N10

membranes two sets of lines are clearly visible. One set is
characteristic of a hydrocarbon zone and the other is characteristic of water. Therefore, under these conditions part of
the spin label is localized in the membrane hydrocarbon zones
and part is dissolved in the aqueous medium. The relative
solubility of the spin label in the membrane hydrocarbon
regions and in water determine the relative size of the two
spectral contributions (see Partitioning of Spin Labels between
Dielectric Zones).

The spin label 2N14 results in an A_N value of about 15.0 gauss

2N14

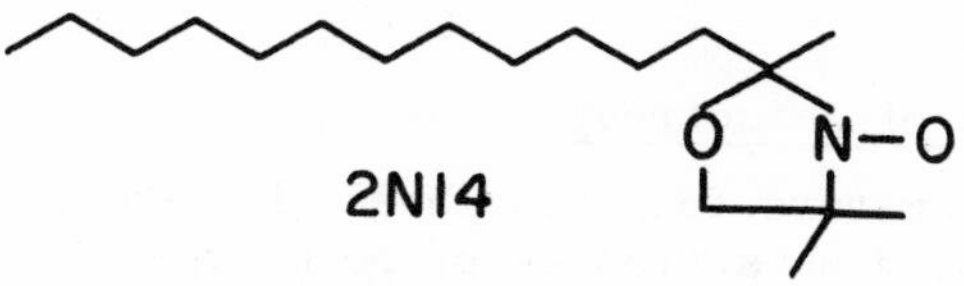

in vesicles of sarcoplasmic reticulum (A_N of 2N14 is 16.1 gauss in water and 14.2 gauss in octadecane). This indicates that the N-oxyl group is either localized at an interface zone which has intermediate polarity or that the N-oxyl group exchanges between two dielectric zones very rapidly. The difference in hyperfine coupling between aqueous and hydrocarbon zones is about 2 gauss or 5.6 megaHz. Therefore the exchange time between the two zones must be on the order of 10^{-7} sec or less in order for such averaging of the A_N to be effective.

The isotropic hyperfine and g-value constants are important measurements in many spin label experiments.

Isotropic rotational diffusion measurements

Rotational diffusion has been the topic of many scientific investigations. Many publications deal with this topic. McConnell, Freed, and Kivelson (6-8) have made important fundamental contributions to the understanding of rotational diffusion in terms of spin label line shape analysis. The Kivelson treatment (8) has been adapted by several workers and forms the basis for the rotational correlation time, τ_c, shown below. K is a constant, h_o and h_i are the mid and high field lines, and W_o is the mid field line width.

$$\tau_c = KW_o \left[\left(\frac{h_o}{h_{-1}} \right)^{\frac{1}{2}} - 1 \right].$$

This equation assumes that rotation is isotropic and that motion is rapid. An empirical method of determining the general validity of this equation is obtained by showing that the τ_c values derived from spin label measurements vary with the same dependencies as τ_c values calculated from Stokes' equation for viscosity. Several small spin labels obey this relationship.

Sometimes it is useful to use the τ_c equation given above without the constant, K. This equation then becomes an empirical motion parameter, R_i, useful for comparative purposes where no numerical accuracy of rotational diffusion is inferred. The integrated intensity under all spectral lines should equal the same value in order for these motion parameters to be valid. Loss of integrated intensity under a single line, where line height ratios are used, introduces large errors.

Measurements of anisotropy

Spectral parameters for spin labels housed in single crystals show pronounced anisotropies of both hyperfine coupling and g-value tensors. Table 1 shows some representative values

(9–12).

Table 1

	di-tert-butyl nitroxide (9)	2N3(10)	3NC(11)	TEMPONE(12)
A_{xx}	7.59	5.9	5.8	6.5
A_{yy}	5.95	5.4	5.8	6.7
A_{zz}	31.78	32.9	30.8	33.0
g_{xx}	2.00881	2.0088	2.0089	2.0094
g_{yy}	2.00625	2.0058	2.0058	2.0061
g_{zz}	2.00271	2.0022	2.0021	2.0021

The basic idea of using oriented multilayers or oriented membranes containing appropriate spin labels is essentially the same as including a spin label in a crystal as an inclusion. The rotational motion must be shown so that the net time spent in one orientation dominates and so that time-dependent averaging of the hyperfine and g-value tensors does not occur extensively. When these requirements are met spin label fatty acids and spin label sterols demonstrate marked changes in hyperfine coupling and g-value constants at different orientations of the sample. Demonstrating spectral anisotropy in this way is direct and shows that these two classes of spin labels orient in oriented lipids (2,3). Several studies using oriented lipid as membrane samples are in good agreement with x-ray diffraction and optical studies which have been carried out on similar samples.

Other important measurements involving anisotropy have been carried out using samples which have no net orientation to the applied magnetic field. This approach, anisotropic motion, has also yielded a variety of important findings. The best known use of anisotropic motion is referred to as the order parameter, S, and is applied to the analysis of the oxazolidine-bearing fatty acids (13,14). This term

$$S_x = \frac{A_{\parallel} - A_{\perp}}{A_{zz} - \frac{1}{2}\left(A_{xx} + A_{yy}\right)}$$

Order parameter, S_{x1} uses known crystal parameters as close approximations of the hyperfine extremes, $A_{\parallel} \simeq A_{zz}$ and $A_{\perp} \simeq \frac{1}{2}(A_{xx} + A_{yy})$. The notation, x, is used to describe the number of carbons between the carboxyl group and the oxazolidine group. The spin label fatty acid, V, shown below has pronounced axial asymmetry and rotates with relative ease on the long axis of the fatty acid. The x-principal axis is along the N-oxyl bond. The y-

$$\underline{\underline{V}}$$

principal axis is perpendicular to the x-principal axis. The
z-principal axis is parallel to the long axis of the fatty
acid chain. The plane of the oxazolidine ring is perpendicular to the long axis of the fatty acid. Rapid rotation about
the long axis of the fatty acid results in rapid motion for
both the A_{xx} and A_{yy} tensors; however, it does not cause the
A_{zz} to enter into the time-averaging process. As the matrix
becomes less viscuous the fatty acid flexes and wobbles caus-
ing some averaging between A_{zz} and $\frac{1}{2}(A_{xx} + A_{yy})$. As the viscosity decreases further $A_{\parallel}$ approaches $A_{\perp}$ and eventually
the spectrum appears isotropic. The restriction to
the time-dependent averaging of $A_{\parallel}$ and $A_{\perp}$ is related to molecular ordering in treatments of the order parameter.

Several spin labels have enhanced motion about the x-principal
axis of the spin label (15). The spin label VI results in
such a signal, particularly in phospholipid or membrane preparations. At x-band (about 9.5 GHz) and in the fast tumbling
range, the low field line has greater amplitude and is more
narrow than the low field line of an isotropic signal (16).
Some spin labels having structure similar to VI, give fairly

$$\underline{\underline{VI}}$$

isotropic appearing signals in isotropic media, but give
signals in membrane preparations with $h_1 > h_0$. For some
experiments the simple measurement, h_1/h_0, is very useful
and shows the degree of ordering of the spin label imposed by
the matrix.

Partitioning of spin labels between dielectric zones

McConnell and associates have carried out extensive experimentation using the small spin label, TEMPO (17).

TEMPO

This spin label is small, has hydrocarbon character, yet still has substantial water solubility. It is a very good probe for measuring the relative solubility between the aqueous medium and the hydrocarbon zones of membrane preparations. TEMPO dissolved in aqueous preparations of membranes often gives a signal which has a two-component high field line. The component marked hydrocarbon originates from hydrocarbon

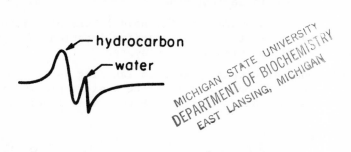

zones. The component marked water originates from aqueous zones. The physical state of membrane hydrocarbon regions markedly affects the size of the hydrocarbon line; therefore, the ratio of the hydrocarbon line height to the aqueous line height is a very useful measurement for determining properties of membrane hydrocarbon zones (17).

Another spin label, 5N10, (structure shown earlier) has similar properties to TEMPO; however, it has greater hydrocarbon solubility and can be used in preparations in which no hydrocarbon contribution would be seen with TEMPO. Both these spin labels and potentially others which partition are very useful in determining the properties of membranes.

Specific site labels

Several alkylating agents have been attached to spin labels. These have varying degrees of group specificity. For example, a spin label with iodoacetate attached to it through an amide or ester bond, has a high degree of specificity for sulfhydryl groups. Different alkylating agents attached to spin labels have specificity for different groups. Labels containing these alkylating agents have been used to study proteins and interactions between lipids and proteins.

The spin label fatty acids, of which V is one, have a certain degree of zone specificity. The carboxyl group anchors to the polar interface. The spacing between the carboxyl group and the oxazolidine group determines the depth with which the oxazolidine group resides in the hydrocarbon zone. These spin labels probe zones of different depths than phospholipid polar zones, depending on the positioning of the oxazolidine on the chain.

Aliphatic hydrocarbons containing spin labels on carbon-2 localize with the N-oxyl group near the interface as determined by measurements of the A_N and g_N values. Such spin labels as these measure perturbations or other events which occur near their local zones probably better than spin labels which locate in several zones.

All these site-specific spin labels penetrate membranes and localize at the site-specific location on both sides of and on all membranes in the system.

Another dimension to spin labeling is the ability to label one side only of a membrane vesicle preparation or the cell surface only of cells. Such a spin label, VII, is shown below.

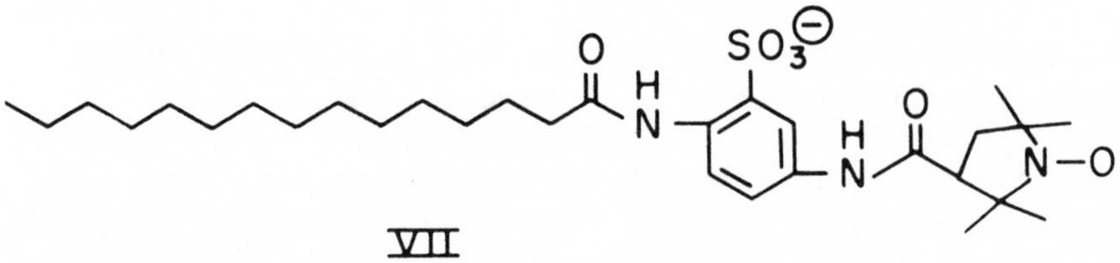

VII

This label was shown to have essentially all of its signal on the outer surface of membrane vesicles and intact yeast cells. (The next section will deal more with this type of label.)

Zone-specific removal of spin label signal

The first such approach was the use of ascorbic acid, or vitamin C, by McConnell and associates to remove part of the aqueous contribution of TEMPO in membrane aqueous

dispersions (19). Ascorbate is a reducing agent and donates
a hydrogen atom to spin labels. The electron of the hydrogen
atom pairs with the unpaired electron of the spin label. The
reduced spin label is a new chemical species which may have
different solubility properties. A modification of the ascorbate approach is to use ascorbyl palmitate, thereby localizing
the spin reduction to membrane surface sites.

Another approach, based on physical rather than chemical prin-
ciples, is to use a second paramagnetic species which has no
visible signal (20). Paramagnetic species interact with each
other in two potential ways. Electron-electron magnetic di-
pole interactions limit line width. The electron-electron
spin exchange interaction causes nitroxide lines to undergo
narrowing and collapse to a single central line, or extensive
broadening, depending on the interacting species. An example
is nickel, which may be used as a salt, such as $NiCl_2$, or as
an organic chelate. $NiCl_2$ causes extensive broadening of
nitroxide lines when the two species are in the same local
environment. Nickel is usually impermeable to membranes.
Nickel does not cause distortion of the nitroxide signal when
the two species are separated by a membrane; therefore, nickel
can be used to remove all the signal on one side of a membrane
without removing any signal from the other side. Membrane
vesicles treated with nickel and a nitroxide result in a
nitroxide signal which originates from zones unavailable to
nickel. These zones are the inside of the membrane enclosure,
deep in the hydrocarbon zones, and the inner membrane surface.
This method has recently been put to use in obtaining information about the aqueous interior of cells (20) and the viscosity profile of sarcoplasmic reticular vesicles (5).

The spin label method has become more versatile with site-
specific and zone-specific labels. These approaches together
with ever increasing sophistication of analysis should lead
to an even more promising future for spin labeling.

Most of the experiments using spin labels which have been
carried out on mitochondrial preparations have been phenomenological. These experiments have been aimed at establishing
a relationship between structure and function. The basic
design of many of these experiments has been to find a correlation between an induced functional change and a simultaneous change in a spin label parameter. Data from these
experiments have often been displayed on Arrhenius plots.

One of the first experiments reported, using spin labels as
probes of membrane structure, was carried out on Neurospora
mitochondria (21). This experiment employed a spin label

fatty acid which was included in the growth medium. The grow-ing Neurospora mycelium incorporated the spin label fatty acid into phospholipids and neutral lipids. The spectra of the spin-labeled mitochondria extracted from the mycelium revealed that the membrane hydrocarbon zones allowed rapid rotational diffusion of the spin label. In addition to the information about spin label motion, the observation that a fatty acid containing an oxazolidine ring on carbon-12 is still treated in approximately the same way by Neurospora as unsubstituted fatty acids is an interesting and revealing observation. This indicates that the structural information on a fatty acid, to satisfy the requirements for incorporation, is small compared to the other major classes of biochemicals such as nucleic acids, carbohydrates, and proteins.

Mammalian mitochondria demonstrate the phenomenon of synchro-nous oscillation (treated in chapter   of the present volume) Mitochondria isolated from rat liver of rats fed on a diet deficient in the so-called essential fatty acids have a damp-ed oscillation and a longer oscillatory period. Spin label data obtained from these mitochondria indicated a hydrocarbon environment which offered greater restriction to molecular rotational diffusion than was seen in the control mitochon-dria (22). These data indicated a direct relationship be-tween mitochondrial fatty acid composition, mitochondrial oscillatory rate, and spin label rotational motion. The spin labels used were localized in mitochondrial membrane lipids. The oscillatory function, which is supposedly controlled by regulatory proteins endogenaous to mitochondria, is modified extensively by an altered lipid composition. Therefore, the modified spin label motion parameters which reflected modi-fied structural properties, helped to demonstrate a relation-ship between molecular structure and function of organelle.

High sucrose concentrations inhibit oxygen-dependent hydrogen transport in submitochondrial and mitochondrial preparations. A spin label study was carried out on submitochondrial preparations using the small hydrocarbon spin label, 5N10 (23). 5N10 and the submitochondrial particles were used at concent-rations such that two high field lines appear. This indi-cates that, as described earlier, 5N10 is partitioned between two environments. One environment would be the hydrocarbon zones of mitochondrial membranes and the other would be the aqueous medium. Adding sucrose to these preparations caused the two high field lines to merge into a single line. The single line was broader than either of the components in the two-component high field signal of the control. A suspension of phospholipid vesicles which demonstrated two high field components, much as the mitochondrial membranes had, also lost the two component signal upon being treated with sucrose. An interesting aspect to this study was the observation that mitochondrial membranes which had been glutaraldehyde-fixed did not lose the two component signal after sucrose treatment.

Earlier in this article some of the considerations about two component signals due to partitioning between dielectric zones are treated. One of the implicit considerations is that the boundary between hydrocarbon and water must be well defined. Any agent which would tend to increase the thickness of the interfacial zone will also create new zones having intermediate polarities. Spin labels residing in these zones of intermediate polarity will give rise to signal contributions which will obscure the resolution of the two component signal. This action would require disorganization of the membrane interface. The loss of a well defined hydrocarbon-aqueous boundary due to sucrose and the loss of oxygen-dependent hydrogen transport simultaneously implies that disorganization of the membrane interface zone causes loss of biological activity. Fixing the membrane with glutaraldehyde before sucrose treatment prevented loss of the two component signal. This implies stabilization of the membrane against the disruptive effects of sucrose.

Other spin label studies carried out on mammalian liver mitochondria revealed that a non-linear event occurred in the neighborhood of $24^{\circ}C$ (24). This was detected by plotting partitioning of the high field lines of 5N10, and by measurements of rotational diffusion of several spin labels. This non-linear event was also detected in the oscillation frequency at the same temperature. Other workers have shown a non-linear event in the activity of several enzymes in mitochondrial preparations at this same temperature (25). The non-linear event has been inferred to be a phase transition.

Current and future investigations in this laboratory concentrate on the use of spin labels which localize on the cell surface (18). These molecules offer the opportunity of studying the physical properties of the cell surface with no interfering signal from other zones.

References:

1. McConnell, H. M. and McFarland, B. G. (1970) Q. Rev. Biophys. 3:91-136.
2. Jost, P., Waggoner, A. S. and Griffith, O. H. (1971) in Structure and Function of Biological Membranes (Rothfield, L., ed.), pp. 84-142, Academic Press, New York.

3. Smith, I.C.P. (1972) in Biological Applications of ESR Spectroscopy (Bolton, J. R., Borg, D. and Swartz, H., eds.), pp. 483-539 , Wiley-Interscience, New York.
4. Keith, A. D., Sharnoff, M. and Cohn, G. E. (1973) Bioch. Biophys. Acta (Biomembranes) 300:379-419.
5. Morse, P. D. II, Ruhlig, M., Snipes, W. and Keith, A. D. (1975) Arch. Biochem. Biophys. 168:40-56.
6. McConnell, H. M. (1956) J. Chem. Phys. 25:709-711.
7. Freed, J. H. and Fraenkel, G. K. (1963) J. Chem. Phys. 39:326-348.
8. Kivelson, D. (1960) J. Chem. Phys. 33:1094-1106.
9. Libertini, L. J. and Griffith, O. H. (1970) J. Chem. Phys. 53:1359-1367.
10. Jost, P. and Griffith, O. H. (1972) in Methods in Pharmacology (Chignell, C., ed.), Vol. II, pp. 223-276, Appleton-Century-Crofts, New York.
11. Hubbell, W. L. and McConnell, H. M. (1969) Proc. Natl. Acad. Sci. U.S. 64:20-27.
12. Snipes, W., Cupp, J., Cohn, G. E. and Keith, A. D. (1974) Biophys. J. 14:20-32.
13. Seelig, J. (1970) J. Am. Chem. Soc. 92:3881-3887.
14. Hubbell, W. L. and McConnell, H. M. (1971) J. Am. Chem. Soc. 93:314-326.
15. Williams, J. C., Mehlhorn, R. J. and Keith, A. D. (1971) Chem. Phys. Lipids 7:207-230.
16. Nordio, P. L. (1970) Chem. Phys. Lett. 6:250-252.
17. Gaffney, B. J. and McConnell, H. M. (1974) J. Magnetic Resonance 16:1-28.
18. Lepock, J. R., Morse, P. D., Hammerstedt, R. H., Snipes, W., Mehlhorn, R. J. and Keith, A. D., in manuscript.
19. Hubbell, W. L. and McConnell, H. M. (1968) Proc. Natl. Acad. Sci. U.S. 61:12-16.
20. Keith, A. D. and Snipes, W. (1974) Science 183:666-667.
21. Keith, A. D., Waggoner, A. S. and Griffith, O. H. (1968) Proc. Natl. Acad. Sci. U.S. 61:819-826.
22. Williams, M. A., Stancliff, R. C., Packer, L. and Keith, A. D. (1972) Bioch. Biophys. Acta 267:444-456.
23. Zimmer, G., Keith, A. D. and Packer, L. (1972) Arch. Biochem. Biophys. 152:105-113.
24. Tinberg, H. M., Packer, L. and Keith, A. D. (1972) Bioch. Biophys. Acta 283:193-205.
25. Lyons, J. M. (1973) Ann. Rev. Plant Physiol. 24:445-466.

NMR STUDIES ON THE PRECISE LOCALIZATION OF DRUGS IN THE MEMBRANE. CHARACTERIZATION OF THE REACTION SITE

Jorge Cerbón

INTRODUCTION

During the past few years there has been an increased interest in membrane biology and considerable progress has been made in the identification and study of receptors (specific sites) for peptide hormones, non-peptide hormones, drugs, antibiotics, etc. The general approach in these studies has been to measure the interaction (binding) of a radioactively-labeled chemical (hormone, drug, etc.) with intact target cells, isolated organelles or specialized cell fractions. A number of problems can be encountered in these kinds of studies, and caution must be exercised in the interpretation of the data (1). Virtually all the compounds used as binding ligands exhibit some non-specific binding properties to a variety of non-receptor biological and inert materials. The binding data must therefore be evaluated and compared to the biological activity of the drug, and since all disruptive methods alter to some extent the natural conformation of the membrane constituents and the binding of many of the drugs is reversible, it is desirable that binding and biological response be measured in the same prepartion before disruptive procedures are performed on the system.

Moreover, the characteristics of the specific binding site or the binding site identified as such after fractionation of the system may not be representative of the original situation. In order to tackle the problem from another point of view, a non-destructive technique, nuclear magnetic resonance (NMR) that allows the detection of changes in the micro-environment around a given nuclei (protons, C^{13}, P^{31}, etc.), was employed to characterize the interaction of local anaesthetics with the membrane. Numerous attempts to elucidate the mode of interaction of local anaesthetics with the nerve membrane, using different techniques and various types of natural and model membranes, have been made and all kinds of possible explanations offered: The interaction is of an electrostatic nature, since interaction was easily detected with acidic phosphatidylserine, but not with the switterionic lecithin (2); there is a limited penetration into the hydrophilic head group region (3,4); there is a primary hydrophobic-interaction (5,6), but there were doubts as to whether the C_4 hydrophobic chain of tetracaine (pantocaine) would be long enough to penetrate the hydrophobic fatty acid region of the phospholipid film (7).

303

Therefore, procaine, lidocaine and other local anaesthetics that do not have such a hydrophobic contribution will be unable to show hydrophobic interaction. The local anaesthetic behaves as a hydrogen donor (X-ray analysis of the crystal structure of procaine) (8) and without any experimental evidence; the local anaesthetic interacts with a protein or lipoprotein receptor (9).

Nuclear Magnetic Resonance Studies of Intermolecular Complexes

In the past few years, the possibility of using the differential changes in the proton magnetic resonance spectra of organic molecules to derive information on the nature of binding sites has received increasing attention in the chemical literature. In particular, the reversible interaction of local anaesthetics with model membranes seemed to be a suitable model for exploring this possibility. As a first step, the interaction of these compounds with lipid membrane models was studied. The first objective of our investigation was to determine which portions of the local anaesthetic molecule are involved in the binding to membrane lipids, and if it was the same portion in all of them. A priori, several alternatives exist: at physiological pH, there is a cationic group which could be important in electrostatic binding to the anionic residues of phospholipids. Also, there is a potential hydrogen bond acceptor at the C=O group in the local anaesthetic derivates of the p-amino benzoic acid, and there is also a benzene ring that in some instances, as in tetracaine, have a small n-hydrocarbon side chain; this portion of the molecule could be involved in hydrophobic binding. The usual methods for the study of binding, such as equilibirum dialysis and ultrafiltration measure only the extent of binding, but fail to give information on the mechanism of interaction between ligand and receptor. The findings obtained in our studies (10-13) have shown that such information can be obtained not only qualitatively, but quantitatively from both the spectral changes (alterations in the signals, width and position) and the changes in relaxation time that accompany drug-membrane interaction (A discussion of the theoretical basis of high resolution NMR is given elsewhere (14)).

In Fig. 1, the possible interaction of a local anaesthetic molecule (tetracaine) with a lipid membrane is represented considering only the electrostatic interactions. Since by NMR spectroscopy the different regions of tetracaine can be clearly identified (Fig. 2, upper line), it is possible to measure the spectral changes that occur upon the interaction of the drug with a lipidic membrane (Fig. 2, middle line). It can be seen

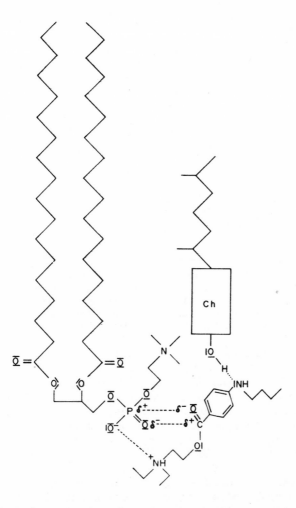

*Fig. 1. Schematic representation of the possible electrostatic inter-
actions of tetracaine with a lipidic membrane.*

that all the protons became broadened and that those in the
$(CH_3)_2$-N group (signal h) are less affected, but clearly identified. The butylene residue (signals a,b and c) as well as
the aromatic ring (signals f and g) which according to Fig. 1
would also be facing the aqueous environment were, however,

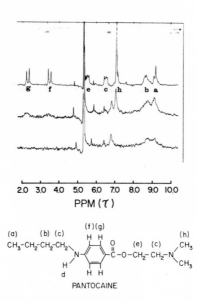

Fi. 2. The NMR spectra of pantocaine-HCl. Upper line: (2% w/v) in
^2H_2O. Middle line: pantocaine-HCl (2% w/v) plus egg phospho-
lipids-membrane (5% w/v) in ^2H_2O (note the broadening of the
NMR signals of the local anaesthetic). Lower line: NMR spectrum
of the lipidic membrane. Temperature 33°C.

almost unidentifiables. These results suggest that the interaction would be as shown in Fig. 3.

In order to test the validity of this model, the interaction
of butacaine with a lipid membrane was studied. This local
anaesthetic has two butylene residues, but at the polar end of
the molecule. The upper line of Fig. 4 shows that again it is
possible to identify all the different parts of the local
anaesthetic molecule, and that upon interaction of butacaine
with the lipid phase (Fig. 4, middle line), the majority of the
signals become broadened except the CH_3 protons of the butyl
residues. Hence, there is a clear difference in the degree of
broadening observed in the spectrum of the CH_3 protons of the
butyl residues that depends upon their localization in the
local anaesthetic molecule. Those located at the non-polar
part of the molecule (tetracaine) are drastically affected,
while those at the polar part of the molecule (butacaine) remain as if they were in aqueous solution.

Once it was shown that local anaesthetic molecules interact
with the lipid membrane in a way that suggests a hydrophobic
interaction plus an electrostatic one (see Fig. 3), it was
necessary to determine whether a net charge of the lipid

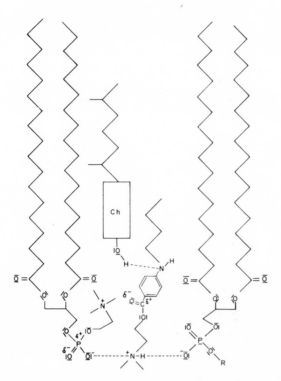

Fig. 3. Schematic representation of the interactions (hydrophobic plus electrostatic) of tetracaine with a lipid membrane.

phase is necessary for the interaction. Procaine and lido-caine, which bear a less hydorphobic contribution than tetra-caine and butacaine, were studied to explore if they behave according to the proposed scheme of interaction. Purified switterionic egg lecithin was employed to prepare the model membrane. The interaction of tetracaine-HCl with lecithin liposomes gave rise to modifications of the spectrum identical to those observed upon their interaction with the lipid mixture (Fig. 2, middle line), indicating that even in the absence of a net charge, tetracaine-HCl is able to interact with the lipid membrane. On the other hand, procaine-HCl and lidocaine-HCl do not seem to interact with purified lecithin liposomes, while butacaine-HCl showed only a certain degree of interaction. At the same time, aqueous dispersions of phosphatidylserine were precipitated by all these local anaesthetics. Thus, the pre-sence of a negative charge in the phospholipids is required for the interaction of those local anaesthetic molecules that do not possess an extra hydrophobic residue at the non-polar aromatic part of the molecule.

307

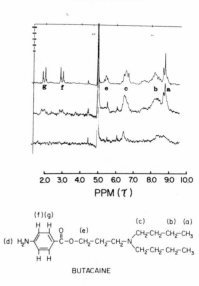

$$2.0 \quad 3.0 \quad 4.0 \quad 5.0 \quad 6.0 \quad 7.0 \quad 8.0 \quad 9.0 \quad 10.0$$

PPM (τ)

Fig. 4. The NMR spectra of butacaine-HCl. Upper line: (2% w/v) in ^2H_2O. Middle line: butacaine-HCl (2% w/v) plus egg phospholipids membrane (5% w/v) in ^2H_2O (note the broadening of the NMR signals of the local anaesthetic with the exception of the signal a). Lower line: the NMR spectrum of the lipidic membrane. Temperature $33^\circ C$.

In mixed phosphatidylserine-lecithin (1/10, w/w) lipid membranes, a differential broadening of the aromatic protons of all local anaesthetic molecules was observed, suggesting that once local anaesthetics are attracted by the opposite charge at the membrane surface, they interact hydrophobically through their aromatic region with the upper part of the lipid carbon chain. As a consequence of their small film penetration as compared to tetracaine, they can be washed out more easily in vivo. The length of duration of the anaesthetic effect and their potency may be related to the strength of the hydrophobic interaction, i.e., tetracaine > butacaine > lidocaine $\simeq$ procaine.

Relation Between Interaction and Mechanism of Action

Studies on the voltage-clamped squid giant axon (15) and on the myelinated nerve fiber (16) have shown that external calcium is intimately associated with the nerve membrane conductance mechanism, and a competitive action of calcium and local anaesthetics has been proposed as part of the mechanism of action of certain of these drugs (17). The efficiency of local anaesthetics in displacing cations, such as calcium,

from the membrane could be related to their capacity to penetrate into the lipid palisade so that their basic groups are
more effective (for being anchored) in neutralizing the negative charge of the phospholipid.

In order to test the above mentioned property of the local
anaesthetic molecules in our model membrane, experiments were
conducted with the paramagnetic cations, europium ($^{3+}$Eu) and
praseodyum ($^{3+}$Pr). These cations induce chemical shifts in
the NMR signal (change the signal position) which arise from
protons of the trimethylammonium (TMA) groups of lecithin
facing the external surface of the lipid membrane. If, indeed,
tetracaine and butacaine interact with lecithin and the results
suggest that their positively-charged polar end faces the membrane surface, it would be expected that they would displace
the polyvalent cations and restore the original position of the
shifted NMR signal. The signal shift can be followed quantitatively, and the efficiency of different local anaesthetics
on modifying this shift can be taken as a measure of their
capacity to remove polyvalent cations from the artificial
membrane.

Figure 5 shows the position of the NMR signal of TMA groups
in a lecithin membrane in the absence of $^{3+}$Pr (upper line).
In the presence of increasing concentration of $^{3+}$Pr (lower
line), the TMA peak splits into two components: peak A arising
from the TMA groups of the lecithin facing the external aqueous
phase and in contact with the cation and shifted low field;
peak B arises from the TMA groups of lecithin facing the internal aqueous phase of the liposomes and maintaining their
original position because they do not have contact with $^{3+}$Pr.
The addition of tetracaine or butacaine reverses the $^{3+}$Pr-
induced splitting of the NMR signal of $-N(CH_3)_3$ protons of
lecithin. Upon increasing the amount of local anaesthetic,
signal A returns to its original position indicating that
tetracaine and butacaine were capable of displacing the polyvalent cation of the lipid membrane. On the other hand, procaine and lidocaine which do not seem to interact hydrophobically with purified lecithin membranes, were unable to reverse
the splitting caused by $^{3+}$Pr.

It was thought that additional insight into the interfacial
behavior of anaesthetics with small hydrophobic contribution
could be obtained through studies on their interaction with
anionic detergent micelles of sodium dodecylsulphate (SDS).
This is a negatively-charged system in which the possibility
of using higher concentrations of the lipid could probably
render it more resistant to precipitation than in the case of
phosphatidylserine and, therefore, accessible to NMR spectroscopy. It was found that 90 mM lidocaine acted similarly to

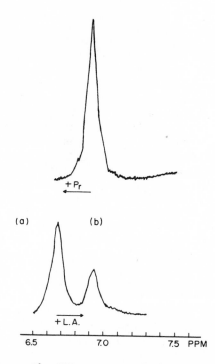

(a)   (b)

6.5   7.0   7.5 PPM

Fig. 5. Upper line: the NMR spectra of the N-(CH₃)₂ protons of egg lecithin membranes (5% w/v). Lower line: splitting of the N-(CH₃)₂ NMR signal upon the addition of 10 mM PrCl₃. The arrows indicate the chemical shift induced by PrCl₃ and the local anaesthetic. Temperature 33°C.

60 mM tetracaine in inducing the formation of highly viscous phases upon interaction with 5% SDS. This property resides in the capacity of local anaesthetics to interact polarly and hydrophobically at the same time. The uncharged form of the local anaesthetic molecule, as well as the cationic residue alone (trimethyl or triethylammonium), are incapable of inducing such changes (12). Also, there was a large broadening and loss of intensity of the SDS methylene signals. This could be indicative of a strong immobilization of the hydrocarbon chains through an extra hydrophobic interaction with the anaesthetic molecule. However, this broadening could also be attributed to the large increment in the viscosity of the system.

To clarify this last point, the measurement of relaxation times of protons from the local anaesthetic as well as protons of the lipid phase were determined by Fourier Transform pulse NMR; the theory and methods of pulse and Fourier Transform

NMR can be found elsewhere (18). With this technique, relaxation times can be determined for all the resonance lines in a complex system, and thus extend greatly the applicability of NMR techniques to the study of dynamic molecular processes in complex biological systems. That is, it is possible to characterize quantitatively the interaction of a drug with the membrane by following the changes in the individual values of T_1 for each group of protons in the different parts of the drug molecule when passing from solution to receptor sites. In the present case (simple lipid-water interphase), the changes of the receptor site during the interaction (Table I) were followed also.

TABLE I

Effect of interaction between tetracaine and SDS micellar systems $1/T_1$ sec^{-1} values

| | Protons of the Local Anaesthetic | | |
	60 mM Tetracaine	60 mM Tetracaine + 173 mM SDS	Δ
N-$(CH_3)_2$	0.89	2.84	1.95
CH_2-N$(CH_3)_2$	2.07	7.55	5.48
$\overset{H}{\underset{H}{\big|}}$ $\overset{O}{\overset{\|}{C}}$ - R	0.76	3.25	2.49
N-$\big\langle\overset{H}{\underset{H}{}}$	1.01	3.9	2.89

| | Protons of the Receptor Site (SDS) | | |
	173 mM SDS	173 mM SDS + 60 mM Tetracaine	Δ
CH_2-S	2.9	3.3	0.4
CH_2-CH_2-S	2.8	2.9	0.1
$(CH_2)_n$-CH_2-CH_2-S	2.9	3.0	0.1
CH_3-$(CH_2)_n$-S	2.04	2.15	0.1

The mobility of the local anaesthetic protons is inhibited in the following sequence suggesting again a type of polar plus hydrophobic interaction, similar to that shown in Fig. 3. This means that upon addition of the local anaesthetic to the lipid-water interphase, the cationic group of the drug neutralizes the negatively-charged interphase. This neutralization

$$CH_2-N(CH_3)_2 > N-\overset{H}{\underset{H}{C}} > \overset{H}{\underset{H}{C}}-\overset{O}{C} > N(CH_3)_2$$

allows a closer proximity of the SDS molecules and, in addition, the apolar part of the anaesthetic (aromatic) region would favor hydrophobic interactions. The mobility of all the protons of the local anaesthetic molecule below the charged protion (going from the polar to the non-polar part of the molecule) is much more inhibited than that of the protons above the charge. Therefore, there is no general effect that can be attributed to the increment in the macroscopic viscosity of the system. High resolution NMR studies of the local anaesthetic plus SDS afford further data that reinforce the above mentioned conclusions on the precise localization of the drug in the lipid model membrane (12). There are upfield displacements of the local anaesthetic NMR signals when passing from the water to the lipid system, and it is known that chemical shifts in a hydrophobic (paraffinic) medium are displaced upfield in relation to their positions in water (19). The larger upfield shifts observed were those arising from the aromatic protons ortho to the amino group in tetracaine and the aromatic protons of lidocaine; i.e., those regions of the local anaesthetic molecules that according to the proposed interaction lie in a more internal region of the lipid membrane. Furthermore, the signals arising from the $N^+(CH_3)_2$ protons are broadened, but practically no shifts were observed. Preliminary results obtained by studying the interaction of local anaesthetics with the intact biological receptor site (cell membrane) suggest a similar type of interaction qualitatively; also some quantitative differences were observed.

By examining the extensively studied field of mitochondria it is possible to see that, at least during the past seven years, people have tried hard to characterize the sites on interaction of numerous drugs that affect in some way the mitochondrial respiratory chain. Some of the most interesting drugs interact non-covalently with "specific" sites and present the same type of problems that I mentioned before. At this time, the "conclusion" is that the specific binding of rotenone and piericidin seem to involve both lipid and protein (20-23).

FUTURE PERSPECTIVES

According to the results obtained in these studies, I believe that: a) it has been possible to characterize the precise location of the drug at the lipid membrane; b) it has been possible to explain how local anaesthetics, in spite of being monovalent cations, can compete with polyvalent cations at the membrane surface; c) it has been established that a negative charge in the active site is necessary to favor the hydrophobic interaction of some of these drugs (procaine and lidocaine), and d) based on both the qualitative and quantitative estimation of their hydrophobic interaction, a logical explanation can now be offered for the difference in potency of serveral local anaesthetics.

Encouraged by the results obtained in these studies, I would suggest that a similar approach be attempted in other areas of scientific research such as mitochondrial electron transfer systems, mitochondrial ATPase, etc.

REFERENCES

1. Cuatrecasas, P. and Bennett, V. (1974) *in* Perspectives in Membrane Biology (Estrada-O, S. and Gitler, C., eds.), pp. 439-453, Academic Press, New York.
2. Hauser, H., Penkett, S.A. and Chapman, D. (1969) Biochim. Biophys. Acta, 183, 466.
3. Gershfeld, N.J. (1962) Phys. Chem. 66, 1923.
4. Feinstein, M.B. (1964) J. Gen. Physiol. 48, 357.
5. Skou, J.C. (1954) Acta Pharmacol. Toxicol. 10, 325.
6. Seeman, P. and Kwant, O.W. (1969) Biochim. Biophys. Acta 183, 512.
7. Hauser, H. and Dawson, R.M.C. (1968) Biochem. J. 109, 909.
8. Sax, M. and Pletcher, J. (1969) Science 166, 1546.
9. Cutting, W.C. (1967) *in* Handbook of Pharmacology, 3rd ed., p. 496, Appleton Century Crofts, New York.
10. Cerbón, J. (1972) Biochim. Biophys. Acta 290, 51.
11. Fernández, M.S. and Cerbón, J. (1973) Biochim. Biophys. Acta 298, 8.
12. Fernández, M.S. and Cerbón, J. In preparation.
13. Cerbón, J., Fernández, M.S. and Pryce, N. In preparation.
14. Becker, E.D. (1969) *in* High Resolution NMR-Theory and Chemical Applications, Academic Press, New York.
15. Frankenhaeuser, B. and Hodgking, A.L. (1957) J. Physiol. 137, 218.
16. Frankenhaeuser, B. (1957) J. Physiol. 137, 245.

17. Blaustein, M.P. and Goldman, D.E. (1966) J. Gen. Physiol. 49, 1043.

18. Farrar, T.C. and Becker, E.D. (1971) *in* Pulse and Fourier Transform NMR. Introduction to Theory of Methods, Academic Press, New York.

19. Tokiwa, F. and Aigami, K. (1971) Kolloid Z. UZ. Polymere 246, 688.

20. Horgan, Y.D., Ohno, H. and Singer, T.P. (1968) J. Biol. Chem. 243, 5967.

21. Gutman, M. and Singer, T.P. (1970) J. Biol. Chem. 245, 992.

22. Gutman, M., Singer, T.P., Beinert, H. and Casida, J.E. (1970) Proc. Nat. Acad. Sci. (US) 65, 763.

23. Yagushinsky, L.S., Smirnova, E.G., Ratmikova, L.A., Kolesova, G.M. and Krasinskaya, I.P. (1973) J. Bioenerg. 5, 163.

THE USE OF APOLAR AZIDES TO MEASURE PENETRATION OF PROTEINS
INTO THE MEMBRANE LIPID BILAYER

Amira Klip and Carlos Gitler

INTRODUCTION

The current membrane models (1,2) propose that proteins are
embedded in a lipid bilayer; these proteins can bear one region exposed to the water medium and one buried in the lipid
apolar environment. Suggestive evidence for this disposition
stems from diverse sources (3-5. By means of water-soluble
chemicals it has been possible to label those regions of membrane proteins which are exposed to the aqueous environments,
both from the outer or extracellular space and from the inner
or cytoplasmic medium (6). Those portions of membrane proteins
which are not labelled by this technique are said to be buried
in the lipid core (7). A more direct approach to determine
which regions of membrane proteins penetrate the lipid core of
the bilayer would be to label these with hydrophobic probes.
Furthermore, a probe tightly bound to the hydrophobic regions
of membrane proteins could detect changes in their structure
and mobility. These changes have been proposed to occur in
membranes performing diverse functions (8,9).

The ideal hydrophobic probe should fulfill the following
requirements: a) to have a large partition towards apolar
phases, b) to be able to react covalently with the rather
inert chemical groups that proteins and lipids expose to non-
polar environments, c) to be reactive when it is located inside the membrane, but not on its way to the apolar core.

These requirements are met by hydrophobic photochemical
probes. Such reagents are generally aromatic azides or diazo
derivatives which are chemically unreactive. However, upon
illumination these compounds generate nitrenes or carbenes
respectively (10) (see Fig. 1), which are very reactive. This
property would allow the compounds to reach the membrane core
in an unreactive state, and then be activated at will by a
flash of light. Carbenes and nitrenes react covalently by
insertion into C-H bonds and by addition to C=C bonds. Since
these bonds are abundant in the protein and lipid portions of
the membrane core, they should easily react with the activated
label.

The present study describes the use of a hydrophobic photo-
active acompound, 1-Azidonapthalene (AzN) to label the regions
of the Ca^{2+}-ATPase of sarcoplasmic reticulum membranes (SR)
that are in close contact with the lipid phase.

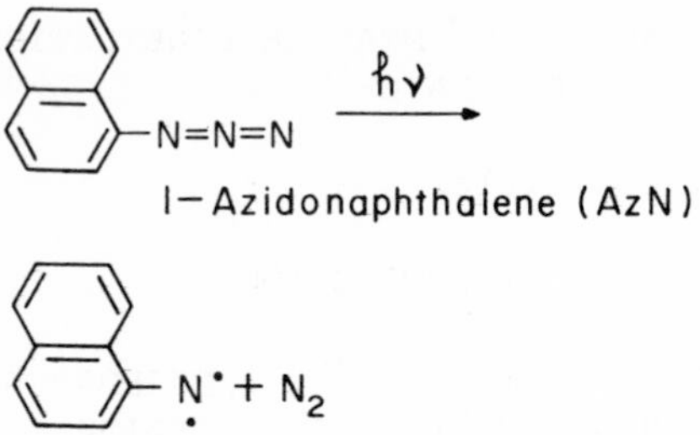

Fig. 1. *Photoactivation of 1-Azidonaphthalene (AzN).*

METHODS

AzN was prepared by the general method of synthesis of aromatic azides (26): 8.25 mmole of 1-aminonapthalene in 50 ml 6 N HCl were exposed to 10 mmole of $NaNO_2$ to form the diazonium salt at $0°C$. After removal of the excess HNO_2 with urea, 13.8 mmole NaN_3 were added. The AzN formed separated as an oil which was extracted from the reaction mixture into ethyl ether. The reaction was performed under dim light. The AzN obtained was identified by IR and UV spectroscopy, and its purity was checked by NMR and thin layer chromatography.

Tritium was incorporated into the aromatic rings in a non-exchangeable way as follows (27): 1 ml [^{3}H]-H_2O (5 Ci/ml) was agitated overnight in benzenethiol to allow the SH hydrogen to exchange with tritium. The thiol was dried over Na_2SO_4 and 5 mmole 1-aminonaphthalene were dissolved in 10 ml of it (1.45 X 10^{14} dpm/mole), together with 10 mg azodiiso-butyronitrile. The solution was incubated in a sealed ampule at $80°C$ for 10 hr. The amine was then precipitated from the thiol as the hydrochloride salt and washed repeatedly with benzene. The powder was crystallized from methanol and checked for purity. The specific activity thus obtained was 5.57 x 10^{12} dpm/mole aminonapthalene. This amine was used to synthesize [^{3}H]-AzN as described above.

Sarcoplasmic reticulum membranes from rabbit muscle (11) (4 mg protein/ml) were exposed to 1 x 10^{-4} M of [^{3}H]-AzN (2.5 Ci/mole). The compound was allowed to penetrate into the membranes in the dark at $25°C$ and the suspension was centrifuged at 45,000 x g for 30 min. The membranes loaded with [^{3}H]-AzN were resuspended in 100 mM KCl, 10 mM imidazole, pH 7.1; one half of this suspension was irradiated at $37°C$ for 30 min to 1 hr at $\lambda=340$ nm with three 275 watt Sylvania sun lamps. Under these conditions, the azide is completely

photoactivated yet no damage of membrane protein occurs (12).
Irradiation of AzN produces 1-nitrenenaphthalene which, in
addition to the insertion and addition reactions, can also
abstract hydrogen atoms or react with other nitrene molecules.
The products of these reactions are not covalently attached to
membrane elements and should be removed from the membranes.
For this purpose, the membranes were washed repeatedly with an
albumin solution until no more radioactivity was liberated
from them. Non-illuminated control membranes lost over 98%
of the [^{3}H]-AzN while 25-30% of the label remained in the
irradiated membranes.

In order to determine which elements of the membranes contained the label, SDS disc gel electrophoresis of membrane
proteins (14) and lipid extractions (28) were carried out.
Enzymatic cleavage of membrane proteins and fluorescence
measurements were also performed to determine the distribution
of the label in the membrane elements. The latter were performed in isotonic phosphate buffer (pH 7.0) to avoid interference of the imidazole-containing buffer used in all other
determinations.

RESULTS

Membrane Location of AzN: AzN is a highly apolar reagent.
It was obtained as an oil which is sparingly soluble in water.
The partition coefficient of AzN between 1-octanol and water
was 166.

Fluorescence quenching experiments were carried out to test
the penetration of AzN into the membrane lipid bilayer. Anthracene is a hydrophobic fluorophore which distributes preferentially towards apolar phases (15). The fluorescent
emission of anthracene can be quenched by molecular collisions
with azides and several other substances. The fluorescence of
anthracene within micelles and SR membranes was scarcely affected by ionic water-soluble quenchers such as iodide. On
the other hand, AzN and the lipid-soluble 1,3-dinitrobenzene
were effective encounter quenchers of anthracene fluorescence
in the membranes. These results demonstrate that AzN and
anthrcene meet in the same locus, i.e., the hydrophobic domain
of the membrane.

Reaction of AzN with Lipids and Proteins: Figure 2 shows
the electrophoretic profile of the proteins of SR membranes
irradiated in the presence of [^{3}H]-AzN and washed as described
under Methods. The radioactivity distribution in the gel is
also shown. The Ca^{2+}-ATPase appears as the major band with a
molecular weight of about 95,000. It contains 15% of the total
radioactivity bound to the membranes. The large radioactive

peak in the running front of the gel presumably represents labeled lipids.

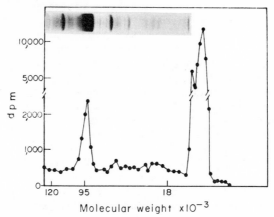

Fig. 2. *Electrophoretic pattern of labelled SR membranes. Sodium dodecylsulfate disc gel electrophoresis was performed in 10% polyacrylamide gels according to Laemmli (14). The radioactivity in the gels was determined as previously described (12).*

The amount of label incorporated into the lipid fraction of membranes was also analyzed by extraction into solvents. When 91.3 ± 8.3% (mean ± SEM) of the lipids were extracted from labelled SR membranes into chloroform-methanol, only 38.9 ± 4.1% of the total radioactivity appeared in the solvent phase. The extracted lipids were then hydrolyzed in methanolic KOH and the fatty acids were protonated with acid and extracted into ether. In this ether extract 86.9 ± 3.8% of the total radioactivity of the lipids was found. This suggests that the label on the lipids is preferentially attached to their fatty acyl moieties.

Pronase Treatment: Membranes are impermeable to proteins. It is considered that when membranes are exposed to proteases, only the regions which the membrane proteins expose to the water medium can be digested (16). This will be so until an appreciable fraction of membrane protein has been cleaved off the membrane; at this point the whole membrane will disintegrate and additional portions of the protein will be available for proteolysis.

SR membranes were labelled with [^{3}H]-AzN and were then exposed to varying concentrations of pronase. After centrifugation of the membranes, the protein and radioactivity content of the membrane pellet and the supernatant were determined. When 25% of the protein was removed from the membranes, the

radioactivity liberated amounted to less than 5% of the total
label in the membranes. Considering that 40% of the label on
the membranes was attached to the lipids, the remaining 60%
should be on proteins. Thus, when 25% of the membrane protein
was cleaved off by pronase, only 8% of the radioactive protein
was liberated. Accordingly, when 50% of the protein was
rendered soluble, only 10% of the total label (17% of the
radioactive protein) was liberated. This implies that the
label is not homogeneously distributed on the proteins and,
furthermore, that the labelled portions are not appreciably
accessible to pronase.

Fluorescence Measurements: The reaction of AzN within the
membrane produces naphthylamines covalently bound to proteins
and lipids. Both the mono- and dialkyl derivatives of aminonaphthalene are produced; these compounds are fluorescent.
When the fluorescence of free aminonaphthalene was measured
in dioxane-water mixtures of variable polarity, the emission
maximum shifted to lower wavelengths as the dielectric constant decreased (Fig. 3a). If the fluorescence of the naphthylamines attached to the membrane elements is considered to
vary accordingly with the dielectric constant of the medium,
it is possible to predict a value for the polarity of the
environment in the vicinity of the label. The emission maximum of labelled SR membranes was 410 nm (Fig. 3b). This value
corresponds to a dielectric constant < 10. This low polarity
can be found in the highly apolar domain of the membrane core.
The polarity in the membrane-water interface is similar to
that of ethanol (17) (dielectric constant 25). Since the
polarity near the naphthyl derivatives is lower, these labels
are apparently not located in the membrane interface to an
appreciable extent.

Labeling of Water-soluble Proteins: It could still be proposed that AzN approaches the membrane proteins from the
aqueous side and reaches an apolar crevice in the protein.
To test this possibility, several water-soluble proteins
bearing hydrophobic sites (18) were exposed to [^{3}H]-AzN.
Neither trypsin nor ribonuclease incorporated any label when
irradiated in the presence of the azide. The label is not
capable of approaching the amino acyl chains that these proteins expose to the water phase. The azide therefore stays
in the aqueous environment and upon illumination can only
react with water. This finding can be extrapolated to the
reaction of AzN with membranes, indicating that the label
which is not incorporated into the apolar phase is unlikely to
react with membrane elements.

Of the water-soluble proteins analyzed, only bovine serum
albumin was able to incorporate the label covalently. This

is not surprising, since albumin binds effectively a large variety of hydrophobic molecules and fatty acids. This finding suggests that if, indeed, some of the label on the membrane proteins were in a non-polar crevice not in contact with the lipid, this crevice should be as hydrophobic as the one in serum albumin.

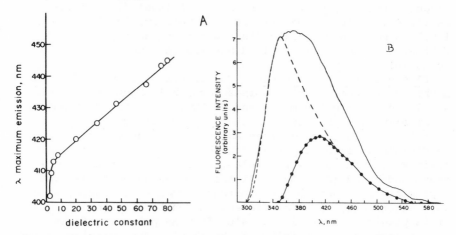

Fig. 3. Fluorescence measurements of SR membranes. A. Emission maxima of 1-aminonaphthalene as a function of dielectric constant. The fluorescent emission of 1-aminonaphthalene was measured in dioxane-water mixtures and dielectric constant values were calculated according to Turner and Brand (18). The excitation wavelength was 350 nm. B. Emission spectra of SR membranes labelled with [³H]-AzN. The membranes were exposed to [³H]-AzN in the dark or under the irradiation lamps. Non-covalently bound label was eliminated as described under Methods. Fluorescent emission spectra of the irradiated (———) and non-irradiated (----) samples were recorded at 285 nm excitation. The emission maximum at 345 nm is due to the membrane trytophan residues. The differential spectrum (—•—) has a maximum at 310 nm.

In order to determine if a lipid environment was necessary to bring the label close to the proteins, the following experiments on proteoliposomes were preformed. Gitler and Montal (19) have shown that water-soluble proteins can bind to lipids by means of Ca^{2+} bridges. The resulting proteolipids can be extracted into an organic solvent and remain soluble therein. When the solvent is evaporated and the proteolipids are sonicated in an aqueous medium, protein-containing membranes (proteoliposomes) are obtained.

These proteoliposomes were prepared with trypsin and phospholipids. When the vesicles were exposed to [³H]-AzN, the

enzyme incorporated a significant fraction of the label. This
result, compared with the inability of water-soluble trypsin
to bind any label, suggests that an apolar environment is
necessary to bring the label close to the protein.

<u>Localization of the Labelled Regions of Membrane Proteins</u>:
The accurate description of the labelled regions of membrane
proteins requires their thorough peptide analysis. This is
difficult to achieve with a protein which has not been se-
quenced yet, as is the Ca^{2+}-ATPase. However, preliminary
attempts to solve this question have provided the following
findings. When SR membranes are treated with low concentrations of trypsin, the Ca^{2+}-ATPase splits into two fragments
of molecular weights 60,000 and 55,000 (20). These fragments
sediment with the membrane indicating that they are attached
to it, probably by means of a hydrophobic anchoring region,
as is the case for other membrane proteins (7,21). If both
fragments of the Ca^{2+}-ATPase contain a hydrophobic site
anchored in the apolar core of the membrane, they should both
be labelled with $[^3H]$-AzN. When $[^3H]$-AzN labelled SR membranes were trypsinized, centrifuged and then analyzed by gel
electrophoresis, both fragments of the ATPase appeared
labelled.

DISCUSSION

Sarcoplasmic reticulum membranes contain a high protein-to-
lipid ratio (28). In addition, the variety of proteins they
contain is fairly small, the Ca^{2+}-ATPase constituting approximately 80% of the protein content. This feature makes SR
membranes very suitable for the study of membrane structure.
In addition, the function of the Ca^{2+}-ATPase, namely the translocation of calcium ions against a concentration gradient
coupled to the hydrolysis of ATP, is well documented.

This paper describes the use of a hydrophobic marker to
label the apolar regions of SR membrane proteins which are in
contact with the lipid core. The reagent is a hydrophobic
aryl azide, a photoactivable compound.

Photochemical reagents have been used before for photoaffinity labeling (22). This technique was developed to label
specific sites of enzymes, especially those involved in substrate recognition and catalytic activity. The reagents
employed were analogs of natural substrates adequately modified
to become photoactivable; they were meant to come close to the
active site of the enzyme and to react covalently with it when
exposed to light. However, this approach has proved inaccurate
on the basis of the reaction of the substrate analogs with

sites other than the catalytic one (23). This becomes important when the lifetime of the activated label is longer than
the time constant of the dissociation of the label from the
active site.

This disadvantage is not met in the approach described in
this chapter. Here we have attempted to label indiscriminately
all membrane elements present in the apolar phase. The apolar
label should not have steric selectivity for any membrane
element, but rather a partition selectivity towards apolar
phases. In this regard, it is interesting that a sterically
different compound, 1-azido-4-iodobenzene, was as effective
as AzN to label the membrane elements (12,13). Both azides
share a preference for highly apolar phases; this stresses the
tenet that the labeling is determined by the partition characteristics of the label.

We consider that the label reacted with membrane elements
present in the apolar domain since: a) the label bound to the
lipids is almost entirely on the fatty acyl chains; b) the
labelled portions of the proteins are relatively inaccessible
to the hydrolytic activity of proteases; c) the fluorescence
emitted by the label covalently attached to membrane elements
reports a highly apolar surrounding; d) water-soluble proteins
such as trypsin and ribonuclease do not get labelled with
$[^3H]$-AzN. However, when trypsin is surrounded by lipids it is
significantly labelled by the azide.

As pointed out before, the label could in principle detect
conformational or topological changes of membrane proteins
associated to their function. When SR membranes were irradiated with $[^3H]$-AzN in the presence or absence of Ca^{2+} and
ATP, the total label incorporated into both membrane preparations was identical. Furthermore, the label distribution on
lipids and proteins was indistinguishable in both cases. This
does not necessarily rule out the possibility that the ATPase
would sink to variable depths in the membrane when actively
transporting calcium. The changes might be too small to be
detected by our method. Moreover, a quantitative agreement
between the labelling with and without Ca^{2+} and ATP does not
mean that no qualitative changes are taking place. The label
could be located on different amino acid residues under both
conditions, but the overall amount of label incorporated could
be the same. This question will be solved when a thorough
peptide analysis of the labelled proteins is carried out. It
is interesting to add that the ATP hydrolytic activity of the
membrane enzyme was unaltered by the labeling with $[^3H]$-AzN.

One further data deserves comment: the total radioactivity
extracted by chloroform-methanol accounted for 40% of the
label on the membranes, when practically all the lipids were
extracted. However, when the same labelled membranes were

analyzed by gel electrophoresis, the labelled peak in the
front of the gel accounted for almost 70% of the total label.
This peak contains the lipid plus a low molecular weight protein (24). This protein is thought to be located in the core
of the membrane and should therefore be highly labelled by
our compound. Alternatively, the discrepancy could be the
result of an incomplete extraction of the labelled lipids
into the organic phase.

At present, the data do not allow a determination of
which proteins of the SR do not penetrate the lipid core.
Proteins other than the ATPase and the low molecular weight
proteolipid are present in such small amounts that the label
incorporated by them, if any, could be too low to be detected.
Nevertheless, it is noteworthy that MacLennan et al. (24)
have proposed that only the Ca^{2+}-ATPase and the proteolipid
penetrate significantly into the lipid phase.

FUTURE PERSPECTIVES

1-Azidonaphthalene penetrates into the membrane apolar core.
This core is about 30Å thick and the label is not more than
6 Å long. Hence, the label can move in a wide region and
react with all C-H and C=C bonds present in that zone. This
makes it difficult to define the accurate depths of the membrane at which the different elements are labelled. Therefore, it would be desirable to anchor the label to a long
amphipathic chain that could orient perpendicularly to the
plane of the membrane. Similar reagents, bearing nitroxides
instead of azides, have been used as monitors of membrane
viscosity (25). The aromatic azides could be attached at different levels to the hydrocarbon chain of an ionic detergent.
Thus, the azide would react only with the elements present at
a known depth in the membrane.

Further information on membrane structure could be provided
by apolar aromatic diazides. These bifunctional compounds
could react simultaneously with two different membrane elements,
crosslinking neighboring molecules in the membrane core.
Water-soluble divalent reagents have been used to crosslink
the elements present in the membrane surface. Crosslinking by
these agents requires the presence of specific reacting groups
on the membrane components. Thus neighboring molecules lacking the reactive group are not detected. In contrast, the
aromatic diazides are capable of reacting with practically all
the groups present in the membrane core. Furthermore, once the
first azide on the molecule is activated, the reacting efficiency of the second azide is enhanced. This would allow a
very effective, fast clipping of neighboring molecules. The

knowledge of the exact nearest neighboring molecules in the membrane could provide further insight into membrane structure and function.

ACKNOWLEDGMENTS

We are indebted to Dr. Gabriel Gojon for his help in the introduction of tritium into aromatic rings, and to Miss Eunice Zavala for excellent technical assistance.

REFERENCES

1. Singer, S.J. and Nicolson, G.L. (1972) Science 175, 720.
2. Gitler, C. (1972) Ann. Rev. Biophys. Bioenerg. 1, 51.
3. Choules, G.L., Sandberg, R.G., Stegall, M. and Eyring, E. M. (1973) Biochemistry 12, 4544.
4. Helenius, A. and Soderlund, H. (1973) Biochim. Biophys. Acta 307, 287.
5. Branton, D. (1969) Ann. Rev. Plant Physiol. 20, 209.
6. Mueller, T.J. and Morrison, M. (1974) J. Biol. Chem. 249, 7568.
7. Segrest, J.P., Kahane, I., Jackson, R.L. and Marchesi, V.T. (1973) Arch. Biochem. Biophys. 155, 167.
8. Blasie, J.K. (1972) Biophys. J. 12, 191.
9. Pang, D.C., Briggs, F.N. and Rogowski, R.S. (1974) Arch. Biochem. Biophys. 164, 332.
10. Knowles, J.R. (1972) Accounts Chem. Res. 5, 155.
11. Martonosi, A. (1968) J. Biol. Chem. 243, 71.
12. Klip, A. and Gitler, C. (1974) Biochem. Biophys. Res. Commun. 60, 1155.
13. Gitler, C. and Klip, A. (1974) in Perspectives in Membrane Biology (Estrada-O, S. and Gitler, C., eds.) pp. 149-178, Academic Press, New York.
14. Laemmli, U.K. (1970) Nature 227, 680.
15. Pownall, H.J. and Smith, L.C. (1974) Biochemistry 13, 2594.
16. Spatz, L. and Strittmatter, P. (1973) J. Biol. Chem. 248, 793.
17. Gitler, C. and Rubalcava, B. (1971) in Probes of Structure and Function of Macromolecules and Membranes, Vol. 1, (Chance, B., Lee, C.P. and Blasie, J.K., eds.) pp. 311-323, Academic Press, New York.
18. Turner, D.C. and Brand, L. (1968) Biochemistry 7, 3381.
19. Gitler, C. and Montal, M. (1972) FEBS Letters 28, 329.
20. Thorley-Lawson, D.A. and Green, N.M. (1973) Eur. J. Biochem. 40, 403.
21. Strittmatter, P., Rogers, M.J. and Spatz, L. (1972) J. Biol. Chem. 247, 7188.

22. Shafer, J., Boronowsky, P., Laursen, R., Finn, F. and Westheimer, F.H. (1966) J. Biol. Chem. 241, 421.
23. Ruho, A.E., Keifer, H., Roeder, P.E. and Singer, S.J. (1973) Proc. Nat. Acad. Sci. 70, 2567.
24. MacLennan, D.H., Yip, C.C., Iles, G.H. and Seeman, P. (1972) Cold Spring Harbor Symp. Quant. Biol. 37, 469.
25. Jost, P., Waggoner, A.S. and Griffith, O.H. (1971) in Structure and Function of Biological Membranes (Rothfield, L., ed.) pp. 84-144, Academic Press, New York.
26. Smith, P.A.S., and Brown, B.B. (1951) J. Amer. Chem. Soc. 73, 2438.
27. Gojon, G., Personal Communication.
28. Meissner, G. and Fleischer, S. (1971) Biochim. Biophys. Acta 241, 356.

NANOSECOND FLUORESCENCE SPECTROSCOPY OF BIOLOGICAL MEMBRANES

P.A. George Fortes

> *Caminante son tus huellas*
> *el camino y nada más.*
> *Caminante no hay camino*
> *se hace camino al andar*
> *caminante no hay camino*
> *sólo estelas en la mar.*
> ANTONIO MACHADO

INTRODUCTION

The first reports on the use of fluorescent probes in membrane studies appeared in 1968-69 (1-4). Since then the literature has grown enormously, as can be judged by a number of reviews on the subject (5-7). Fluorescence spectroscopy has been used to study several membrane functions including fluidity, microenvironment, membrane potentials, conformation changes, etc. (cf. 5-7). However, since several factors affect fluorescence and probes may be located in different membrane sites, it is often difficult to interpret measurements obtained with conventional techniques.

Nanosecond fluorescence spectroscopy offers the advantage that different types of probe binding regions can be distinguished and the contributions of each region to the total intensity can be separated. Thus for a system with n different types of binding sites, the decay of fluorescence intensity $F(t)$, following excitation by a short light pulse, is given by:

$$F(t) = \sum_{i=1}^{n} a_i e^{-t/\tau_i}$$

where τ_i is the lifetime of the probe in the ith site. The coefficient a_i is defined by $a_i = A \cdot k_f \cdot \varepsilon_i \cdot X \cdot C_i$, where A is an instrumental constant, X is the optical path length and k_f, ε_i and C_i are the rate of fluorescence emission, the extinction coefficient and the concentration of the probe bound to the ith site, respectively.

For certain fluorescent probes, such as 1-anilino-8-naphthalene sulfonate (ANS), τ_i depends on the properties of the ith site, which allows identification of different binding sites from the lifetimes of the fluorescence decay; ε_i and k_f are not significantly dependent on the environment. Thus, nanosecond fluorescence data yield values of τ_i and a_i which, respectively, identify and quantify the probe concentration in each binding site.

Experimentally, the duration of the light pulse, L(t), and the response of the detectors alter the decay curve. The recorded fluorescence decay, R(t), is related to the actual decay, F(t), by the equation:

$$R(t) = \int_0^t L(T)F(t-T)dT$$

F(t) is obtained from analysis of the experimental R(t) using a method of moments (8). A detailed description of the theory, instrumentation and methods of analysis is available (9).

In this chapter, a nanosecond fluorescence study of the ANS binding regions in human erythrocyte membranes is presented. Since ANS inhibits anion transport (10) and changes the cell shape (11,12), it was of interest to see if an anion transport site and the mechanism of the perturbations induced by the probe could be identified. In addition, some questions on the usefulness of ANS as a membrane probe are studied.

The ANS Environment in Model Systems

The fluorescence parameters such as emission spectrum, quantum yield and lifetime depend on the composition and properties of the environment surrounding the fluorophore. Since a variety of factors can affect these parameters, studies of the properties of the probe in model systems can be useful in the interpretation of measurements in complex systems such as biological membranes.

Figure 1 shows fluorescence decays of ANS solutions in alcohols of increasing chain length. In these homogeneous systems the decay follows a single exponential and the lifetime (τ) increases as the polarity of the medium decreases. The change in τ results from varying rates of depopulation of the excited state, which decays by fluorescence and nonradiative processes (i.e., internal conversion, intersystem crossing, quenching, etc.). The rates of fluorescence, k_f, and nonradiative decay, k_n, can be calculated if the quantum yield, Φ, and τ are known using the equations:

$$\Phi = \frac{k_f}{k_f + k_n} \qquad \tau = \frac{1}{k_f + k_n}$$

so that $k_f = \Phi/\tau$ and $k_n = k_f - 1/\tau$

Table I shows Φ, τ and derived values of k_f and k_n for ANS in a variety of solvents. Whereas the rate of light emission remains relatively constant, the rate of nonradiative decay increases with solvent polarity. This indicates that the

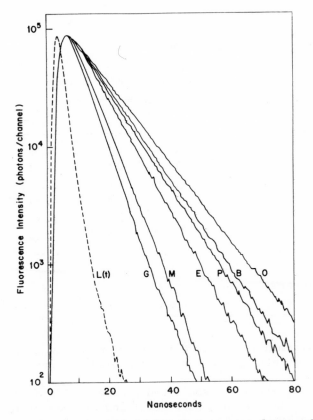

Fig. 1. Effect of solvent on ANS fluorescence decay. All curves are normalized to the same peak count. In this and the next figures: L(t) = exciting pulse. G = glycerol, M = methanol, E = ethanol, P = proponal, B = butanol, O = octonal. ∿ 1 μM ANS. Temperature 23°C.

main effect of polar solvents is to increase the rate of nonradiative decay - probably intersystem crossing (13) - and thus decrease the quantum yield. But solvent polarity is not the only variable. A viscous environment with hindered movement of the solvent dipoles that quench the excited state results in an increased lifetime and quantum yield, even if it is polar (13-15). When the viscosity of glycerol (relaxation time ∿ 1 nsec, similar to the ANS lifetime) is increased by lowering the temperature from 23° to 4°C, τ increases from 4.9 to 8.1 nsec (Table I). In less viscous solvents like ethanol, the lifetime changes only fractions of a nsec in this temperature range. Thus, ANS in a highly structured and

TABLE I

Fluorescence Parameters of ANS in a Series of Solvents

Solvent	Φ	τ (nsec)	$k_f(\text{sec}^{-1})$ $\times 10^{-7}$	$k_n(\text{sec}^{-1})$ $\times 10^{-7}$
Octanol	0.646	12.3	5.23	2.86
Butanol	0.516	10.9	4.73	4.44
Propanol	0.476	10.2	4.65	5.12
Ethanol	0.361	8.85	4.08	7.22
Methanol	0.216	6.05	3.57	12.9
Dioxane	0.505	11.8	4.28	4.19
H_2O	--	0.55	--	--
D_2O	--	1.0	--	--
Glycerol (23°C)	--	4.9	--	--
Glycerol (4°C)	--	8.1	--	--

viscous environment, such as a membrane, may show increased lifetime and quantum yield even if its binding site is not hydrophobic.

Detergent micelles and sonicated phospholipids offer a good model system for the interfacial and bilayer regions of biological membranes. Table II shows the fluorescence lifetime of ANS in cetyltrimethylammonium bromide (CTAB) micelles and egg lecithin liposomes with and without cholesterol. In all these systems the decay is a single exponential and τ is between 5 and 7 nsec, which corresponds to an environment between methanol and ethanol in the alcohol scale (Table I).

Sonicated red cell membrane lipids bind ANS with a much lower affinity than lecithin, probably because they contain negatively charged phospholipids, which repel the anionic ANS. The fluorescence decay of ANS in red cell lipid vesicles is shown in Fig. 2, compared with that of lecithin-cholesterol vesicles. Although the decay in the red cell lipids is multi-exponential, indicating a variety of environments, the lifetimes range from 5 to 8 nsec and the average lifetime is similar to that in lecithin-cholesterol vesicles.

These results indicate that the lifetime of ANS in phospholipids has a value in the range of 5 to 8 nsec and is relatively insensitive to the composition of the lipid or the presence of cholesterol, but is monitoring mainly the lipid-water

TABLE II

ANS Lifetimes in Various Lipid Systems

Lipid	τ (nsec)
CTAB (10 mM)	5.0
Lecithin Vesicles	6.7
Lecithin Vesicles in 84% D_2O	10.6 (2 exponentials)
Lecithin Vesicles + 2 mM $CaCl_2$	6.4
Lecithin-Cholesterol (1:1) Vesicles	5.9
RBC Membrane Lipid Vesicles	5.5 average (range 4-8)

Temp. 23°C. Medium: 20 mM Tris-Cl, pH 7.5. The lifetimes varied less than 0.3 nsec with 10-60 μM ANS.

interface. If water is substituted by D_2O a large increase in lifetime is observed (Fig. 3). The large effect of D_2O suggests that there is structured water in the region where ANS binds to the lecithin vesicle since the difference in lifetime of ANS solutions in H_2O and D_2O is only 0.5 nsec (cf. Table I).

The ANS Environment in the Red Cell Membrane

Fluorescence decays of ANS in ghost suspensions, measured at two different time scales, are shown in Figs. 4 and 5. The curvature of these semilogarithmic plots indicates that the decay is not a single exponential. Analysis of the decays with the method of moments (8,9) yields a sum of two exponentials with lifetimes of about 7 nsec for the short component (τ_1) and 19 nsec for the long component (τ_2). The accuracy of the analysis can be seen in the excellent fit of the theoretical decay curves (dashed lines) which are superimposed on the experimental curves and follow them for several decades, down to the noise level (Fig. 5). We interpret these lifetimes as characteristic of two different environmental domains in the membrane monitored by ANS. A relative measure of the ANS concentration in each membrane region, as characterized by its lifetime, is given by the coefficient of each exponential. The values of these coefficients indicate that 60 to 75% of the ANS molecules in the membrane are in the short lifetime region (τ_1) and the remainder occupy the long lifetime region (τ_2).

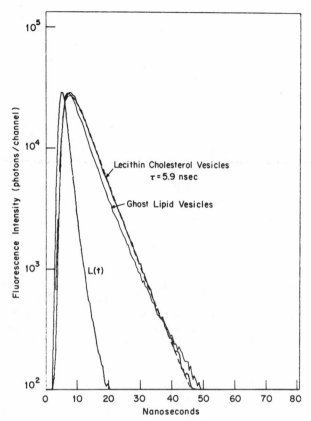

Fig. 2. ANS fluorescence decays in lecithin-cholesterol and ghost lipid vesicles, in 20 mM Tris-Cl, pH 7.5; 33 μM ANS. Temp. 23°C.

τ_1 is similar to that measured in the extracted membrane lipids and the lecithin vesicles, suggesting that it represents the phospholipid bilayer portion of the membrane. τ_2 is much longer than τ in any of the model systems tested, including the heme binding site of apomyoglobin where τ = 16.4 nsec (16). This suggests that τ_2 represents membrane protein or regions of lipid-protein interactions.

The lifetimes in both membrane regions show a strong temperature dependence (Fig. 6) suggesting that their environment is highly viscous and has relaxation times of the order of the ANS lifetime. The behavior of τ_1 is similar to τ of lecithin-cholesterol vesicles, supporting the conclusion that it is composed by a phospholipid bilayer. The changes in τ_2 above 40°C are irreversible, suggesting protein denaturation. The ratio of ANS molecules in regions 1 and 2 (a_1/a_2 in Fig. 6) also changes with temperature, indicating that the activation energy for ANS binding is different for the two regions.

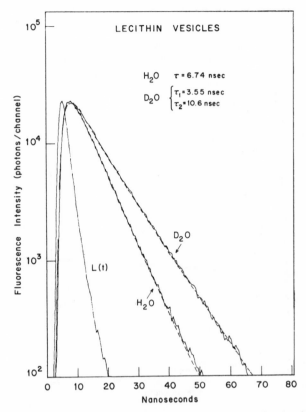

Fig. 3. Effect of D_2O on ANS fluorescence decay in lecithin vesicles. ANS, 10 μM; D_2O, 84%. Temperature 23°C.

As in the case of phospholipid vesicles, substituting the water in the suspension medium by D_2O changes the ANS lifetimes in ghosts significantly. τ_1 increases from 7.5 nsec in H_2O to 10.5 nsec in D_2O, and τ_2 from 19.4 to 22.1 nsec. These results indicate that both ANS binding regions in ghosts are superficial, accessible to water, but that the properties of the water in these regions are different from free solvent H_2O. Indirect effects of D_2O on the membrane are also possible.

The dependence of the coefficients and lifetimes on ANS concentration are shown in Fig. 7. As the medium ANS increases, the amount bound to regions 1 and 2 (a_1 and a_2 in Fig. 7) follows a saturation curve.

This binding behavior is similar to that observed with conventional steady state fluorescence measurements (17), although in this case it is possible to resolve the binding at each

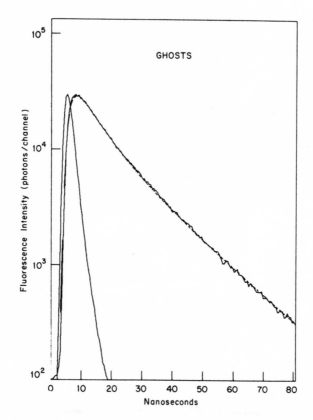

Fig. 4. *ANS fluorescence decay in ghosts. Medium: 20 mM Tris-Cl, pH 7.5; 30 μM ANS; 0.83 mg ghost protein/ml. Temp. 23°C. The dashed line shows the two exponential fit using F(t) = 0.3 exp (-t/7.2) + 0.16 exp (-t/18.7). See Fig. 5 for full scale.*

region. Surprisingly, although the lifetimes indicate quite different environments, the two regions appear to have the same affinity for ANS, since the ratio (a_1/a_2) of the amount bound to the two sites remains relatively constant through the titration.

In contrast with the solvents and lipid systems where τ is independent of the amount of added ANS, the two lifetimes in ghosts vary with concentration (Fig. 7).

The change in lifetimes with increasing ANS concentration may represent a perturbation of the membrane by ANS itself. Evidence that ANS perturbs the membrane at the same concentrations used in these fluorescence measurements is given by its

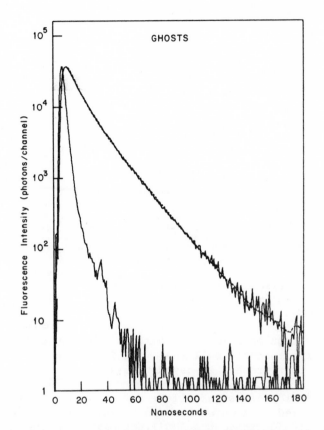

Fig. 5. *ANS fluorescence decay in ghosts. Medium: 20 mM Tris-Cl, pH 7.5; 20 μM ANS; 1 mg ghost protein/ml. Temp. 22°C. The full scale is presented to show the noise level and the theoretical fit (dashed line) of the two exponential decay analysis. The experimental values were F(t) = 0.3 exp (-t/7.1) + 0.13 exp (-t/18.9). Note the fit even at less than 1% of the initial intensity.*

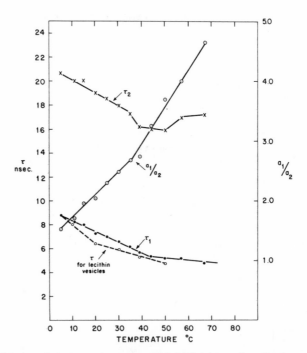

Fig. 6. *Effect of temperature on ANS lifetimes in ghosts and lecithin vesicles. Ghosts (.96 mg protein/ml) or lecithin vesicles in 20 mM Tris-Cl, pH 7.5; 20 μM ANS. Decay curves were collected at each temperature (increasing). a_1/a_2 is the ratio of ANS concentration in sites of lifetime τ_1 and τ_2 in the ghost, respectively.*

inhibitory effect on anion transport (10) and by it echino-cytogenic effect on red cell shape (11,12) which has been interpreted to represent an asymmetric expansion of the membrane (12).

The lyotropic anions salycilate and thiocyanate which, like ANS, inhibit red cell anion transport (18) and crenate red cells (19) also decrease the ANS lifetimes in ghosts, particularly τ_2 (Fig. 8). Thus, it is possible that the decrease in lifetimes caused by ANS and the lyotropic anions represents a change related to the functional alterations. However, the lifetimes are also decreased in salts of other anions (Fig. 8) including citrate and p-amino hippurate, which are impermeable in red cells. Thus, this effect is not specific for inhibitors of anion transport.

A second possibility is that energy transfer between bound ANS molecules causes the decreased lifetimes. A test of this is the dependence of τ_1 and τ_2 on bound ANS (Fig. 9). At low ionic strength τ_1 and τ_2 decrease with increasing ANS binding. At high ionic strength they are lower, but independent of the

336

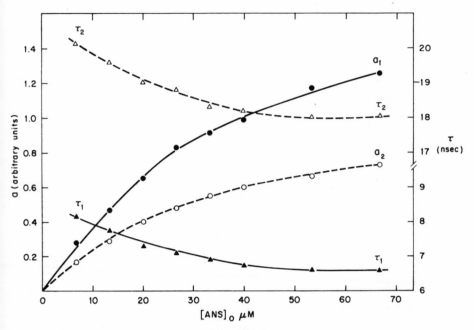

Fig. 7. *Dependence of ANS binding and lifetimes in ghosts on ANS concentration. Ghosts (1.8 mg protein/ml) in 20 mM Tris-Cl, pH 7.5. Temp. 22°C. Each point was determined from a decay curve at the indicated ANS concentrations as in Figs. 4 and 5. a_1 and a_2 are proportional to the concentration of bound ANS at regions of lifetime τ_1 and τ_2.*

extent of ANS binding. This rules out the possibility of energy transfer as the cause of the decrease in lifetimes, since this mechanism depends on the distance between ANS molecules, which are closer when more ANS is in the membrane.

An alternative interpretation of these changes is that they reflect some heterogeneity of ANS lifetimes and affinities in each membrane region. If shorter lifetime sites in each region have a lower affinity for ANS, higher ANS concentrations are required to populate them. Then, the measured lifetimes will decrease with increasing ANS concentrations. This predicts that an increase in the affinity of shorter lifetime sites for ANS, or displacement of ANS from longer lifetime sites will decrease the average lifetime of the population. The observations agree with this hypothesis since various conditions that increase the membrane affinity for ANS, e.g., high ionic strength, divalent and trivalent cations, and local

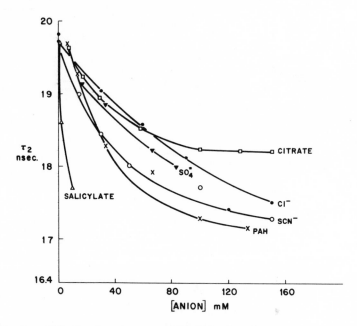

*Fig. 8. Effect of anions on ANS lifetime in ghosts. Ghosts (0.7-
1.1 mg protein/ml) in 20 mM Tris-Cl, pH 7.5; 33 μM ANS; and the
Na salts of the indicated anions. Each point corresponds to a
decay curve. PAH = p-amino-hippurate. Similar changes were
observed in τ_1.*

anesthetics (2-4, 17) decrease τ_2 (Figs. 9,10) suggesting that
the increased binding is at sites of shorter lifetime. The
lyotropic anions have been shown to compete with ANS for membrane binding sites (17) suggesting that the competition is
at longer lifetime sites. Figure 10 shows the effect of Na^+,
Ca^{2+} and SCN^- on τ_2. A similar response is observed for τ_1.

Changing the pH of the medium does not affect the lifetimes
significantly (Fig. 11) although it alters substantially the
steady state fluorescence. This indicates that the ANS environment is not sensitive to the state of ionization of the
membrane and the changes are due to variations in binding
affinity only.

The use of hydrolytic enzymes offers the possibility of
selective perturbation of the protein and lipid components of
the membrane. Incubation of red cells with pronase under
conditions that hydrolyze the anion transport protein and

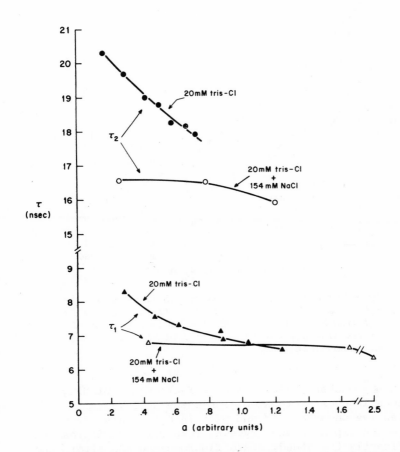

Fig. 9. *Dependence of ANS lifetimes in ghosts on bound ANS and ionic strength. Ghosts (1.3 mg protein/ml) suspended in the indicated media at pH 7.5 (22°C) were titrated with ANS. "a" is proportional to the concentration of bound ANS at each membrane region.*

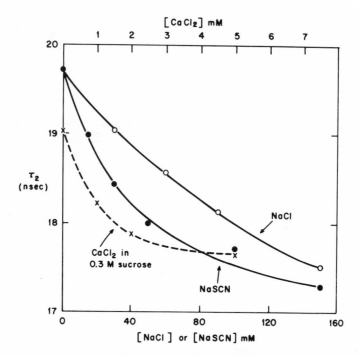

Fig. 10. *Effect of anions and cations on ANS lifetime in ghosts.*
Ghosts (1.05 mg protein/ml) in 20 ml Tris-Cl, pH 7.5; 33 µM ANS;
titrated with NaCl or NaSCN; for the CaCl$_2$ titration, 0.3 M
sucrose was also present as a control of the increasing osmolarity
with the monovalent ions.

inhibit $SO_4^=$ exchange (20) has no effect on ANS binding or
lifetimes in ghosts prepared from these cells.

Extensive phospholipid and protein hydrolysis of ghosts
with phospholipase C and trypsin, respectively, decreases
significantly the steady state fluorescence and alters the
binding kinetics of ANS (Fig. 12). Surprisingly, the decay
parameters are not affected significantly by either enzyme
treatment, particularly when the ghosts are at low ionic
strength (Fig. 13, left). The differences in response to ion
addition (Fig. 13, center and right) result mainly from decreased ANS binding at the phospholipid region in the phospholipase C-treated ghosts, since τ_1 and τ_2 are within 1 nsec
of the control.

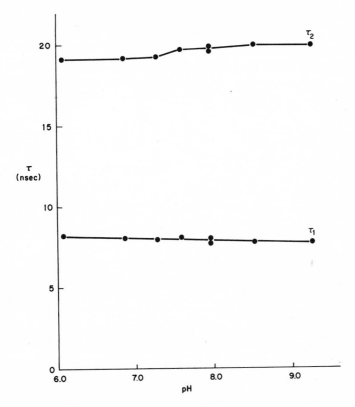

Fig. 11. *Effect of pH on ANS lifetimes in ghosts. Ghosts (1.05 mg/ml)*
suspended in 20 mM Tris-Cl; 33 μM ANS. The pH was adjusted with
HCl or NaOH.

Surface Potential and ANS Binding to the Membrane

The previous experiments show that ANS is in two environ-
mentally different types of membrane sites. In spite of the
difference in environment of these sites, the proportion of
ANS bound at sites 1 and 2 remains relatively constant even
if the total amount bound varies several-fold under different
ionic incubation conditions, ANS concentrations and enzyme
treatments. Figure 14 shows an ANS titration at regions 1
and 2 (a_1 and a_2, respectively), at high and low ionic
strength. Adding NaCl causes increased binding to both regions
and the apparent saturation level is increased. The increase
in region 1 is proportionally larger than in region 2, since
the ratio a_1/a_2 increases from 1.6 at low ionic strength to
2.1 in isotonic NaCl. However, at constant salt concentration,

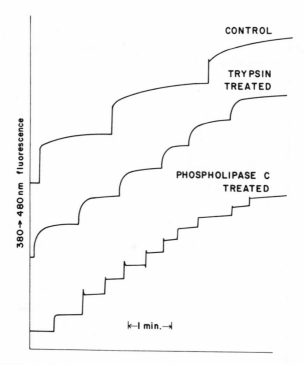

Fig. 12. Effect of phospholipase C and trypsin on steady state ANS fluorescence in ghosts. Enzyme-treated or control ghosts (0.25 mg protein/ml) in 20 mM Tris-Cl, pH 7.5; titrated with ANS (each addition is 4 μM ANS).

the binding to both regions shows a similar dependence on ANS concentration, since the ratio a_1/a_2 is constant (within 15 to 20%) throughout the titration. Since these ghosts are freely permeable to ions, no gradients or transmembrane potentials are present which might produce these effects.

To account for these observations, we propose that ANS binding depends on and alters the surface potential of the membrane. The ANS ions that bind to the membrane make the surface potential more negative, which opposes the binding of additional ANS ions. At high ANS concentrations, the electrostatic potential becomes sufficiently large so that the binding appears to saturate. In addition, we assume that this apparent saturation occurs far below the true saturation level of the available sites. Under these conditions the shapes of the curves in Fig. 14 are determined essentially by the surface potential and not by the number of available sites. Since this potential is the same for regions 1 and 2, their binding curves should have the same shapes, accounting for the constancy of a_1/a_2 observed experimentally. Furthermore,

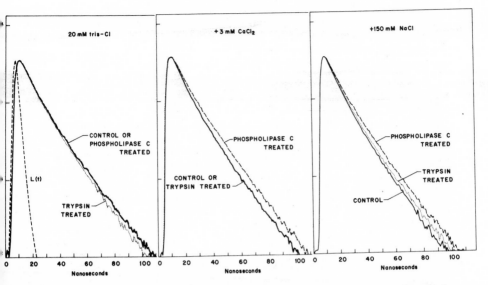

Fig. 13. *Effect of ions on ANS fluorescence decays in enzyme-treated ghosts. Ghosts of Fig. 12 in 20 mM Tris-Cl, pH 7.5; 33 μM ANS (left), plus 3 mM $CaCl_2$ (center) or 150 mM NaCl (right). The curves are normalized to the same peak count.*

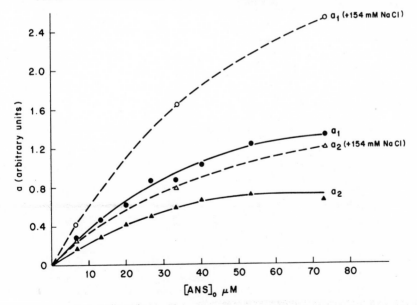

Fig. 14. *Effect of ionic strength on ANS binding to ghosts at regions 1 and 2. Experimental conditions as in Fig. 9. a_1 and a_2 are proportional to the concentration of ANS bound to membrane regions 1 and 2, respectively.*

increasing the ionic strength reduces the surface potential by screening the negative charge on the membrane as predicted by double layer theory. Since the sites are far from true saturation, charge screening allows more ANS ions to bind to the membrane, accounting for the data in Fig. 14.

The agreement between the above hypothesis and experimental observations is not only qualitative, but quantitative. Figure 15 shows experimental and theoretical titration curves (points and solid lines, respectively) of ANS binding at region 1 in ghosts at two ionic strengths. The theoretical curves were derived from a model, described in detail elsewhere (21,22), in which the ANS-dependent surface potential is calculated with the Gouy-Chapman equation. The estimated ANS surface potential is used to correct the dissociation constant; the curve is generated obtaining values for the ANS concentration necessary to give assigned values of ANS bound to region 1, according to the equation in Fig. 15. Similarly, a comparison between theory and experiment for ANS binding at region 2, and

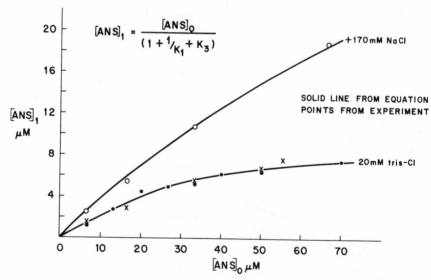

Fig. 15. *Predicted and experimental ANS binding curves for region 1. In the equation, K_1 is the dissociation constant for binding at site 1 and K_3 is the ratio a_2/a_1 obtained experimentally. Values of $[ANS]_1$ are assigned, the ANS surface potential (ψ_{ANS}) is calculated with the Gouy-Chapman equation to correct K for the change in ψ, and corresponding values of $[ANS]_0$ are obtained from the equation to generate the curve. Experimental values at low ionic strength show two different experiments (x and •).*

the estimated surface potentials produced by ANS are shown in Table III. Note that in this model, the membrane is far from saturation with ANS at the concentrations in Fig. 15. The only factor that makes the curves bend is the surface potential. This is consistent with the effects of ANS on ion transport and cell shape, which increase with ANS concentration up to the millimolar range (10,12). The excellent agreement between theory and experiment supports the validity of the model, which suggests a method, to be described elsewhere, to measure surface potentials in biological membranes with ANS titrations (23).

Furthermore, since the negatively-charged phospholipids appear to be mainly on the internal half of the red cell membrane (24), ANS binding must be asymmetric, being less favored at internal sites. This asymmetric binding causes the shape changes of intact red cells (11,12) confirming the predictions of the bilayer couple hypothesis (25).

TABLE III

Theoretical Fit for ANS Binding at Region 2 in Ghosts

Experimental		Calculated		
$[ANS]_0$ (μM)	$[ANS]_2$ (μM)	$[ANS]_2$ (μM)	ψ_{ANS} (mV)	
6.6	.9	.9	− 6.7	Low
16.6	1.6	1.8	−13	Ionic
33.3	2.8	2.7	−24.4	Strength
50	3.4	3.4	−29.2	(C = 20 mM)
6.6	1.1	1.1	− 3.4	High
16.6	2.2	2.6	− 7.1	Ionic
33.3	4.8	4.6	−14.5	Strength
66.6	6.2	6.9	−23.4	(C = 190 mM)

Conditions as in Fig. 15. Note that the effect of ionic strength is to decrease the surface potential produced by bound ANS.

CONCLUSIONS AND PERSPECTIVES

The data presented indicate that it is possible, using nanosecond fluorescence spectroscopy, to distinguish different types of probe binding sites. Thus, the interactions of ANS with the phospholipid bilayer and the protein regions of the membrane can be distinguished and evaluated quantitatively. Unfortunately ANS binding is not specific enough to distinguish a possible anion carrier site from other sites in the protein region of the membrane. Although ANS is not a specific probe, the approaches presented in this chapter indicate that, given an appropriate fluorescent molecule, studies of the properties of its binding sites and the distinction between different binding sites are feasible with these methods. Thus, at the present the main problem is to design appropriate probes, specific for the membrane function to be studied. At least in the case of the galactoside transport system of bacteria, specific probes are already available (26).

Despite the large differences in fluorescence lifetime of ANS in solvents and in the two regions of red cell membranes, relatively small differences (< 2 nsec) are observed between a variety of lipid systems of different composition (cf. Table II), and with significant membrane perturbations such as extensive enzymatic hydrolysis. D_2O and temperature were the only variables tested which caused lifetime changes that can alter significantly the steady state fluorescence. Thus, changes in ANS binding, not on quantum yield, are the cause of most fluorescence changes, unless water structure or the viscosity of the membrane is altered.

The main factor that controls ANS binding is the surface potential, which is made more negative by bound ANS. The surface potential will affect the interaction of all ions with the membrane. If the ANS binding sites are near ion transport sites the ANS surface potential should alter ion transport. ANS does inhibit anion transport (10), but does not increase cation transport, except at high (> 0.5 mM) concentrations (12), suggesting that the cation transport sites are separated at least 5 to 10 $\overset{o}{A}$ from the ANS binding sites. Estimates of ANS-dependent surface potential from Cl^- permeability data (27) agree well with the surface potentials estimated from fluorescence measurements in ghosts (Table III), suggesting that at least part of the inhibition of anion transport may be due to electrostatic factors.

ACKNOWLEDGMENTS

Drs. J.F. Hoffman and J. Yguerabide collaborated in various aspects of this work. This work was done at the Department of Physiology, Yale University, School of Medicine, and was supported by grants HE-09906 (NIH) and GB-18924 (NSF) to J.F. Hoffman. I thank Dr. L. Stryer for allowing the use of his instruments, and the American Heart Association for support during the writing of this paper.

REFERENCES

1. Tasaki, I., Carnay, L., Sandlin, R. and Watanabe, A. (1968) Proc. Nat. Acad. Sci. 61, 883.
2. Azzi, A., Chance, B., Radda, G.K. and Lee, C.P. (1969) Proc. Nat. Acad. Sci. 62, 612.
3. Rubalcava, B., Martinez de Muñoz, D. and Gitler, C. (1969) Biochemistry 8, 2742.
4. Vanderkooi, J. and Martonosi, A. (1969) Arch. Biochem. Biophys. 133, 153.
5. Brand, L. and Gohlke, J.R. (1972) Ann. Rev. Biochem. 41, 843.
6. Radda, G.K. and Vanderkooi, J. (1972) Biochim. Biophys. Acta 265, 509.
7. Azzi, A. (1975) Quart. Rev. Biophys. 8, 237.
8. Isenberg, I. and Dyson, R.D. (1969) Biophys. J. 9, 1337.
9. Yguerabide, J. (1972) in Methods in Enzymology (Hirs, C. H.W. and Timasheff, S.N., eds.)Vol. 26, p. 498, Academic Press, New York.
10. Fortes, P.A.G. and Hoffman, J.F. (1974) J. Membrane Biol. 16, 79.
11. Yoshida, S. and Ikegami, A. (1974) Biochim. Biophys. Acta 367, 39.
12. Fortes, P.A.G. and Ellory, J.C. (1975) Biochim. Biophys. Acta, in press.
13. Seliskar, C.J. and Brand, L. (1971) J. Amer. Chem. Soc. 93, 5414.
14. Seliskar, C.J. and Brand, L. (1971) Science 171, 799.
15. Chakrabarti, S.K and Ware, W.R. (1971) J. Chem. Phys. 55, 5494.
16. Tao, T. (1969) Biopolymers 8, 609.
17. Fortes, P.A.G. and Hoffman, J.F. (1971) J. Membrane Biol. 5, 154.
18. Wieth, J.O. (1970) J. Physiol.(London) 207, 581.
19. Deuticke, B. (1970) Naturwiss. 57, 172.
20. Passow, H. (1971) J. Membrane Biol. 6, 233.

21. Fortes, P.A.G. (1972) Ph.D. Thesis, University of Pennsylvania. Diss. Abstr. Int. B 33, 2961.
22. Yguerabide, J. (1973) *in* Fluorescence Techniques in Cell Biology (Thaer, A.A. and Sernetz, M., eds.) p. 325, Springer-Verlag, Berlin.
23. Fortes, P.A.G., Yguerabide, J. and Hoffman, J.F. (in preparation).
24. Zwaal, R.F.A., Roelofsen, B. and Colley, C.M. (1973) Biochim. Biophys. Acta 300, 159.
25. Sheetz, M.P. and Singer, S.J. (1974) Proc. Nat. Acad. Sci. 71, 4457.
26. Shuldiner, S., Kerwar, G.K., Weil, R. and Kaback, H.R. (1975) J. Biol. Chem. 250, 1361.
27. Fortes, P.A.G. and Hoffman, J.F. (1973) *in* Erythrocytes, Thrombocytes, Leukocytes: Recent Advances in Membrane and Metabolic Research, (Gerlach, E., Moser, K., Deutsch, E. and Wilmanns, W., eds.) p. 92, Georg Thieme Publishers, Stuttgart.

THE STRUCTURE OF THE MITOCHONDRIAL INNER MEMBRANE: CHEMICAL
MODIFICATION STUDIES

Harold M. Tinberg and Lester Packer

INTRODUCTION

The mitochondrial inner membrane, and other membranes as
well, can be best envisioned as an array of protein molecules
embedded to varying degrees in a bilayer of amphipathic lipids.
These proteins can be localized in three loosely defined membrane domains. They may be: a) at one or both surfaces in
contact with the bulk aqueous phase; b) buried within the
hydrophobic core; c) transmembranous, extending from surface
to surface through the apolar zone. Some information concerning the localization of proteins in membranes has been obtained by exposing sealed membrane systems to polar, water-
soluble probes, which permeate poorly through the membrane
permeability barrier (1-3). The addition of alkylating potential to these probes makes it possible to covalently label
surface-situated proteins.

Surface-Localized Proteins

Surface probes can be classified into two major groups:

Small probes: The impermeability of these molecules is
made possible by conferring hydrophilicity to the reagent via
a neutral polar moiety (e.g., carbohydrate) or more frequently
a charged group (e.g., sulfonate, phosphate). It is imperative that these reagents react with proteins under mild conditions of pH, temperature and ionic strength, thereby insur-
ing minimal membrane perturbation during exposure. Groups
fulfilling these requirements include diazonium salts, iso-
thiocyanates, aldehydes, imidates, mercurials and nitrenes.
The two probes utilized in our studies are diazobenzenesulfonate (DABS) and p-mercuriphenylsulfonate (PMPS).

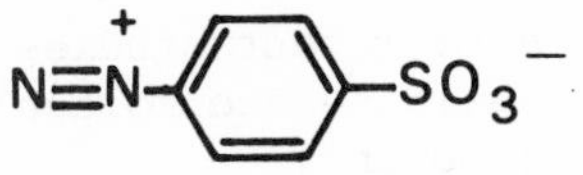

$$N \equiv N^+ - \text{\textcircled{}} - SO_3^- \qquad {}^+Hg - \text{\textcircled{}} - SO_3^-$$

DABS

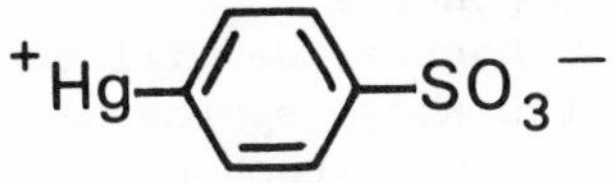

PMPS

Macromolecular probes: The second class of surface
probes includes macromolecules, which because of their size do
not penetrate membranes. These include proteases (e.g., tryp-

sin), and carbohydrases (e.g., neuraminidase). Exterior tyrosines can be labeled with radioiodine with the enzyme lactoperoxidase.

Protein-Protein Associations

The probes described above yield information relevant to
the localization of components in a plane normal to the long
axis of the membrane. Of interest also are the lateral associations between membrane proteins, associations which give
rise to supramolecular functional complexes. These associations, which are generally non-covalent, are destroyed when
membranes are solubilized and constituent proteins analyzed by
conventional approaches. One approach to investigating protein-protein associations in membranes involves the use of
bifunctional alkylating agents to introduce covalent linkages
between neighboring proteins, thereby rendering lateral
associations resistant to solubilization, and therefore amen-
able to analysis by procedures used for monomeric proteins.

The covalent crosslinking of membrane proteins to stabilize
protein associations can be accomplished in several ways.
Treatment of membranes with a chelate complex of 0-phenanthro-
line-Cu results in the production of disulfide bonds between
neighboring proteins (4). It is apparent that this method
can only be applied to sulfhydryl-containing proteins. Bifunctional alkylating agents can also be used to crosslink
membrane components. The use of glutaraldehyde to link adjacent proteins contains many drawbacks. The polydisperse size
distribution of autopolymerized glutaraldehyde in solution
implies that this compound can couple even distant proteins.
In the studies to be described below, we have used dimidates
to crosslink inner membrane proteins. These compounds, which
exist as monomers in solution, react primarily with free
amines to yield cationic amidines (Fig. 1), thereby conserving
the charge characteristics of the alkylation site. Furthermore, biimidates of various chain lengths can be utilized as
molecular rulers to determine intermolecular distances between
membrane components.

In this chapter, we will describe some recent studies
utilizing the above approaches to investigate the molecular
structure of the mitochondrial inner membrane.

Labeling of the Inner Membranes with Impermeant Labels

The analysis of the surfaces of the mitochondrial inner
membrane is facilitated by the fact that one can obtain preparations in which the membrane possesses a normal orientation

(in intact mitochondria or inner membrane-matrix preparations [mitoplasts]) or in an inverted orientation (in sonically-prepared inner membranes, SMPs). It is therefore possible to independently expose either membrane surface to the incubation medium containing the labeling agents. The labeling of both surfaces was carried out as described in Fig. 2. This procedure gave rise to the following preparations: a) <u>outer surface preparations (OSP)</u> obtained from labeled mitochondria; b) <u>inner surface preparations (ISP)</u>, representing SMPs treated directly with DABS.

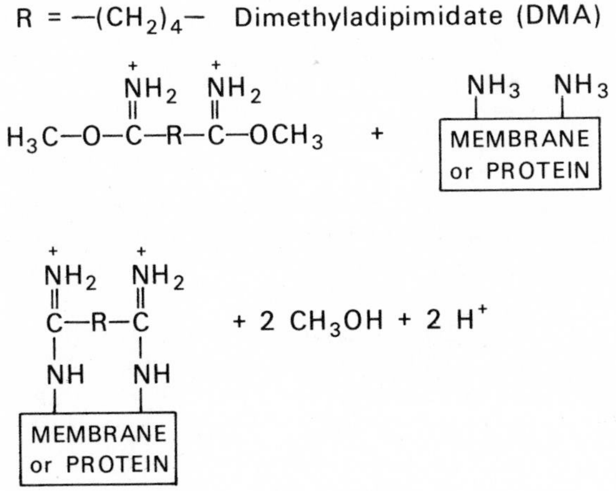

$R = -(CH_2)_4-$ Dimethyladipimidate (DMA)

$$\overset{+}{\underset{\parallel}{NH_2}} \quad \overset{+}{\underset{\parallel}{NH_2}} \qquad\qquad NH_3 \quad NH_3$$

$$H_3C-O-C-R-C-OCH_3 \quad + \quad \boxed{\begin{array}{c} MEMBRANE \\ or\ PROTEIN \end{array}}$$

$$\overset{+}{\underset{\parallel}{NH_2}} \quad \overset{+}{\underset{\parallel}{NH_2}}$$
$$C-R-C \qquad + 2\ CH_3OH + 2\ H^+$$
$$\underset{|}{NH} \quad \underset{|}{NH}$$
$$\boxed{\begin{array}{c} MEMBRANE \\ or\ PROTEIN \end{array}}$$

Fig. 1.   The amidination reaction.

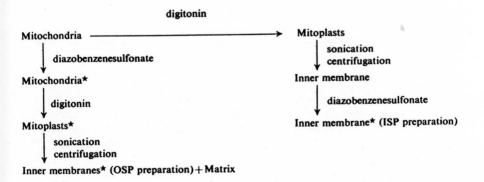

Fig. 2.   Labeling of inner membranes with diazobenzenesulfonate. Labeled preparations are designated with an asterisk (*). (cf. 3)

Although DABS has been shown to be impermeable to several membrane systems (1,3), the reagent does appear to penetrate some membranes (5). It is imperative, therefore, to substantiate the impenetrability of the probe to the membrane system to be investigated. In our system, this can be accomplished by determining the amount of DABS present in the matrix obtained from labeled mitochondria. Under our experimental conditions, the label present in the matrix fraction was 20% of that bound to the SMPs prepared from these mitochondria. Furthermore, a significant percentage of the DABS in the matrix can be attributed to contamination of this fraction by fragmented inner membranes. Similarly, the impermeability of PMPS was determined by measuring the loss of free sulfhydryl groups in the matrix following treatment of intact mitochondria. Although the molar ratio of mercurial to matrix sulfhydryl groups during treatment was approximately eight, only about 10% of these groups were lost. It appears, therefore, that both DABS and PMPS penetrate the mitochondrial inner membrane poorly, indicating the potential usefulness of these reagents to examine the localization of surface components in this membrane system.

Table I gives the distribution of DABS in the various inner membrane prepartions. ISP preparations possesed a specific activity greater than three times that of the OSP preparation suggesting that the inner surface of the membrane possesses substantially more diazoreactive sites than the outer surface.

TABLE I

Incorporation of Diazobenezene-[^{35}S]Sulfonate into Mitochondrial Inner Membrane Preparations

Preparation	Diazobenezene-[^{35}S]sulfonate labeling (cpm/mg protein)
OSP	106 000
ISP	357 000
Matrix	20 000

Intact mitochondria (10 mg protein/ml) and isolated inner membranes (4 mg protein/ml) were labeled with diazobenzene-[^{35}S]sulfonate (2 μmoles/ml, 5·10^6 cpm/μmole) as described in Fig. 2. (cf. 3).

Futhermore, since diazonium salts are relatively nonspecific alkylating agents, reacting with histidyl, tyrosyl, and lysyl residues in proteins (6), the incorporation of DABS into proteins reflect the extent to which these proteins are exposed to the bulk aqueous phase. Only 2-6% of the label present in the inner membrane could be extracted with chloroform:methanol (2:1,v/v), indicating that essentially all the label present represents protein-bound reagent.

The incorporation of PMPS into mitochondrial fractions was determined by measuring the decrease in free sulfhydryl groups (Table II). The labeling protocol was similar to that employed for the DABS studies except that SMPs were prepared by soni-cating intact mitochondria instead of mitoplasts. Treatment of mitochondria with PMPS resulted in SMP preparations which had lost approximately 30% of the free sulfhydryls present.

TABLE II

Free Sulfhydryl Content of Inner Membranes Treated with PMPS

Sample	Sulfhydryl content (n moles/mg protein)
C_{OSP}	73.2
SMP_{OSP}	54.4
C_{ISP}	74.8
SMP_{ISP}	1.5
Matrix C_{OSP}	87.1
Matrix SMP_{OSP}	76.4

Mitochondria (2.5 mg protein/ml) and SMPs (1 mg protein/ml) were treated with PMPS (1 mM) in 100 mM sodium phosphate-0.25 M sucrose (pH 8) for 15 min at $4^{O}C$. Mitochondria were washed once with 0.25 M sucrose and then sonicated to prepare SMPs (OSP, outer surface pre-parations) while treated SMPs (ISP, inner surface preparations) were reisolated by centrifugation. Both preparations were suspended in 0.25 M sucrose for enzyme determinations. Controls were carried through all treatments in the absence of PMPS.

In contrast, nearly all these groups disappeared when SMPs were treated directly with PMPS. Therefore, results obtained with both probes suggest a marked asymmetry of the inner mem-brane with the inner surface containing more exposed protein than the outer surface.

Sodium dodecyl sulfate (SDS)-polyacrylamide gel electro-phoresis was used to identify those polypeptides labeled with [35S]-DABS (Fig. 3). Electrophorograms of SMPs consist of approximately 15 major components ranging in molecular weight from 10000 to 140000 (7). However, labeling in OSP prepara-tions occurred in essentially a single molecular weight class

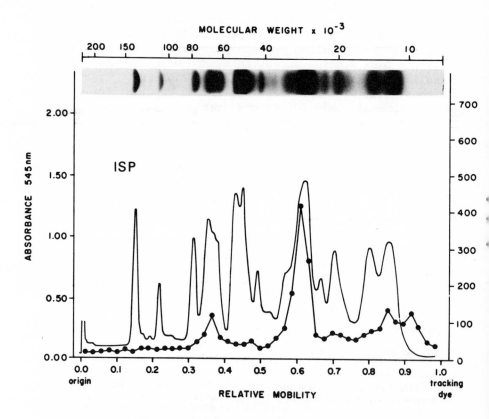

Fig. 3. *Sodium dodecylsulfate-polyacrylamide gel electrophoresis of diazobenzenesulfonate-labeled inner membranes. Labeling with [35S] diazobenzenesulfonate was performed as described in Fig. 2. Inner membranes (2 mg protein/ml) were solubilized in 1% sodium dodecylsulfate and 1% mercaptoethanol and subjected to electrophoresis, in duplicate, in 6% polyacrylamide gels containing 0.1% sodium dodecylsulfate. Each gel contained 45 μg protein. One gel was stained with coomassie blue, while the other was fractionated prior to radioactivity determination. Coomassie blue stain (absorbance 545 nm) (——); cpm [35S] per fraction (●——●) (cf. 3).*

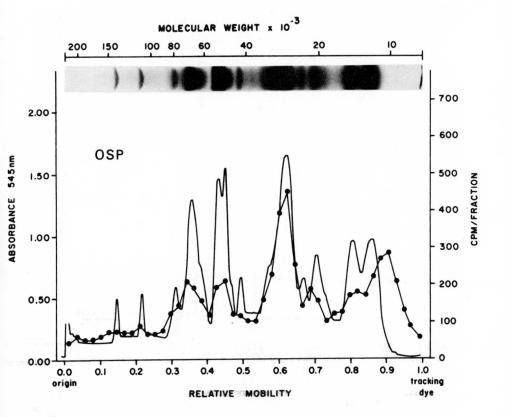

(26000) with small amounts of radioactivity also present in
polypeptides possessing an apparent molecular weight of 66000.
In contrast, several bands were labeled in ISPs, these components possessing apparent molecular weights of 80000, 66000,
51-48000, and 26000. The nature of surface polypeptides was
also investigated by exposing both membrane surfaces to trypsin (8). In these experiments the outer surface was probed
in mitoplasts to insure that the outer membrane posed no
restriction on the access of the proteases to the inner membrane. Following treatment, the inner membranes derived from
these mitoplasts and inner membranes digested directly were
solubilized and analyzed by SDS-polyacrylamide gel electrophoresis (Fig. 4). Digestion of the outer surface resulted
in decreased staining intensity of components possessing
apparent molecular weights of 70-65000 and 26000. Several
bands decreased in intensity when inner membranes were treated
directly with trypsin (mol. wts. 80000, 70-65000, 50000 and
26000. These results agree quite well with those obtained

using DABS labeling, confirming both the impermeability of the probe and the disposition of polypeptides indicated by the probe.

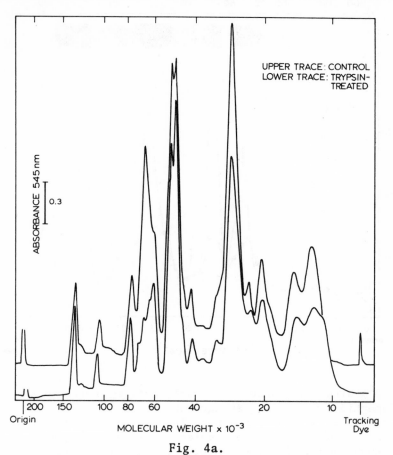

UPPER TRACE: CONTROL
LOWER TRACE: TRYPSIN-
TREATED

ABSORBANCE 545 nm

0.3

200 150 100 80 60 40 20 10
Origin   Tracking
                                          Dye
MOLECULAR WEIGHT x 10^{-3}

Fig. 4a.

Fig. 4. Trypsin treatment of rat liver mitochondrial inner membranes. Mitoplasts (inner membranes plus matrix, 5 mg protein/ml) and SMPs (2 mg protein/ml) were incubated with trypsin (0.05 mg/ml) in 0.05 M sodium phosphate-0.25 M sucrose (pH 8.0) for 30 min at room temperature. After the addition of trypsin inhibitor (0.25 mg/ml), both preparations were isolated by centrifugation. SMPs were prepared from digested mitoplasts and both SMP samples adjusted to 2 mg protein/ml. Samples were solubilized in 1% SDS and 1% mercaptoethanol and subjected to polyacrylamide gel electrophoresis in 6% gels. Proteins were visualized with coomassie blue. 4a) SMPs derived from treated and control mitoplasts. 4b) SMPs treated directly with trypsin (cf. 8).

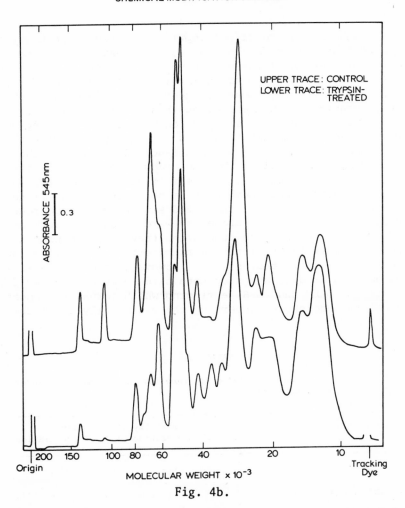

Fig. 4b.

Although a 26000 mol. wt. component appeared to be labeled
in both preparations, it is unlikely that this component
exists in a transmembrane orientation considering the molecular size of this component in monomeric form. Furthermore,
it is possible that this component represents two or more different species possessing similar molecular weights. Our
results on the distribution of inner membrane polypeptides
agree well with those obtained using other surface probes.
Lactoperoxidase-iodination has been used to demonstrate little
or no labeling of the outer surface and extensive iodination
of the inner surface (9). The differences between these results and ours may be due partly to a paucity of tyrosines at
the outer surface of the inner membrane.

357

Although the nature of polypeptides exposed at each surface can be defined electrophoretically, these data do not aid in the identification of these components. Another approach to the localization of components involves inactivation of these components by the impermeant reagents used. Several points must be considered when employing this approach. Inactivation of a component indicates interaction with the alkylating agent. However, lack of inactivation does not equivocally mean that the component in question was not labeled, but rather that labeling has not involved residues requisite for the proper functioning of the enzyme. The effect of DABS and PMPS labeling on several inner membranes is shown in Tables III and IV.

Several activities were affected similarly by both reagents. ATPase and succinate dehydrogenase were strongly inactivated only at the inner surface suggesting that these components face the matrix in the intact mitochondrion, a localization suggested by other studies (10). However, other activities such as NADH oxidase and succinate oxidase were affected differently by DABS and PMPS, a finding which serves to emphasize the points made above concerning the use of alkylating agents

TABLE III

The Effect of Diazobenzenesulfonate Labeling on Mitochondrial Inner Membrane Enzyme Activities

Enzyme System	Activity (% of control)	
	OSP preparation	ISP preparation
ATPase	114	20
NADH oxidase	19 ± 6	80 ± 24
NADH-dehydrogenase	45	46
NADH-cytochrome c reductase	53	99
Succinate oxidase	14	5
Succinate dehydrogenase	84 ± 12	7 ± 1
Succinate-cytochrome c reductase	28	13
Ascorbate-TMPD oxidase	110	156

Labeling of membranes was carried out as shown in Fig. 2. Mitochondria (10 mg protein/ml) and SMP (4 mg protein/ml) were labeled with 2 mM DABS. Activities have been normalized to unlabeled controls and are given either as the means of two experiments or as the means ± standard deviations (cf 3).

as enzyme inactivators. Some activities such as ascorbate-TMPD oxidase (a measure of cytochrome oxidase) were not in-activated at either surface by either probe (and in fact dis-played unexplained activation). However, other studies indi-cate that components of this system are labeled both by DABS (10,11) and lactoperoxidase (11) and therefore appear to exist at the membrane surface. It is apparent that these portions of the molecule contain neither diazoreactive or sulfhydryl groups required for enzyme activity.

TABLE IV

The Effect of Mercuriphenylsulfonate on Mitochondrial Inner Membrane Enzyme Activities

Enzyme System	Activity (% of control)	
	OSP	ISP
NADH oxidase	198	0
NADH dehydrogenase	26	9
Succinate oxidase	90	0
Succinate dehydrogenase	67	0
Ascorbate-TMPD oxidase	176	185
ATPase	82	28

Activities are expressed as percent controls. (Cf Table II)

In summary, the results described above indicate a marked asymmetry in the mitochondrial membrane. This asymmetry has been suggested by other experimental approaches such as freeze-fracture electron microscopy, which has revealed a significant difference in particle density on the two fracture faces of the inner membrane (12), (Fig. 5). Furthermore, our studies point out the potential use of impermeant enzyme in-activators to localize components in other membrane systems.

Interaction of Inner Membranes with Bifunctional Alkylating Agents

The studies to be described below involve the modification of the mitochondrial inner membrane with bifunctional alkylat-ing agents. These studies were carried out to obtain infor-mation relevant to: a) the nature of protein-protein associa-tions in the inner membrane, and b) the effect of molecular

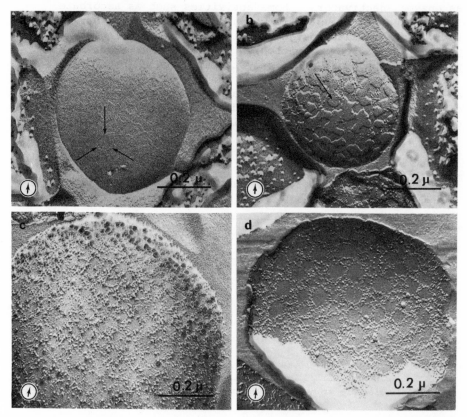

Fig. 5. *Asymmetric distribution of particles seen in fracture faces of outer and inner membranes of rat liver mitochondria (cf. 12). a) Smooth fracture face of outer membrane showing a faint circular pattern (arrows); b) Fracture face showing smooth patches (arrow) overlying the particle-covered inner membrane face; c) Convex fracture face of inner membrane; d) Concave fracture face of inner membrane.*

crosslinking on inner membrane function.   Bifunctional imidates were chosen for these experiments because previous studies have demonstrated that amidination is associated with minimal structural alteration of proteins (2,13).   The extent of polypeptide crosslinking was ascertained using SDS-polyacrylamide gel electrophoresis.   In this system, crosslinked components would possess higher molecular weights and lower mobilities than constituent monomers.   Therefore, the disappearance of monomer bands and the appearance of new polymer bands can be taken as evidence of polypeptide crosslinking. The use of gel electrophoresis to demonstrate crosslinking is shown in Fig. 6. In these experiments, isolated oligomycin-insensitive ATPase (F_1) was treated with biimidates of varying chain lengths (14). In this gel system (3% acrylamide-0.4% agarose) untreated F_1 consists of two major bands corresponding to apparent molecular weights of 80000 and 40000. These values are somewhat higher than those reported (15) and may be attributed, in part, to the gel system used. Imidate treatment resulted in the appearance of bands of higher molecular weight, apparently representing crosslinking polypeptides. The extent of crosslinking was dependent on the chain length of the imidate, with greater coupling occurring with the longer reagents, dimethyladipimidate (DMA) and dimethylsuberimidate (DMS). Insignificant interoligomer crosslinking was observed since bands greater than the reported molecular weight of F_1 (384000) (15) were not present.

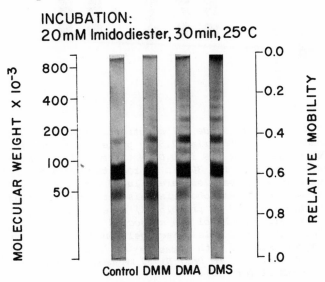

INCUBATION:
20 mM Imidodiester, 30 min, 25°C

Control DMM DMA DMS

Fig. 6. *Crosslinking of F_1 with bifunctional imidoesters. F_1 (0.34 mg protein/ml) was reacted with imidoester (20 mM) for 30 min at 25°C. At the end of the incubation, sodium dodecylsulfate and mercaptoethanol were added and the samples subjected to electrophoresis in 3% acrylamide-0.4% agarose gels. Each gel contained 17 μg protein. DMM, dimethyl malonimidate; DMA, dimethyl adipimidate; DMS, dimethyl suberimidate (cf. 14).*

Mitochondria

The effect of DMS on the structure and function of rat liver mitochondria was investigated (16). Incubation of rat liver mitochondria with increasing concentrations of DMS resulted in

the progressive loss of primary amino groups. SDS-gel electrophoresis indicated that treatment with 1 mM DMS resulted in
the disappearance of several bands (mol. wts. 130000, 60000,
and 40000) (Fig. 7). Since the 130000 mol. wt. component has
been shown to be located in the matrix (7), it appears that
DMS is permeable to the inner membrane. Polypeptides crosslinked with DMS also failed to enter 4% acrylamide gels (mol.
wt. exclusion ~ 500000) indicating the large size of these
components.

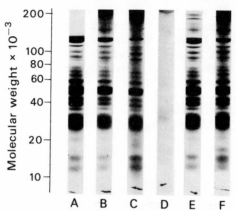

Fig. 7. SDS-polyacrylamide gel electrophoresis of amidinated mitochondria. Mitochondria (2.5 mg protein/ml) were incubated
with either EA or DMS in 0.05 M triethanolamine-HCl-0.25 M
sucrose (pH 8.5) for 30 min at room temperature. Following
incubation, the samples were diluted with ice cold 0.25 M
sucrose and the mitochondria reisolated by centrifugation.
Mitochondria (2.0 mg protein/ml) were solubilized in 1% SDS
and 1% mercaptoethanol and subjected to electrophoresis in
6% polyacrylamide gels. Each gel contained 30 μg protein.
Gels were stained with comassie brilliant blue. A) Control;
B) 1 mM DMS; C) 5 mM DMS; D) 50 mM DMS; E) 2 mM EA;
F) 100 mM EA (cf. 16).

Mitochondria treated with 2-5 mM DMS failed to swell when
placed in deionized water, while preparations amidinated with
the monoimidate, ethylacetimidate (EA) displayed swelling properties similar to untreated samples. The lack of osmotic
sensitivity serves to underscore the effectiveness of DMS as
a crosslinking agent.

The effect of DMS treatment on electron transport was studied. Preparations amidinated by EA were also examined to
ascertain the effect of chemical modification per se. Ascorbate-TMPD oxidase, which involves the transfer of electrons
from cytochrome c to oxygen was measured because this system
does not require permeation of the substrates prior to

oxidation. Effects, therefore, can be attributed to modification of the electron transport system directly and not to a permease.

Treatment of mitochondria with increasing concentrations of DMS resulted in the progressive loss of oxidase activity (Fig. 8). Nearly 60% inhibition was observed when less than 20% of mitochondrial amines were amidinated. Nearly complete inhibition was obtained at approximately 50% amidination. The inhibition of electron transport was not due to chemical modification of free amines since EA had little effect on activity, with only 10% inhibition occurring at high levels of amidination (65%). Although DMS did inhibit electron transport, treatment with low concentrations (1-5 mM) resulted in the retention of considerable activity. Since these mitochondria were unable to respond osmotically, this system should be of value in studying function in the absence of gross morphological changes.

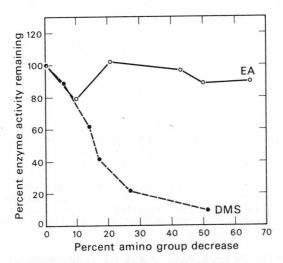

Fig. 8. *The effect of amidination on mitochondrial electron transport. Amidinated mitochondria were prepared as described in Fig. 7. Ascorbate-TMPD oxidase activity was determined polarographically at room temperature.* ●--- *DMS;* O—— *EA (cf. 16).*

Submitochondrial Particles

The effect of crosslinking on membrane function was also investigated in SMP preparations (17). The use of SMPs obviates such difficulties as the crosslinking of matrix protein to membrane proteins as suggested in mitochondrial experiments.

Treatment of SMPs with DMS also resulted in the cross-linking of membrane polypeptides into polymers too large to enter 6% polyacrylamide gels. Due to the complex polypeptide profile of SMP membranes, however, it is difficult to determine the origin of crosslinked components as has been achieved in simpler membrane systems (4,18).

Treatment of SMPs with DMS resulted in the marked inhibition of both electron transport (Table V) and ATPase activity (Fig. 9).

TABLE V

The Effect of Amidination on Electron Transport

	Activity (% of control)			
	%NH$_2$ loss	NADH oxidase	Succinate oxidase	Ascorbate-TMPD oxidase
Control		100	100	100
2 mM DMS	16	58	62	61
10 mM DMS	41	10	22	28
4 mM EA	16	78	61	73
20 mM EA	46	56	45	64

Inner membranes (1 mg protein/ml) prepared from mitoplasts were treated with imidates in 0.05 M triethanolamine-HCl-0.25 M sucrose (pH 8.5) for 30 min at room temperatures. Following incubation, the membranes were reisolated by centrifugation at 39000 rpm (45 min, 4°C, spinco #40 rotor) and resuspended in 0.25 M sucrose. NADH oxidase was measured spectrophotometrically while the other activities were determined polarographically. The values shown are from two separate experiments. Free amines were assayed using the Fluram technique. (cf 17)

Amidination of approximately 40% of inner membrane primary amines was accompanied by a loss of 70-90% of activity. Monofunctional modification by EA also caused inhibition of enzyme activity, with succinate oxidase exhibiting the greatest sensitivity to EA treatment. However, at comparable levels of amidination, DMS was more inhibitory to all activities than EA, indicating that amidination per se cannot totally account for the inhibition resulting from DMS treatment. We suggest that the loss of enzyme activity in DMS-treated membranes is a consequence of both chemical modification and molcular cross-linking. That EA and DMS inhibit activity via different mechanisms is also evident when one examines the kinetic parameters of ATPase in membranes treated with both reagents (17). In untreated preparations, ATPase possessed an apparent km of 3.7×10^{-4} M, a value which agrees closely with that reported in the literature. Treatment with EA did not change the km. However, a decrease in the km to 2.6×10^{-4} M did result from amidination by DMS.

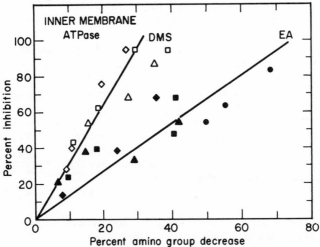

Fig. 9. *The effect of amidination of inner membrane ATPase activity. Inner membranes (1 mg protein/ml) were treated with DMS (1-10 mM) and EA (2-100 mM) in 0.05 M triethanolamine-HCl-0.25 M sucrose (pH 8.5) as described in Table V. After reisolation by centrifugation and suspension in 0.25 M sucrose, aliquots were removed for protein, primary amine and ATPase activity determinations. The ATPase activity (μmole hydrolyzed/min/mg protein) of treated samples was normalized to that of untreated controls. The results of several experiments are presented. DMS-treated membranes (open symbols); EA-treated membranes (closed symbols). In each experiment, membranes were treated with both imidates. In one experiment (●), membranes were treated with high concentrations of EA to determine the extent of amidination required for complete inhibition of activity (cf. 17).*

The above studies demonstrate that crosslinking of the mitochondrial inner membrane results in the inhibition of several membrane-associated activities. Crosslinking in membranes may involve protein-protein, protein-lipid (19), and lipid-lipid coupling. At the present time, it is difficult to ascertain the relative affect of these processes on the activity of membrane enzymes. However, crosslinking irrespective of mode, may prevent or restrict molecular interactions between components necessary for proper functioning, thereby resulting in loss of activity.

REFERENCES

1. Berg, H.C. (1969) Biochim. Biophys. Acta 183, 65.
2. Whitely, N.M. and Berg, H.C. (1974) J. Mol. Biol. 87, 541.
3. Tinberg, H.M., Melnick, R.L., Maguire, J. and Packer, L. (1974) Biochim. Biophys. Acta 345, 118.
4. Steck, T.L. (1972) J. Mol. Biol. 66, 295.
5. Amar, A., Rottem, S. and Razin, S. (1974) Biochim. Biophys. Acta 352, 228.
6. Riordan, J.F. and Vallee, B.L. (1972) in Methods in Enzymology (Hirs, C.H.W. and Timasheff, S.N., eds.) Vol. XXV, p. 521, Academic Press, N.Y.
7. Melnick, R.L., Tinberg, H.M., Maguire, J. and Packer, L. Biochim. Biophys. Acta 311, 230.
8. Tinberg, H.M. and Packer, L. (1975) in Membrane Bound Enzymes (Martonosi, A., ed.) Plenum Press, N.Y. In press.
9. Astle, L. and Cooper, C. (1974) Biochemistry 13, 154.
10. Schneider, D.L., Kagawa, Y. and Racker, E. (1972) J. Biol. Chem. 247, 4074.
11. Eytan, G.D. and Schatz, G. (1975) J. Biol. Chem. 250, 767.
12. Wrigglesworth, J.M., Packer, L. and Branton, D. (1970) Biochim. Biophys. Acta 205, 125.
13. Dutton, A., Adams, M. and Singer, S.J. (1966) Biochem. Biophys. Res. Commun. 23, 730.
14. Tinberg, H.M., Melnick, R.L., Maguire, J. and Packer, L., (1974) in BBA Library Vol. 13, Dynamics of Energy Transducing Membranes (Ernster, L. Estabrook, R.W. and Slater, E.C., eds.) p. 539, Elsevier, N.Y.
15. Catterall, W.A. and Pedersen, P.L. (1971) J. Biol. Chem. 246, 4987.
16. Tinberg, H.M., Lee, C. and Packer, L. (1975) J. Supra-molecular Structure. In press.
17. Tinberg, H.M., Nayudu, P.R.V. and Packer, L. (1975) Submitted for publication.
18. Wang, K. and Richards, F.M. (1974) J. Biol. Chem. 249, 8005.

FLUIDITY OF MITOCHONDRIAL LIPIDS

P.M. Vignais

INTRODUCTION

Mitochondria play a specialized role in cellular metabolism. Their main function is to supply the cell with ATP
and reducing equivalents. Another function, perhaps over-
looked, may be to create compartments within the cell where
some metabolites can be processed, preserved or used for
definite purposes. These functions imply numerous exchanges
of metabolites between the mitochondria and the other compartments of the cell. Controlled exchanges of metabolites via
specific transport systems are exerted by the inner membrane
which constitutes a selective permeability barrier. The inner
mitochondrial membrane also contains specialized enzymes involved in energy transducing reactions: the electron transport
carriers and the ATPase complex. These protein constituents
of the inner membrane have been studied extensively during the
last two decades in relation to their functions. The membrane
lipids have more recently attracted the attention of investigators, leading physical chemists to take the inner mitochondrial membrane as an example of biological membrane for
analyzing structure-function relationships.

This chapter contains studies on the molecular dynamics of
lipids in the inner mitochondrial membrane. The molecular
motion of the membrane lipids was assessed by the spin-label
technique. The lipid-protein interactions between a specific
integral protein, the ADP/ATP carrier, and its natural lipid
environment were probed by the use of spin-labeled amphipathic
molecules able to recognize selectively the carrier in the
inner mitochondrial membrane.

Molecular Basis of Membrane Fluidity

The Fluid Mosaic Model: The concept of a phospholipid bilayer forming the basic structure of biological membranes came
from the early studies of Gorter and Grendel (1). The striking
similarities of results obtained, by different physical techniques including X-ray diffraction, with aqueous phospholipid
dispersions and a wide variety of biological membranes confirmed the occurrence of a phospholipid bilayer as a predominant structural component of many biological membranes (2).
The technique of freeze-etching allowed a more direct view of
the structural organization of a biological membrane in which
globular hydrophobic proteins are embedded in a continuous

bilayer of phospholipids. The fluid mosaic model of Singer
(3) illustrates clearly the current trend of ideas (Fig. 1).
Observations of freeze-fractured mitochondrial membranes (see
Tinberg and Packer, this volume) indicated that the fluid
mosaic model was applicable to mitochondrial membranes.

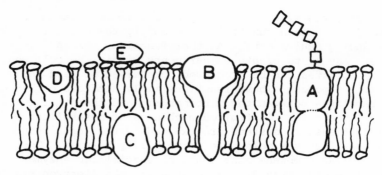

*Fig. 1. The fluid mosaic model of membrane (3) A and B intrinsic
or integral proteins (spanning); C intrinsic, inwardly exposed
protein; D intrinsic, outwardly exposed protein; E extrinsic
or transit protein.*

Lipid Phase Transition: In the fluid mosaic model the
lipids matrix constitutes a "solvent" for intrinsic proteins.
The fluidity of the lipids depends on the temperature. At
a characteristic transition temperature (T_t) bilayers of pure
phospholipids undergo a transition from a semi-crystalline to
a liquid-like phase (4,5). T_t depends on the chemical nature
of the phospholipids (chain length, number of cis double bonds,
polar head groups). In living cells, the T_t of membrane
lipids is usually lower than the physiological temperature
(6-8). Below T_t the phospholipids in the bilayer are in a
gel-like ordered structure with the hydrocarbon chains fully
extended, lying parallel and closely packed in an all-trans
configuration. When the temperature is increased in the region
of T_t, hindered rotation is possible around C-C bonds giving
rise to transgauche rotational isomers (kinks) (9,10). Highly
mobile kinks are formed at the order-to-fluid transition of
the lipid phase observed at T_t. This endothermic process also
called "chain-melting" process, is accompanied by a shortening
of the chains producing a reduction of the bilayer thickness
(6 to 7 Å for dipalmitoyllecithin (DPL)) and a lateral expansion of the bilayer (4-8). The phase transition is a cooperative effect. Träuble (8) has estimated that for the order-to-
fluid transition of DPL about 30 molecules act cooperatively.
The cooperativity of the transition increases with the chain
length of the phospholipids.

In a heterogeneous system such as a real biological membrane, the transition from the rigid to the fluid state will be spread over a range of temperatures (7). Within the temperature range there will be a coexistence of rigid and fluid regions (11). This phase separation accompanied by an increase in lateral compressibility of the phospholipids can favor the insertion of an exogenous protein in the bilayer (e.g., phospholipase A_2 (12)) or facilitate a carrier-mediated transport (e.g., sugar transport through E. coli membrane (13)).

Most of the living organisms being homothermic, the structural changes observed at the phase transition would not have any physiological interest if produced solely by changes of temperature. However, the lipid phase transition may be induced at constant temperature. The density of packing of the hydrocarbon chains decreases above T_t, and varies with the relative arrangement of the polar groups. Modification of the polar group interactions may lead to changes in chain packing and hence in T_t. It has been shown (14-16) that changes of pH, ionic strength, concentration of divalent cations or interactions with basic proteins able to bind the negative charges of the head groups may alter T_t. Preferential interaction with intrinsic proteins by increasing the ordering of vicinal lipids or decreasing the cooperativity of the lipid molecules may shift the melting point of boundary lipids towards higher temperatures or make the transition broader. Possibly triggering an order-disorder transition locally by interaction with the charged groups of the phospholipids in the vicinity of some specific enzymes (e.g., a metabolite carrier) may contribute to the adaptability of a biomembrane.

Motion of the Hydrocarbon Chains: The membrane fluidity results from the configurational freedom and the kinetic motions of the phospholipids in the bilayer and of the hydrocarbon chains in particular.

The motions of hydrocarbon chains have been investigated by the use of spin-label (electron spin resonance, ESR) and nuclear magnetic resonance (NMR) techniques. These two techniques allow the fluidity to be described quantitatively in terms of correlation times (roughly defined as the time required for a detectable reorientation of the molecule). In the liquid-crystalline state, the molecules are endowed with a high degree of rotational and translational motion.

a) Rotational motion. The rotational isomerization process interconverting the methylene groups between trans and gauche forms is very fast (correlation time $\sim 10^{-10}$ sec) and roughly uniform over much of the acyl chain (17). It is considered to be the main factor responsible for the fluidity in the bilayer (10). A second type of motion (10 to 100 times slower) is the rotation of the whole molecule around its long molecular axis. The rate of rotational reorientation of the whole molecule has been determined by the use of cholestane spin label intercalated in lecithin multibilayers and found to be approximately 10^8 sec^{-1} (18).

b) Translation motion or lateral diffusion - also occurs at a high rate in the plane of the bilayer. A diffusion constant of $D=10^{-8}$ cm^2/sec was found when the lateral diffusion of spin-labeled phosphatidylcholine (PC) molecules was studied in artificial bilayers (19,20) as well as in biomembranes (see below). This corresponds to an exchange rate between PC molecules of approximately 10^7 sec^{-1}.

c) Transverse motion - is the flipping of a phospholipid moelcule from one side of the bilayer to the other. This so-called "flip-flop" mechanism is a very slow process in synthetic bilayers. A half-time of 6.5 hr at 30° was found for the flipping of PC molecules in egg PC liposomes (19). This corresponds to a rate some 10 to 11 orders of magnitude smaller than rotational motion in the same type of preparation.

d) Mobility gradient along the hydrocarbon chain. The use of spin-label ESR and NMR indicated that the hydrocarbon chains are more ordered near the polar head groups and more mobile near the center of the bilayer. A quantitative term, the order parameter S, was defined to express the mobility gradient along the hydrocarbon chain (S gives the average orientation of the chain segment at the position of the label, e.g., a nitroxide radical or deuterium) (21,22). In recent deuterium magnetic resonance studies (10) a constant order parameter was found in the DPL systems for the first nine carbon atoms even above T_t. In other words, even in the so-called disordered state, the nine first carbon atoms remain closely packed in a nearly parallel fashion. This indicates that gauche conformations in the region of constant order parameter can only occur pairwise (kink, jog) so that the disordered chains leave the bilayer still well-ordered (10). In the central region, the kink contribution is greater. There may be from 3 to 6 kinks per chain in total (8,10).

As discussed by Seelig and Seelig (10) the order parameter gives static information on the ordering of the molecular system. The fluidity of the bilayer is a dynamic attribute. It depends upon the rate of formation of kinks (lifetime < 10^{-6} sec) and their diffusion rate.

Membrane Proteins: The occurrence of lateral diffusion of the phospholipids in biological membranes implies that membrane proteins are able to move laterally, unless they form an ordered array as the acetylcholine receptor of the electric

organ of Torpedo marmorata (23). There are other known examples (24) of long range protein-protein aggregation in
natural membranes but these are, however, probably exceptional
cases. In the fluid mosaic model, membrane proteins are able
to move laterally and to rotate and have already been found
to do so experimentally in natural membranes (24-26).

Fluidity of the Mitochondrial Lipids

The mitochondrion is surrounded by two membranes. The outer
membrane in many ways resembles the endoplasmic reticulum, in
particular as far as lipid composition is concerned (27). The
inner membrane is clearly distinct from the outer membrane.
It holds the enzymes of oxidative phosphorylation and exhibits
a selective permeability towards cations and cellular metabolites. The studies reported below concern the lipids of the
inner mitochondrial membrane.

Isolation of Inner Membrane Vesicles: Several methods have
been devised to isolate the inner mitochondrial membrane. A
chemical method (28,29) exploits the presence of cholesterol
in the outer membrane using treatment with digitonin to cause
fragmentation of the membrane. Digitonin particles surrounded
by the inner membrane and containing most of the matrix material ("mitoplasts") are then released; they have properties
very similar to those of mitochondria. It is also possible to
break the mitochondria by sonication and to isolate the inner
membrane by centrifugation on a sucrose gradient. The submitochondrial particles obtained form vesicles which are mostly
inside-out (30). This method is more suitable for heart mitochondria very rich in cristae. A third method (31) consists
of a hypotonic phosphate treatment that breaks open the outer
membrane which is then removed by differential centrifugation.
This method gives particles which still contain some matrix
material and are surrounded by the inner membrane ("inner
membrane particles"). It is possible to extract further these
particles from their matrix content by repeated phosphate
treatment and centrifugation on sucrose gradient. Inner membrane vesicles which are still inside-in are then obtained.
This last method, better adapted for liver mitochondria, is
probably the less deleterious for the inner membrane. It
gives preparations freed from most contaminants as assessed by
the use of marker enzymes and electron microscopy (27).

Lipid Composition of the Inner Membrane: The inner mitochondrial membrane is characterized by the presence of cardiolipin, a highly unsaturated phospholipid which makes up about
20% of the total membrane phospholipids. The other two main
phospholipids (roughly 40% each) are phosphatidylcholine and

phosphatidylethanolamine. The inner mitochondrial membrane is probably devoid of cholesterol (27).

The lipid composition (high content of polyunsaturated fatty acids, absence of cholesterol) favors a high fluidity of the membrane.

Use of Spin-labels to Study the Fluidity of the Inner Mito-chondrial Membrane: The general formula of the spin-labeled fatty acids, (m,n)FA, and their derivatives used in these studies are given in Fig. 2. (m,n)FA and (m,n)PC were used to probe the lipid bilayer; whereas specific molecules (e.g., (m,n)acyl-CoA and (m,n)acyl-atractylate) able to recognize a specific intrinsic protein, the ADP/ATP carrier, have been used to probe the lipid environment of that particular protein.

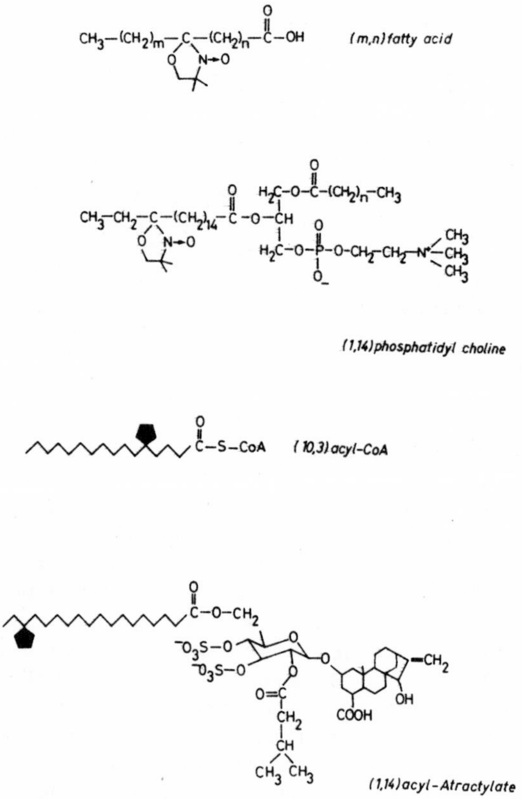

Fig. 2. Structure of the spin-labeled lipids used to probe the inner mitochondrial membrane.

a) Lipid ordering. From the ESR spectra given by (m,n) fatty acids incorporated in heart mitochondria or inner membrane particles, the order parameter S corresponding to the different positions of the nitroxide group was calculated (32). In Fig. 3, values of S are plotted versus the position of the nitroxide radical along the hydrocarbon chain. At $0°$ a curve was obtained with a plateau up to n=10, a rapid fall for n>10 indicating a high fluidity in the central region of the bilayer. At $20°$, the plateau was shorter and for (5,10)FA, S was already as low as 0.3. This means that at $0°$, (5,10)FA interacts with an ordered lipid environment and has a low rotational motion while at $20°$ the environment detected by (5,10)FA is fluid. The (5,10)FA is, therefore, a good probe for the study of the structure of the bulk of lipids.

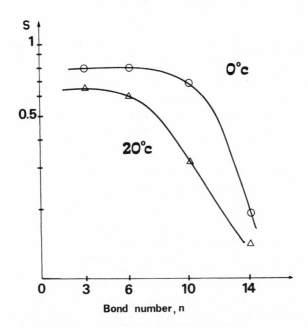

Fig. 3. Logarithmic plot of the order parameter S as a function of n determined with (m,n)FA in heart mitochondria.

b) Lateral diffusion of phospholipids. The lateral diffusion of spin-labeled PC incorporated by fusion of spin-labele PC liposomes with inner membrane particles has been measured with (1,14)PC. The resulting lateral diffusion constant D (D $\sim$ 5×10^{-8} cm^2/sec at $37°$) determined by Devaux (33) is of the same order of magnitude as that found in PC

multilayers (20) and in sarcoplasmic reticulum (34) indicating
a high diffusion rate of spin-labeled PC in the inner mitochondrial membrane.

c) Transverse motion of phospholipids. The transverse
motion (or flip-flop) of the phospholipds across the bilayer
randomizes in both layers the distribution of spin-labeled PC
molecules originally present in one-half of the bilayer (19).
In the inner membrane particles, the anisotropy of labeling
was obtained in two ways (35). The spin-labeled PC molecules
were first incorporated on both sides of the bilayer by
fusion with liposomes of (10,3)PC. The signal arising from
the molecules in the outer layer of the particles was then
immediately removed by ascorbate treatment at 0° (19) (in the
conditions used, ascorbate did not penetrate inside the particles). The second method involving the use of PC exchange
protein (36) afforded a direct preferential labeling of the
outer layer of the membrane. The labeled particles were maintained at 22° in isotonic conditions for different periods of
time and then chilled at 0°. The amount of ESR signal removable by ascorbate was then determined. The anisotropy of
labeling was found to remain stable for at least 3 hr. (The
incubation periods were limited to 3 hr in order to avoid
structural alterations of the membrane which could possibly
be produced by the mitochondrial phospholipase A_2 (37)).

It is concluded that, in physiological conditions, the rate
of outside-inside and inside-outside transitions is very slow
(half-life time greater than 24 hr). Therefore, contrary to
lateral diffusion which is very rapid, the flip-flop of phospholipids in the inner mitochondrial membrane is a negligible
event.

Enzyme Activity and Membrane Fluidity: A very large number
of reports have associated thermal transitions in the membrane
lipids with changes in the activity of membrane-bound enzymes
(38). Actually very good correlations between lipid phase
transitions and carrier-mediated transport processes have been
demonstrated with the use of mutants of E. coli that were
unsaturated fatty acid auxotrophs (13,39). Manipulation of
the acyl chains of mitochondrial phospholipids has been achieved by using the fatty acid desaturase mutant of Saccharomyces
cerevisiae, strain KD 115, and supplementing the growth medium
with a well-defined unsaturated fatty acid (40-42). Breaks
in the Arrhenius plots for cytochrome oxidase and ATPase activities and for ADP transport were observed at the temperature
producing a phase transition in the extracted mitohcondrial
phospholipids (42). This indicates that the membrane proteins
have little effect on the long range order of the bulk membrane
lipids or, in other words, that only a small fraction of the

phospholipids are interacting with membrane proteins; on the other hand, the physical state of the bulk lipids has a long range effect on the embedded proteins. The activity of membrane associated enzymes could also be modulated locally by an order-disorder transition of the boundary lipids whose contribution in calorimetric experiments is probably too small to be detected.

Probing of the Lipid Environment of an Intrinsic Protein of the Inner Mitochondrial Membrane, the ADP/ATP Carrier

To probe the immediate environment of the ADP carrier in situ, the strategy has been to attach a spin-labeled fatty acid to a molecule having a high affinity for the carrier, namely a specific inhibitor. The acyl chain affords the anchorage of the label in the membrane and the specific inhibitor provides the specificity of recognition of the carrier-protein. By placing the spin-label at different positions along the chain, information on the protein may be obtained concerning its interaction with the surrounding lipids, its position in the lipid matrix, and possibly its shape or size.

Use of Spin-labeled Acyl-CoA: Long-chain acyl-CoA esters have a high affinity for the ADP carrier ($K_i^{ADP} \sim 10^{-7}M$) and inhibit it competitively with ADP (43) (see P.V. Vignais et al., this volume). Spin-labeled acyl-CoA esters bearing a nitroxide radical at the positions shown in Fig. 4 have been synthetized and their inhibitory properties on the ADP transport determined (32). They still have a high affinity for the ADP carrier ($K_i \sim 10^{-7}M$) to which they bind reversibly. They are displaced from their binding site by the substrates of the carrier (ADP or ATP) or by specific inhibitors such as carboxyatractylate. Reversibility by specific ligands demonstrates unambiguously the specificity of binding.

The ESR spectrum of (10,3) acyl-CoA incorporated in heart mitochondria is shown in Fig. 5. It corresponds to a rather immobilized probe. Spectra of the corresponding (10,3)FA incorporated into mitocondria show a higher degree of motion of the probe at temperatures between $0°$ and $30°$. Similar differences in the spectra of (10,3) acyl-CoA and (10,3)FA were observed with inner mitochondrial membrane preparations from rat liver but not with preparations of outer mitochondrial membrane. Addition of ligands competing for binding (ADP, ATP or carboxyatractylate) displaced (10,3) acyl-CoA from its site and rendered the spectrum more mobile, similar to that given by (10,3)FA, indicating that (10,3) acyl-CoA is now free to diffuse away from the protein and conversely that the ADP-carrier is accessible to acyl-CoA from the lipid core of the membrane.

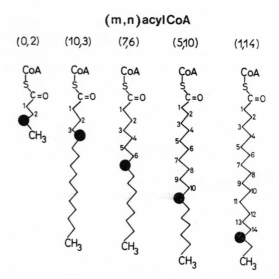

Fig. 4. *Schematic representation of the set of (m,n)acyl-CoA esters used to probe the ADP/ATP carrier. The position of the nitroxide radical relative to the polar head group is shown.*

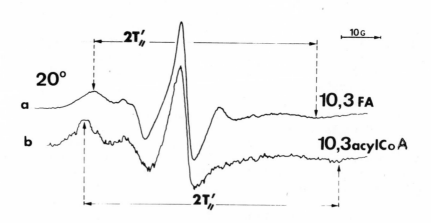

Fig. 5. *ESR spectra of (10,3)FA and (10,3)acyl-CoA incorporated in pigeon heart mitochondria.*

The nitroxide-protein interactions could be detected with (10,3)acyl-CoA and (7,6)acyl-CoA (between 0° and 30°) but tended to disappear when the nitroxide was moved towards the end of the chain. No interaction could be detected with (5,10)acyl-CoA or (1,14)acyl-CoA although they were bound to the carrier as revealed by the measure of the inhibition of ADP transport. However, at 0° a somewhat greater immobiliza tion of the nitroxide was observed with (5,10)acyl-CoA than with (5,10)FA. The abrupt changes in the spectra of the (5,10) acyl-CoA observed around 10° may indicate a phase transition of the lipid microenvironment of the ADP-carrier. A discon- tinuity in the activation energy for ADP transport has also been observed in the same range of temperature (44).

It must be noted that the lack of immobilization of the nitroxide ring when the probe is down the hydrocarbon chain of acyl-CoA is a characteristic feature of the probing of the lipid environment of the ADP-carrier. It did not occur when (1,14)acyl-CoA was intercalated in microsomal membrane. It was not observed either when (1,14)FA was incorporated in partially delipidated inner mitochondrial membrane (33).

These data are interpreted as follows: 1) The fact that once displaced from the protein, (10,3)acyl-CoA gives an ESR spectrum similar to that of (10,3)FA intercalated in a lipid bilayer, means that the ADP-carrier protein is in direct con- tact with a lipid bilayer in the membrane. 2) The loss of characteristic interactions when the probe is moved beyond the 8th carbon atom indicates that the carrier protein is neither of the type schematized as A nor as C in Fig. 1. However, the ESR data do not allow a distinction between the B type or the D type.

Use of Spin-labeled Acyl-atractylate: Because of the struc- tural resemblance between CoA and ADP, it was assumed that ADP and acyl-CoA were competing for the same binding site. Atrac- tylate and carboxyatractylate also compete with ADP for binding to the ADP-carrier, but probably by binding to an allosteric site (cf. P.V. Vignais, this volume). It may be hoped that another region of the protein can be monitored by this second spin-labeled inhibitor. (m,n)acyl-atractylate binds to mito- chondria with high affinity ($K_i < 10^{-7}M$) and competitively inhi- bits ADP transport as atractylate does. Work is in progress in view of using this probe for exploring the carrier protein in situ with regard to its shape (G. Lauquin, personal communi- cation).

The tight binding of (m,n)acyl-atractylate to the ADP- carrier should also enable one to estimate, by saturation- transfer sepctroscopy (45), the rate of rotational motion of the protein in the membrane. Preliminary data indicate a correlation time of approximately 10^{-5} to 10^{-4} sec for rotational motion of the ADP-carrier in the mitochondrial membrane.

REFERENCES

1. Gorter, E. and Grendel, F. (1925) J. Exp. Med. 41, 439.
2. Levine, Y.K. (1972) Prog. Biophys. Mol. Biol. 24, 3.
3. Singer, S.J. and Nicolson, G.L. (1972) Science 175, 720.
4. Luzzati, V. (1968) in Biological Membranes - Physical Fact and Function (Chapman, D., ed.) pp. 71-123, Academic Press, London.
5. Phillips, M.C., Williams, R.M. and Chapman, D. (1969) Chem. Phys. Lipids 3, 234.
6. Phillips, M.C. (1972) Prog. Surface and Membrane Sci. 5, 139.
7. Oldfield, E. and Chapman, D. (1972) FEBS-Letters 23, 285.
8. Träuble, H. and Eibl, H. (1975) in Functional Linkage in Biomolecular Systems (Schmitt, F.O., Schneider, D.M. and Crothers, D.M., eds.) pp. 59-101, Raven Press, New York.
9. Träuble, H. (1971) J. Membrane Biol. 4, 193.
10. Seelig, A. and Seelig, J. (1974) Biochemistry 13, 4839.
11. Shimshick, E.J. and McConnell, H.M. (1973) Biochemistry 12, 2351.
12. Op Den Kamp, J.A.F., De Gier, J. and Van Deenen, L.L.M. (1974) Biochim. Biophys. Acta 345, 253.
13. Linden, C.D., Wright, K.L, McConnell, H.M. and Fox, C.F. (1973) Proc. Nat. Acad. Sci., US 70, 2271.
14. Hegner, D., Schummer, U and Schnepel, G.H. (1973) Biochim. Biophys. Acta 307, 452.
15. Verkleij, A.J., De Kruyff, B., Ververgaert, Ph.J.Th., Tocanne, J.F. and Van Deenen, L.L.M. (1974) Biochim. Biophys. Acta 339, 432.
16. Träuble, H. and Eibl, H. (1974) Proc. Nat. Acad. Sci., US 71, 214.
17. Horwitz, A.F., Michaelson, D.M. and Klein, M.P. (1973) Biochim. Biophys. Acta 298, 1.
18. Smith, I.C.P. (1971) Chimia 25, 349.
19. Kornberg, R.D. and McConnell, H.M. (1971) Biochemistry 10, 1111.
20. Devaux, P. and McConnell, H.M. (1972) J. Amer. Chem. Soc. 94, 4475.
21. Seelig, J. (1970) J. Amer. Chem. Soc. 92, 3881.
22. Hubbell, W.L. and McConnell, H.M. (1971) J. Amer. Chem. Soc. 93, 314.
23. Dupont, Y., Cohen, J.B. and Changeux, J.P. (1974) FEBS-Letters 40, 130.
24. Singer, S.J. (1974) Ann. Rev. Biochem. 43, 805.

25. Gitler, C. (1973) Ann. Rev. Biophys. Bioengineering 1, 51.
26. Edidin, M. (1974) Ann. Rev. Biophys. Bioengineering 3, 179.
27. Colbeau, A., Nachbaur, J. and Vignais, P.M. (1971) Biochim. Biophys. Acta 249 462.
28. Levy, M., Toury, R. and André, J. (1967) Biochim. Biophys. Acta 135, 599.
29. Schnaitman, C. and Greenawalt, J.W. (1968) J. Cell. Biol. 38, 158.
30. Malviya, A.N., Parsa, B., Yodaiken, R.E. and Elliott, W.B. (1968) Biochim. Biophys. Acta 162, 195.
31. Parsons, D.F. and Williams, G.R. (1967) Methods in Enzymology 10, 443.
32. Devaux, P.F., Bienvenüe, A., Lauquin, G., Brisson, A.D., Vignais, P.M. and Vignais, P.V. (1975) Biochemistry 14, 1272.
33. Vignais, P.M., Devaux, P.F. and Colbeau, A. (1974) in Biomembranes - Lipids, Proteins and Receptors (Burton, R.M. and Packer, L., eds.) pp. 313-332, Bi-Science Publ. Division, Webster Groves, Mo.
34. Scandella, C.J., Devaux, P. and McConnell, H.M. (1972) Proc. Nat. Acad. Sci, US 69, 2056.
35. Rousselet, A., Colbeau, A., Zachowski, A., Vignais, P.M. and Devaux, P.F. (1975) 10th FEBS Meeting, Paris, Abst. # 1043.
36. Wirtz, K.W.A. (1974) Biochim. Biophys. Acta 344, 95.
37. Nachbaur, J., Colbeau, A. and Vignais, P.M. (1972) Biochim. Biophys. Acta 274, 426.
38. Raison, J.K. (1973) J. Bioenergetics 4, 285.
39. Overath, P., Schairer, H.U. and Stoffel, W. (1970) Proc. Nat. Acad. Sci., US 67, 606.
40. Haslam, J.M., Cobon, G.S. and Linnane, A.W. (1974) Biochem. Soc. Transactions 2, 207.
41. Watson, K., Houghton, R.L., Bertoli, E. and Griffiths, D. E. (1975) Biochem. J. 146, 409.
42. Lunardi, J. and Lauquin, G. (1975) 10th FEBS Meeting, Paris, Abst. #1041.
43. Morel, F., Lauquin, G., Lunardi, J., Duszynski, J. and Vignais, P.V. (1974) FEBS-Letters 39, 133.
44. Duée, E.D. and Vignais, P.V. (1969) J. Biol. Chem. 244, 3920.
45. Thomas, D.D., Seidel, J.C., Hyde, J.S. and Gergely, J. (1975) Proc. Nat. Acad. Sci., US 72, 1729.

THE INTERACTION OF IONIC DETERGENTS WITH SUBMITOCHONDRIAL MEMBRANES

R.J. Mehlhorn

Considerable research activity is focused on the problem of how membrane structure, particularly the relationship between lipids and proteins, correlates with membrane functions. Despite intensive efforts of many laboratories, this problem remains essentially open, except perhaps for a few simple model systems. Complex membranes, like those of mitochondria, containing a large variety of lipid and protein components have proven particularly difficult to study and a number of workers have incorporated selected protein complexes into model membranes in an effort to simplify the structural problem.

An alternative to the reconstitution approach is the perturbation approach. In the latter technique, some structural component of a complex membrane is altered, or new molecules are introduced into the membrane to modify the structure in some known way and the consequences of this modification on membrane functions are studied. An example of the utility of this approach is provided by yeast lipid mutants, where the fatty acid composition is varied to study the lipid requirements for mitochondrial functions.

Detergents are another useful method for perturbing membrane structure. They are particularly useful because: a) they are often highly water soluble so their incorporation into membranes is rapid; b) they are usually simple molecules whose properties in a number of solvents are well characterized; c) they are generally chemically inert, so their effects can safely be ascribed to their physical effects; d) their amphiphylic character resembles that of membrane components so their disposition among these components can, to a limited extent, be predicted.

As elaborated in a recent review (1), detergents can be divided into two classes and three concentration categories on the basis of their disruptive effects. Thus ionic detergents are regarded as more disruptive than non-ionic ones, and it is convenient to interpret detergent data on membranes in terms of three concentration ranges: low, intermediate, and high.

Low detergent levels with many membranes exhibit high affinity binding, i.e., binding has a saturation value at some detergent to protein ratio. At low detergent levels, a number of membrane effects have been observed: an increase in membrane area (erythrocytes), stabilization against lysis, reduction of buoyant density, uncoupling of phosphorylation (mitochondria and chloroplasts), and, often, a marked increase in enzyme activities.

At intermediate detergent levels membrane lysis occurs, accompanied by a dramatic increase in detergent binding, and intrinsic membrane proteins are released from the membrane as lipoprotein-detergent complexes.

At high detergent to membrane ratios, lipid is removed from intrinsic proteins (except perhaps for some tightly bound lipid), and intrinsic proteins can be segregated from lipid by physical techniques, e.g., density centrifugation. Protein-protein interactions generally seem to be maintained by nonionic detergents, while they are disrupted by ionic ones. Moreover, extensive protein denaturation is common for high levels of ionic detergents.

Studies with SDS and CTAB on Submitochondrial Membranes

We have been using the ionic detergents sodium dodecyl sulfate (SDS) and cetyl trimethyl ammonium bromide (CTAB) to interact with sonicated rat liver mitochondrial preparations (SMP's). Unlike many of the nonionic detergents, these two detergents have the advantage of chemical homogeneity. Also the alkyl group is similar (except for length) so that any differences between these two detergents are probably due to charge effects. Functional studies were those of electron transport using the substrates NADH, succinate, and ascorbate-TMPD with oxygen as the terminal electron acceptor. Structural studies were carried out with lipid soluble spin probes: a CTAB analogue, denoted by CDTAB,

a stearic acid derivative, A12NS,

and a hydrocarbon label, 5N9.

Functional Studies

The inhibition of substrate oxidation with SDS and CTAB
was assayed polarographically. The respiratory rates were
diminished to 10% of the control values by the following
detergent levels (μMoles/mg protein):

	NADH	Succinate	Ascorbate TMPD
SDS	0.7	0.9	2.0
CTAB	0.3	0.6	0.7

The inactivation of respiration for all three substrates
by CTAB occured in three stages: initially, there was a
progressive decrease in respiratory rates with increasing
CTAB levels. The numbers in the table refer to this range.
At higher CTAB levels, there was a range of concentrations
for which respiratory activity was slightly stimulated by
increasing levels of the detergent. Beyond this range, increasing levels of CTAB again inactivated the respiratory
rate until total inactivation occured.

On the other hand, SDS produced a stimulation of respiration at the lowest detergent levels considered (0.1 μMole/mg
protein). The extent of stimulation was similar to that
observed for an uncoupler, e.g., FCCP. Beyond this uncoupling range, SDS progressively inactivated the oxidation of
all three substrates.

Inactivation brought about by SDS or CTAB could be rapidly
reversed by adding detergent of the opposite charge. Optimal
reactivation generally required about 30% more SDS than CTAB.
Reactivation was observed for all three substrates, but was
most marked for succinate. In the case of CTAB inactivation,
the addition of 1.2 equivalents of SDS led to 230% of the
control rate. This was considerably higher than could be
accounted for by uncoupling effects.

The rate of reactivated respiration decreased as the concentration of the detergent mixture increased. At sufficiently high levels of detergent, no reactivation was observed. This is to be expected from the known denaturing
effects of charged detergents at high concentrations.

Structural Studies

The structural effects of SDS and CTAB on submitochondrial
membranes were assayed with the spin labels CDTAB and A12NS.

The former probes the structure of the polar-nonpolar interface, while A12NS is localized in the hydrophobic membrane interior.

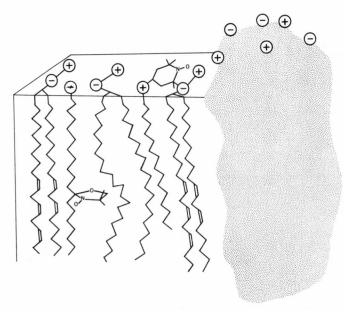

Fig. 1. Cutaway view of an SMP membrane showing the presumed intercalation of spin labels among lipid and protein components.

Spin resonance experiments were carried out at much higher membrane concentrations than the respiratory studies. The probe CDTAB was used to show that both experiments may be compared on the basis of µMoles of detergent/mg protein. The critical micelle concentration of CDTAB was found to be 460 µM by ESR methods. This is close to the known CMC of CTAB (0.92 mM). This information, taken together with the structural similarity of the two molecules was used to deduce binding properties of CTAB from the CDTAB studies. An ESR spectrum of 0.25 µMoles of CDTAB/mg protein of submitochondrial particles displays both aqueous and membrane-bound components at high membrane concentrations (15 mg protein/ml). The aqueous label concentration can be estimated directly from line height measurements of the spectrum. Similar spectral analyses were carried out for a series of dilutions of submitochondrial particles in suspension medium to give the dependence of bound versus free CDTAB upon the membrane concentration. This study showed that the majority of CDTAB molecules are membrane-bound over a broad concentration range, including concentrations used for the respiration assays.

The effects of the detergents on membranes were analyzed in terms of the rotational correlation time, τ_0; to a first approximation this parameter is inversely proportional to the fluidity of the environment of the spin label.. The effect of detergents on τ_0 for the two spin labels in SMP membranes is shown in Fig. 2.

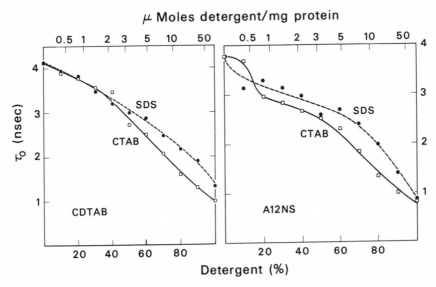

Fig. 2. *The effect of SDS and CTAB on the mobility of two spin labels in submitochondrial membranes.*

There is a progressive increase in the fluidity of the label microenvironment as the level of detergents increase. Since both CDTAB and A12NS are affected to about the same extent, the disruption of the membrane is presumed to extend from the surface into the hydrophobic membrane interior for both SDS and CTAB. It is also seen that significant alteration in membrane fluidity occur at the detergent levels which result in the loss of respiratory functions.

To study the structural aspects of reactivation, a similar analysis of rotational correlation times was carried out with CTAB-treated membranes (o.25 μMoles/mg protein) as a function of SDS additions. A partial reversal of the fluidizing effect of CTAB was noted which reached a maximum at one equivalent of SDS to CTAB. However, this reversal was not sufficient to account for the restoration of respiratory function. Thus, it appears that the loss of respiratory rates caused by charged detergent disruption is not due to the disordering of membrane structure as reflected in the membrane fluidity, both at the membrane surface and in the hydrophobic interior.

To further elucidate the structural aspects of reactivation, SDS and CTAB mixtures were studied in the absence of membranes. X-ray studies of the equimolar mixture in water showed a crystalline pattern characteristic of a fatty acid salt. The spin label 5N9 which partitions into aqueous and hydrophobic domains for a membrane preparation of 15 mg/ml, has virtually no solubility in an equimolar detergent mixture of 15 mg/ml. This suggested to us that SDS and CTAB in a membrane might also form a crystalline complex which might separate from the membrane components by a process of lateral diffusion.

To assess the possibility of detergent crystallization in the membrane, we exploited the phenomenon of the exchange interaction between spin labels. An equimolar mixture of SDS and CDTAB produces a single spectral line which is easily distinguished from the ESR signal of CDTAB in dilute dispersion in the membrane. However, the addition of SDS to CDTAB labeled SMP membranes (0.25 µMoles/mg protein) led to no detectable signal change so it was concluded that no aggregation of CDTAB into exchange broadened crystalline domains occurred.

The conclusion of the structural studies then is that the detergent molecules are fairly randomly dispersed in the membrane, even in the reactivated state. Moreover, neither inactivation nor reactivation are primarily due to effects of the detergents on membrane fluidity. It thus seems that SDS and CTAB effects are primarily due to an alteration of membrane surface charge caused by the detergents.

Crosslinking Studies (collaboration with L. Packer)

Rat liver mitochondria were treated with dimethyl suberimidate (DMS) at sufficient levels to prevent osmotic swelling of the organelles. Respiratory rates of these crosslinked mitochondria were several times smaller than those of untreated controls. Inactivation of the oxidation of ascorbate-TMPD of the treated mitochondria occurred at about the same detergent levels as for controls. However, about one-fifth the detergent concentrations were required to inactivate the succinate activity as compared with control mitochondria. Reactivation patterns for the crosslinked organelles were similar to those found for control mitochondria.

Thus it is clear that the charged detergents are interacting with crosslinked mitochondria. This suggests that
local changes in membrane area by the insertion of detergent
molecules can occur even though the limiting membrane area is
fixed (as inferred from the absence of osmotic changes).

REFERENCE

1. Helenius, A., Simons, K. (1975) Biochim. Biophys. Acta,
    415, 29-79.

LIGHT INCREASES THE ION PERMEABILITY OF RHODOPSIN-PHOSPHOLIPID BILAYER VESICLES

A. Darszon* and M. Montal

INTRODUCTION

Photoreceptor cells in both vetebrate and invertebrate species are capable of converting the energy of a single photon into a neural impulse. The membrane-bound visual pigment, rhodopsin, in vertebrate and invertebrate photoreceptors, triggers the photoresponse in these cells by altering the plasma membrane ionic conductance (1,2).

The molecular mechanisms of the function of vetebrate rhodopsin remain obscure, but appear to involve a direct regulation of ion flow across disc membranes: Rhodopsin, itself, may be the ion translocator, or it might act by regulating an internal transmitter.

We have approached the problem of the mechanism of action of rhodopsin by studying the properties of rhodopsin in re-constituted experimental membranes (3). The results obtained on the planar bilayer system have been discussed before (3). We will restrict ourselves on this account to the studies on bilayer vesicles (cf. 2,3).

MATERIALS AND METHODS

All of the procedures were performed under dim red light (Kodak series 1 filter) unless otherwise stated. Rod outer segments from dark adapted bovine retinas (Hormel, Co.,) were isolated by sucrose flotation and purified in a discontinuous sucrose gradient (4). Cetyltrimethylammonium bromide (CTAB) was used to extract rhodopsin (5). Rhodopsin is incorporated into spherical bilayers as follows: the detergent CTAB is partially removed from Rhodopsin-lipid-detergent micelles with the use of Bio Beads SM-2 (Bio Rad Laboratories), a neutral porous styrene divinylbenzene copolymer (6,7). Following detergent removal, bleached rhodopsin can be regenerated in the dark with 11-cis or 9 cis-retinal (7). These micelles are then used to form spherical bilayers containing rhodopsin by cosonication with lipids in the presence of a salt medium (3,8). A typical protocol is: Soybean phospholipids, partially purified (9) containing 37% phosphatidylcholine, 37% phosphatidylethanolamine and 8% cardiolipin, are mechanically dispersed in a vortex during 15 min (15 mg/1.5 ml) in a

---

*CONACYT Fellow

solution buffered with 0.01 M imidazole, pH 7. Thereafter,
rhodopsin (0.5 mg of absorbance at 500 nm, assuming an extinction coefficient of 4×10^4 and a molecular weight of 4×10^4
(2)) in 0.25 ml of 50 mM CTAB, 66 mM phosphate buffer, pH 7.0 -
which was incubated for 50 min at room temperature with 0.12
gr of Bio Beads - is cosonicated with the lipid for 3.5 min
at 4°C by immersion of the test tube in a Bransonic (Heat
Systems Ultrasonics, Inc., Plainview, N.Y.; ultrasonic
cleaner-power, output 100 watts). In the permeability studies
reported here, the sonication step is carried out in the presence of radioactively-labeled ions (Amersham/Searle, Corp.).
     Two methods have been used to measure the permeability of
the vesicles. In the first one, the sonication step is
followed by centrifugation at 105,000 x g for 30 min. The
pellet (P_1) is washed in a cold-salt medium by resuspension
and centrifugation. The supernatant is centrifuged again for
1 hr, which yields another pellet that is subsequently washed
(P_2). Thereafter, the particles are resuspended and the
appearance of label in the solution is measured. The second
method is an adaptation of that used by Michaelson and Raftery
to study the permeability of vesicles reconstituted with
purified acetylcholine receptor (10). The method consists of
loading a DEAE column with the vesicles and washing exhaustively
the untrapped label with a cold-salt medium. Thereafter, the
trapped label is eluted from the column by means of Triton
X-100 detergent. Both methods yield similar results, but the
latter has been used more often for its convenience, ease of
operation and rapidity. Radioactivity was measured in a
Packard Scintillation Counter, utilizing detergent-based
toluene liquid scintillation solution (11) or in a Nuclear
Chicago well type counter. Absorption spectra were recorded
in a Cary 14 spectrophotometer (Applied Physics Corp.,
Monrovia, CA); phospholipid phosphate was analyzed according
to Dawson (12).

                              RESULTS

     We have examined the rhodopsin-lipid recombinants by
means of freeze-fracture electronmicroscopy (13). Freeze-
fracture replicas of the recombinants show them to be vesicles
with an average diameter of 2500 Å. The fracture faces of the
vesicles show a random distribution of particles of about
100 Å in diameter (14). Thus, the recombinants formed by this
procedure display a similar structure to those formed by dialysis (15). The labeled ions are retained by the recombinants
and can be released by the addition of the ionophores X-537A
or gramicidin, thus establishing an existing population of

sealed vesicles. This could be expected in view of the lipid-rhodopsin ratio of 10^3, which implies the presence of extensive regions of bilayer. It is important to note that the manipulations required to incorporate rhodopsin into the vesicles result in protein denaturation to a certain extent: after the detergent removal and sonication steps were performed, 73% of the quantity used at the beginning of the procedure is preserved.

Figure 1A illustrates the type of results obtained by the centrifugation method. The distribution of radioactivity in the pellet and supernatant indicates that on bleaching the pellet is depleted and the supernatant enriched; the effect is more striking when Ca^{2+} is present as well. Figure 1B shows that the content of Na^+ in the vesicle decreases after bleaching.

In Table I, it is shown that the extent of Na^+ leakage is greater in the bleached than in the dark recombinants. It is also illustrated that the Na^+ leakage is enhanced when cold $CaCl_2$ or $MgCl_2$ (1 mM) is included in the sonication medium. Having found this favorable effect of divalent cations on Na^+ leakage, Ca^{2+} was thereafter included to obtain optimal differences. It is evident that the number of Na^+ released per bleached rhodopsin (Na^+/Rho) is different in the two methods of determination. There appears to be no simple explanation for the discrepancy since the two methods involve very distinct processes. The data presented in the remainder of this article was collected using the column technique. An important feature of the Table is to show that the effect of light is measurable with different techniques.

The relation between the amount of Na^+ released and the number of rhodopsin molecules bleached is illustrated in Fig. 2. The initial slope indicates that about 184 Na^+ diffuse out of the vesicles per bleached rhosopsin. As the number of bleached rhodopsin increases, the relation departs from linearity suggesting that the Na^+ content of the vesicles becomes a limiting factor. This, indeed, appears to be the case, since the number of ions released per bleached rhodopsin increases with the ionic strength of the sonication medium as shown for Cs^+ in Table II.

To provide further support to the notion that rhodopsin is responsible for the permeability effects observed, the action spectrum illustrated in Fig. 3 was obtained. The measurements are acquired in the region of linearity indicated in Fig. 3. Due care was exercised to deliver the same number of photons in all the wavelengths explored. The agreement between the rhodopsin absorption spectrum and the wavelength dependence of the Na^+ release identifies rhodopsin as the absorbing species.

DISTRIBUTION OF $^{22}Na^+$-IN PELLET (P) AND
SUPERNATANT (S) OF RHODOPSIN VESICLES
(T= 4° C)

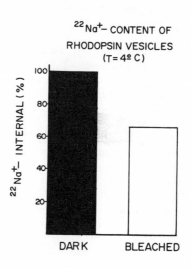

$^{22}Na^+$- CONTENT OF
RHODOPSIN VESICLES
(T= 4º C)

Fig. 1. Na$^+$ efflux from rhodopsin-lipid vesicles. A) Centrifugation method: The vesicles were formed in 0.1 M NaCl with ^{22}NaCl at about 1 x 10^6 cpm/ml. When indicated, 1 mM CaCl$_2$ was also included in the sonication buffer. The vesicles (P$_1$ and P$_2$) were washed in a (0.1 M KCl-0.01 M imidazole, pH 7) buffer. After illumination (Sylvania movie lamp, 150 w output), the aliquots were centrifuged and the radioactivity in the pellet and in the supernatant counted. The plot indicates the distribution of radioactivity between the two compartments; 100% is equivalent to the sum of the radioactivity in pellet and supernatant. All the procedures were performed at 2-4°C.

B) Column method: The vesicles were formed in 0.1 M NaCl with ^{22}NaCl at about 1 x 10^6 cpm/ml and 1 mM CaCl$_2$. 1.5 ml of DEAE-cellulose were vacuum packed in a column, and equilibrated with 0.1 M KCl, 0.01 M imidazole, pH 7. A 0.5 ml sample was carefully layered over the column and introduced. A volume of 5.5 ml of KCl buffer was passed to wash the untrapped external radioactivity. Thereafter, 3.0 ml of 1% Triton-X 100 in 66 mM phosphate buffer, pH 7.0 was introduced to elute the ^{22}Na retained by the vesicles. The bleached sample was introduced into the column within one minute after illumination. Thus, the Na$^+$ content expressed does not represent equilibrium values. The radioactivity was counted in both samples. Aliquots from the detergent sample were used to measure the rhodopsin content spectrophotometrically. Phospholipid phosphate was measured in the washing medium and corresponded to 10-15% of the total amount of lipid used (12). Temperature 2-4°C.

TABLE I

Effect of Divalent Cations on Na$^+$Efflux

	Cations per Rhodopsin		
	Centrifugation		Column
	P$_1$	P$_2$	
Na$^+$	67	191	22
Na$^+$+ Ca^{2+}	308	235	92
Na$^+$+ Mg^{2+}	153	313	--

The vesicles were formed as in 1B with 1 mM CaCl$_2$ or MgCl$_2$ where indicated. Temperature 2-4°C.

As illustrated in Table III, the permeation pathway is not highly selective since Na$^+$, Cs$^+$ as well as Ca^{2+} can pass through. There is apparently a larger affinity for Cs$^+$ over Na$^+$. The ratio for Ca^{2+} should be taken as a lower boundary since it is limited by the low Ca^{2+} content of the vesicles. When the sonication step is performed in the presence of a CaCl$_2$ concentration larger than 2.5 mM, the resultant vesicles remain turbid; this is due to the progressive precipitation induced by Ca^{2+}. Considering that the vesicles formed in 0.1 M solutions of NaCl or CsCl yield ratios within the order of those found for Ca^{2+}, it is suggestive that the Ca^{2+}/rhodopsin ratio may actually be larger than for the other ions. When we consider the results reported with both methods, the light effect observed corresponds to an 80% reproducibility.

Controls: It has been previously reported (16,17) that retinaldehyde increases the diffusion rate of K$^+$ our of phosphatidylethanolamine vesicles. Since our vesicles contain this lipid, it is necessary to rule out any spurious effect of retinal.

All the experiments were carried out at temperatures ranging between 2-4°C. The intermediate in the photo-bleaching pathway that is stable at this temperature range is metarhodopsin II (λ max = 380 nm); in this intermediate, all-trans retinal remains still covalently bound as a Schiff base to the epsilon-amino of a lysine in the protein (cf. 1,2). Thus, no free retinal is available to disrupt the membrane on account of its surfactant capacity. Furthermore, additional controls where vesicles were kept in the dark but supplemented with an amount of all-trans retinal equivalent to that present as 11-cis retinal in rhodopsin, indicate that no significant Na$^+$ leakage from the vesicles resulted from the addition of retinal. In many experiments, the dark versus bleached difference is taken in reference to dark retinal control samples.

DISCUSSION

It has been demonstrated that light increases the cation permeability of rhodopsin-lipid vesicles. The action spectrum as well as the dismissal of the spurious effect of retinal establish that the light response can be accounted for by rhodopsin bleaching.

The permeation pathway appears to be relatively non-selective since Na$^+$, Cs$^+$ and even Ca^{2+} pass through with an apparent equivalent efficiency.

Experiments are at present underway to establish the temporal sequence of events and the correlation of the lifetimes of rhodopsin intermediates and the time course of cation release.

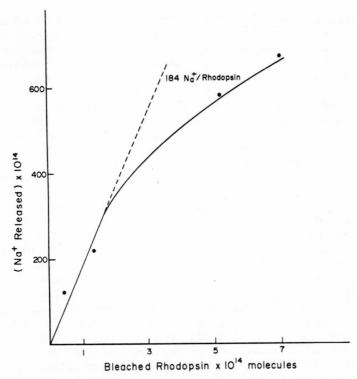

Fig. 2. *The relation between the extent of rhodopsin bleaching and* $^{22}Na^+$ *efflux. The vesicles were prepared as in 1B. Rhodopsin was bleached by illuminating at a wavelength of 500 nm for different exposure periods, and the extent of bleaching measured spectrophotometrically. Temperature 2-4°C.*

TABLE II

Recombinants of Native and Chymotrypsin-Treated Rhodopsin

Cation	Cations/Rhodopsin[†]	
	Native	Chymotrypsin-treated
Na^+	92	72
Cs^+	70	110 (222)*
Ca^{2+}	20	10

[†]*100% bleached rhodopsin.*
Rod outer segment discs were treated with chymotrypsin according to Saari (32). Rhodopsin was solubilized into 50 mM CTAB, 66 mM phosphate buffer, pH 7. Thereafter vesicles were prepared as described in Table II. Temperature 2-4°C.
**The vesicles were formed in 0.25 M CsCl.*

Mechanism of the Ca^{2+} effect: The favorable effect of Ca^{2+} on the permeability of the monovalent cations, can be explained from the well known effects of Ca^{2+} on charged interfaces of biological relevance (18). Ca^{2+} condenses negatively-charged monolayers and raises the transition temperature of negatively-charged visicles (19,20). This condensing effect leads to phase separations in the plane of the monolayer, and thus leaves free space to be occupied by the protein. In addition, it binds to negatively-charged residues in the protein, thus forming ternary complexes - lipid-Ca^{2+} - protein (21,22). The combination of these two effects may be responsible for a deeper penetration of the protein into the bilayer hydrocarbon domain (23).

Mechanism of Permeability Change: The results we have obtained do not permit us to propose a specific mechanism for the observed change in permeability, nevertheless, they do suggest certain alternatives that deserve serious consideration The experiments accumulated up-to-date on the planar bilayers with incorporated rhodopsin reproduce in remarkable parallelism the effects reported on the vesicles: Light increases the membrane conductance; in the presence of Ca^{2+}, at concentrations at which it cannot be the charged carried species, the maximum level of conductance reached is higher than in its absence. In addition, we obtained new information which is unaccessible in the vesicles. The conductance exhibits fluctuations of about 0.4 nmho (in 0.2 M NaCl) which suggest a wide permeation pathway of about 10 Å in diameter. Recent hydrogen-tritium exchange measurements are consistent with a hydrophylic channel of about 10-15 Å in diameter (40). Likewise, the unit conductance of 0.4 nmho indicates a transport velocity in the order of 10^8 ions/sec (cf. 3). This is the transport rate calculated for the unit conductance of the Na$^+$ channel in squid giant axon (24) and the chemically-activated post-synaptic channel (25-27). In addition, this value is the maximum transport rate calculated for a model transmembrane channel (28).

There are at least two models that can account for the results obtained (cf. 3): 1) the rhodopsin molecule itself may be a transmembrane channel; 2) alternatively, rhodopsin may alter the dielectric structure of the lipid bilayer in which it is embedded. We do not know of any evidence that strongly supports one and not the other.

There are some known effects that tend to support the second alternative: Rhodopsin and opsin (the protein moiety left after bleaching and release of retinal) behave as two distinct proteins. For example, the pH range at which rhodopsin is stable is from 3-9, whereas that for opsin is from 6-8 (29);

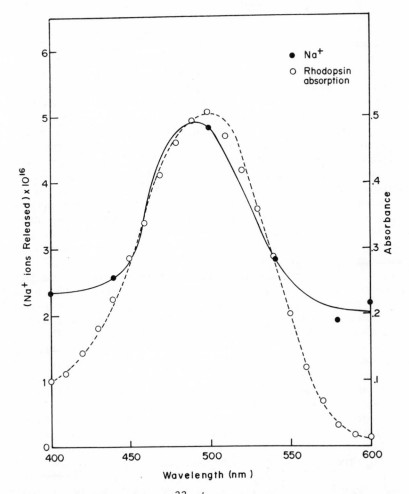

Fig. 3. *Action spectrum of* $^{22}Na^+$ *efflux. The vesicles were prepared as in 1B and exposed to equivalent light intensities at the indicated wavelengths. Narrow band interference filters from Baird Atomic were used. Temperature 2-4°C.*

thermal denaturation indicates that opsin denatures at a temperature 20°C lower than rhodopsin (30). The most relevant difference for this discussion is the strong tendency of opsin to aggregate in aqueous solutions. Thus, it is conceivable that on bleaching, several opsin molecules aggregate and as a result a discontinuity in the bilayer dielectric core appears.

Lateral diffusion of opsin monomers and lipids would lead to dissociation (cf. 3). Rhodopsin is a glycoprotein (cf. 5), thus the effect of Ca^{2+} would be consistent since it has been recognized that Ca^{2+} is required for the aggregation of glyco-proteins that eventaully lead to permeability changes (31).

The models have the attraction of suggesting experiments to distinguish between them, which we are performing at present. The data shown in Table II are an example of the strategy involving a systematic cleavage of the macromolecule. This is feasible now that we have identified a "functional assay" of rhodopsin as a light-dependent increase in ionic per-meability. We know at present that we can cleave rhodopsin from a molecular weight of 40,000 down to two fractions, one of 20,000 and another of 29,000, without loosing activity. On the contrary, in the case of Cs^+ an apparent enhancement of specific activity follows digestion. Since under these conditions the bleached protein residue is regenerable upon incubation in dark with 9-cis retinal (32), the residual poly-peptide preserves photochemical as well as "functional" activity. This approach is leading us to cleave the carbo-hydrate moiety as well, and to continue to resolve the minimum fragment that exhibits activity. We have carried out prelimi-nary experiments to measure the cationic permeability of normal and chimotrypsin digested reinal rod discs. The discs were loaded either by sonication in the presence or absence of lipids or by 8 hr equilibration with the radioactively-labeled cations (^{134}Cs, ^{22}Na). The results obtained on this system, as well as those obtained with the purified rhodopsin liposomes, are on close agreement.

The aggregation scheme is presently being explored by looking at the effect of temperature. If a transmembrane structure is responsible for the permeability changes, it would be expected to be active even in frozen membranes, as is the case of gramicidin A in bilayers (33). In contrast, the aggregation mechanism depends stringently on temperature since the collision of the monomers is limited by the diffu-sion rate in the plane of the monolayer.

Mechanism of Phototransduction: Irrespective of rhodopsin's mode of action at a molecular level, as a monomeric or aggre-gated channel it is evident that the amplification gained on allowing the passage of 10^8 ions/sec per absorbed photon is, indeed, enormous.

Ca^{2+} has been postualted as the transmitter in vertebrate photoreceptors (cf. 1,2). Several recent reports indicate that the Ca^{2+} content of rod disc membranes decreases upon illumination. The stoichiometries range between 1-1000 Ca^{2+}/photon (34-37). The Ca^{2+} hypothesis requires a Ca^{2+} uptake mechanism for the dark restoration process. In this regard, a Ca^{2+}-ATPase has been described in rod outer segments (38) and in purified disc membranes (39). Thus, the apparent selectivity of phototransduction for Ca^{2+} (cf. 1) may not arise from the selectivity of the channel, but from that of the pump that would concentrate one ionic species over others - Ca^{2+} in this case, and on bleaching, this gradient would be dissipated via the non-selective channel (3). There is no evidence at present to extend this scheme to the invertebrate photoreceptors where the situation appears more complex (cf. 1,2).

TABLE III

Cation Efflux from Rhodopsin-Lipid Vesicles

	Cations/Rhodopsin	
Cation	67% Bleached	100% Bleached
Na^+	78	92
Cs^+	84	70
Ca^{2+}	26	20*

The vesicles were prepared in 0.1 M NaCl-^{22}NaCl (about 1 x 10^6 cpm) and 1 mM $CaCl_2$; 0.1 M CsCl-^{134}CsCl (about 2 x 10^6 cpm) and 1 mM $CaCl_2$; or 2.0 mM $CaCl_2$-$^{45}CaCl_2$ (6 x 10^6 cpm). Temperature 2-4°C.

*This experiment has been performed by the centrifugation method on 13 preparations yielding differences between bleached and dark pellets of about 15% (average).

ACKNOWLEDGEMENTS

It is a pleasure to thank Dr. Amira Klip for her collaboration in the rhodopsin cleavage experiments, and Mr. Jorge Zarco for his skillful assistance.

REFERENCES

1. Hagins, W.A. (1972) Ann. Rev. Biophys. Bioengineer. 1, 131.
2. Montal, M. and Korenbrot, J.I. (1975) *in* Membrane Bound Enzymes (Martonosi, A., ed.) Plenum Press, New York, In press.
3. Montal, M. (1975) *in* Molecular Aspects of Membrane Phenomena (Kaback, H.R., Radda, G.K. and Neurath, H., eds.) Springer Verlag/Heidelberg, New York, In press.
4. McConnell, D.J. (1965) J. Cell Biol. 27, 459.

5. Heller, J. (1968) Biochemistry 7, 2906.
6. Holloway, P.W. (1973) Anal. Biochem. 53, 304.
7. Feldberg, N.T. (1974) Invest. Opthamol. 13 155.
8. Racker, E. (1973) Biochem. Biophys. Res. Commun. 55, 224.
9. Kagawa, Y. and Racker, E. (1971) J. Biol. Chem. 246, 5477.
10. Michaelson, D.M. and Raftery, M.A. (1974) Proc. Nat. Acad. Sci. US 71, 4768.
11. Klip, A. and Gitler, C. (1974) Biochem. Biophys. Res. Commun. 60, 1155.
12. Dawson, R.M.C. (1960) Biochem. J. 75, 45.
13. Branton, D. and Deamer, D.W. (1972) Protoplasmatologia, Vol. 11E1.
14. Darszon, A., Martíniez-Palomo, A. and Montal, M. Unpublished observations.
15. Chen, Y.S. and Hubbell, W.L. (1973) Exp. Eye Res. 17, 517.
16. Bonting, S.L. and Bangham, A.D. (1968) in Biochemistry of the Eye (Dardenne, M.M. and Nordman, J., eds.) pp. 493-513, Karger, Basel-New York.
17. Daemen, F.J.M. and Bonting, S.L. (1969) Biochim. Biophys. Acta 183, 90.
18. Montal, M. (1975) Ann. Rev. Biophys. Bioengineer. 6, in press.
19. Träuble, H. and Eibl, H. (1974) Proc. Nat. Acad. Sci. US 71, 214.
20. Verkleij, A.J., Dekruyff, O., Ververgaert, P.H.J.Th., Tocanne, J.F. and VanDeenen, L.M.M. (1974) Biochim. Biophys. Acta 339, 432.
21. Bulkin, B.J. and Hanser, R. (1973) Biochim. Biophys. Acta 326, 289.
22. Gitler, C. and Montal, M. (1972) FEBS Letters 28, 329.
23. Montal, M. and Korenbrot, J.I. (1973) Nature 246, 219.
24. Hille, B. (1970) Prog. Biophys. Mole. Biol. 21, 1.
25. Katz, B. and Miledi, R. (1972) J. Physiol. (London) 224, 665.
26. Andersen, C.R. and Stevens, C.F. (1973) J. Physiol. (London) 235, 655.
27. Sachs, F. and Lecar, H. (1973) Nature (London) 246, 214.
28. Haydon, D. and Hladky, S.B. (1972) Quart. Rev. Biophys. 5, 187.
29. Radding, C.M. and Wald, G. (1955-56a) J. Gen. Physiol. 39, 923.
30. Hubbard, R. (1958) J. Gen. Physiol. 42, 259.
31. Gingell, D. (1973) J. Theoret. Biol. 38, 677.
32. Saari, J.C. (1974) J. Cell Biol. 63, 480.
33. Krasne, S., Eisenman, G. and Szabo, G. (1971) Science 174, 412.

34. Abrahamson, E.W., Fager, R.S. and Mason, W.T. (1974) Exp. Eye Res. $\underline{18}$, 51.
35. Hendricks, Th., Daemen, F.J.M. and Bonting, S.L. (1974) Biochim. Biophys. Acta $\underline{345}$, 468.
36. Liebman, P.A. (1974) Invest. Opthamol. $\underline{13}$, 700.
37. Szuts, E.Z. and Cone, R.A. (1974) Fed. Proc. $\underline{33}$, Abst. 1403.
38. Bownds, D., Gordon-Walker, A., Gaide-Hugnenin, A.C. and Robinson, W. (1971) J. Gen. Physiol. $\underline{58}$, 225.
39. Ostwald, T.J. and Heller, J. (1972) J. Biochemistry $\underline{11}$, 4679.
40. Downer, N.W. and Englander, S.W. (1975) Nature $\underline{254}$, 625.

SUBJECT INDEX

A

A12NS, 383
A23187, 42, 54
Acetaldehyde, 172
Acriflavine, 275
Acyl-atractylate, 372
  spin label of, 377
Acyl CoA, 65, 90, 110, 372
  spin label of, 375
Adenine nucleotide, transport of, 75, 109, 130,
  271, 375
Adenine nucleotide transport, nuclear mutants of,
  112
Adenylate kinase, 132
ADP, binding protein, 112
Affinity label, 269
Alamethicin, 12
Alkylating agent, 359
Alkyguanidine, 25, 172
Aminonaphthalene sulfonate, 225
Aminooxyacetate, 72
Amytal, 172
Anaesthetics, interaction with membrane, 303
1-Anilino-8-naphthalene sulfonate (ANS), 327
Anion translocator, 79
  citric cycle interaction, 89
  model of, 66
Anthracene, 317
Antimycin-A, 152, 253
Arsenazo III, 33
Ascites cells, 86
Ascorbyl palmitate, 299
Aspartate aminotransferase, 89
ATPase, 22, 75, 97, 157, 183, 358
  Ca^{++}-dependent, 315
  complex, 241
    reconstitution of, 189
  mercurial-dependent, 14
  oligomycin-insensitive, 361
  oligomycin-sensitive, 171, 214
Atractyloside, 110, 127, 271
  binding protein, 112
1-Azido-4-iodobenzene, 322

2-Azido-4-nitrophenol (ANP), 226
1-Azidonaphthalene (AzN), 315

B

Bathophenanthroline, 70
Black film, 50
Bongkrekic acid, 110, 136
Butacaine, 306
Butylammonium, 160

C

Ca^{++}
  binding glycoprotein, 51
  energy-dependent uptake, 35
  transport of, 31, 49
Carboxyatractyloside, 110, 136, 147
  binding protein, 112
Cardiac muscle, 36
Cardiolipin, 171, 371
Carnitine, 65
Cation pump models, 10
Cationic dyes, 26
CCCP, 184
CDTAB, 382
Cetyl-trimethyl ammonium bromide (CTAB),
  382, 389
Chemical hypothesis, 38, 155
Chemiosmotic hypothesis, see Mitchell hypothesis
Chloramphenicol, 195, 224, 253
p-Chloromercuriphenyl sulfonate, 12
Cholate dialysis technique, 183
Citrate synthase, 93
Complex I, 159, 167, 188
Complex III, 186
CoQ-cytochrome c reductase, 183
$CoQH_2$:cytochrome c reductase, 213
Crosslinking agent, 362
Cychloheximide, 170, 172, 253

Cyclic AMP, 55
Cytochrome a, 247
Cytochrome a_3, 207, 220, 243
Cytochrome b, 209, 242
Cytochrome c, 86
  synthesis of, 209
Cytochrome oxidase, 183, 213, 242, 359

D

DCCD, 156, 226
Decamethylene diguanidine, 157
Detergent micelles, 309, 330
Detergents, types of, 381
Diazobenesulfonate (DABS) 349
Dibenzo-18-crown-6, 173
Dibucaine, 14
Dictyostelium discoideum, 252
Diethylethylbestrol, 172
5,5-Dimethyl-2,4-oxazolidine, 82
Dimethyl adipimidate (DMA), 361
Dimethylsuberimidate (DMS), 361, 386
1,3-Dinitrobenzene, 317
2,4-Dinitrophenol, 226
Dio-9, 225

E

Electrochemical H^+ gradient, 39
Electron paramagnetic resonance (EPR), 167,
  217, 373, 384
Electroneutral cation/H^+ exchanger, 4
Electrophoresis gel, 50, 147, 216, 317, 353
p-Enolpyruvate carboxykinase, 88
Erythrocyte membrane, 328
Erythromycin, 255
ESR, see EPR
Ethidium bromide, 25, 157, 221, 244, 275
n-Ethylmaleimide, 118, 143, 163
ETP particles, 174
Euflavine, 225, 275

F

F_1-ATPase, 120, 160
Fatty acid
  β-oxidation, 61
  spin label of, 298, 372
Fatty acids, transport of, 64
FCCP, 41, 226
"Flip-flop" phospholipids, 121, 370

Fluorescent probe, 327
Formycine phosphate, 132

G

Gluconeogenesis, 61, 87
Glutamate-aspartate carrier, 74
Glutamate dehydrogenase, 74, 89
Glutamate oxalacetate transaminase, 75
Glyceraldehyde-3-P dehydrogenase, 85
Glycolysis, 22
Glycoproteins, 49
Gramicidin, 8, 390
Guanidine, 25

H

H^+ pump, see Proton pump
Hatefi complex I, 173
Hexylammonium, 160

I

Inheritance, cytoplasmic, 268
Inner mitochondrial membrane, isolation of, 371
Iodination of proteins, 216
Ionophores, 3, 39, 54, 187
Isocitrate dehydrogenase, 95

K

K^+ transport, 21
Ketogenesis, 101
α-Ketoglutarate dehydrogenase, 96
Kluyveromyces lactis, 277

L

La^{3+}, 47
Lactate dehydrogenase, 85
Lactoperoxidase-iodination, 357
Lecithinases, 172
Lidocaine, 304
Lipid bilayers, 65
Lipid phase transition, 368

Liposomes, 17, 50, 171, 183, 307, 330, 374

M

Macrocyclic polyethers, 172
Malate dehydrogenase, 73, 94
  release of, 151
Mersalyl, 118
Metallochromic indicator, 33
n-Methylphenazinium methylsulfate, 190
p-Mercuriphenylsulfonate(PMPS), 349
Mitchell hypothesis, 3, 21, 38, 62, 81, 155, 156, 183
Mitochondrial drug resistant mutants, 266-272
Monovalent cation transport, 4
Murexide, 33, 48

N

NADH CoQ reductase, 183
NADH dehydrogenase, 171, 178, 259
NADH iron-sulfur centers, 168
NADH oxidase, 358
NADH-ubiquinone reductase, *see* Complex I
Naphthoquinone sulfonate, 185
Nernst potential, 50
NiCl$_2$, 299
Nigericin, 8, 10, 42, 71, 185
Nitrosoguanidine, 277
Nitroxide free radicals, 291
NMR, 303
Nonactin, 177

O

Octylguanidine, 25, 157
Oligomycin, 7, 47, 112, 156, 163, 171, 225, 253
Oscillatory mitochondria, 300
Oxidative phosphorylation complex, biogenesis of, 265

P

Palmityl-carnitine, 101
Paramagnetic cations, 309
PCMB, 118
Petite mutant E5, 201
0-Phenantroline, 172

Phenethylbiguanide, 157
Photoaffinity label, 221, 321
Phosphatidyl ethanolamine, 171
Phosphatidylcholine, 171
Photochemical probe, 315
3-Phosphoglycerate kinase, 185
Piericidin-sensitivity, 170
Procaine, 304
Progesterone, 172
Prostaglandins, 55
Proteoliposomes, 320
Proton pump, 21, 62, 168
Pyruvate carboxylase, 87, 89
Pyruvate dehydrogenase complex, 91

R

Rb^{++}, electrophoretic transport of, 39
Redox carriers, 21
Reducing equivalents, transport of, 83
Rhein, 172
Rhodopsin, 389
  spherical bilayer into, 389
Ribosomes of mitochondria, 207
  depletion of, 209
Rifampicin, 261
Rotenone, 70, 171
  specific binding sites, 170
Ruthenium red, 47

S

Sacchromyces cerevisiae, 252
  Cytoplasmic mutants, 247
  KD115-2, 200, 374
  L410 ρ^+, 204
  E5 ρ^o, 204
  ρ^+, 230, 243
  ρ^o, 213, 230, 243
  nuclear mutants, 245
Sarcoplasmic reticulum, 36, 315
Site I, 171
Site III, 171
Spin label, 291, 382
  5N10, 292
  partitioning of, 293
Submitochondrial particles (SMP), 351
  crosslinking of, 363
  detergent interaction, 382
Succinate dehydrogenase, 259, 358
Succinate oxidase, 358
Succinate thiokinase, 97
Swelling of mitochondria, 3, 33, 151, 362

T

TEMPO, 297
Tetraalkylammonium, 157
Tetramethylammonium, 8
Tetraphenylboron, 188

U

Ubiquinone, 168
Ubiquinone reductase, 171
Uncoupler
  binding protein of, 272
  1799, 184, 226
Uncouplers, 7, 112, 171
  K^+ uptake inhibitors, 27
  Ca^{++} release induced by, 40, 53

Urea cycle, 61
Ureogenesis, 88

V

Valinomycin, 3, 39, 81, 177, 184
Venturicidin, 225

XY

X-537 A, 54, 390
Yeast
  auxotrophic mutants, 198
  chemostate culture of, 195
  metabolism of, 22

A 6
B 7
C 8
D 9
E 0
F 1
G 2
H 3
I 4
J 5